DEVELOPMENTS IN BIOCHEMISTRY

BIOLOGICAL AND CLINICAL ASPECTS
OF SUPEROXIDE AND SUPEROXIDE DISMUTASE

BIOLOGICAL AND CLINICAL ASPECTS OF SUPEROXIDE AND SUPEROXIDE DISMUTASE

Proceedings of the Federation of European Biochemical Societies
Symposium No. 62

Editors:

W.H. BANNISTER
*Nuffield Department of Clinical Biochemistry, University of Oxford,
Radcliffe Infirmary, Oxford, England*

J.V. BANNISTER
Inorganic Chemistry Department, University of Oxford, Oxford, England

Organizing Committee:
J.V. Bannister (England), B. Babior (U.S.A.), E.M. Fielden (England),
I. Fridovich (U.S.A.), J. McCord (U.S.A.), A.M. Michelson (France),
G. Rotilio (Italy)

ELSEVIER/NORTH-HOLLAND
NEW YORK • AMSTERDAM • OXFORD

Published by:

Elsevier North Holland, Inc.
52 Vanderbilt Avenue, New York, New York 10017

Sole distributors outside USA and Canada:

Elsevier/North-Holland Biomedical Press
335 Jan van Galenstraat, P.O. Box 211
Amsterdam, The Netherlands

Library of Congress Cataloging in Publication Data

International Symposium on Superoxide and Superoxide
    Dismutase, Malta, 1979.
    Biological and clinical aspects of superoxide and superoxide dismutase.
    (Developments in biochemistry; v. 11B ISSN 0165-1714) (Proceedings of the
    Federation of European Biochemical Societies symposium; no. 62)
    Papers presented at the symposium held Oct. 1-5, 1979 and sponsored by
    the Federation of European Biochemical Societies.

    Bibliography: p.
    Includes index.
    1. Superoxide—Metabolism—Congresses. 2. Superoxide dismutase
    —Congresses. I. Bannister, W. H.   II. Bannister, Joe V., 1945-
    III. Federation of European Biochemical Societies. IV. Title. V. Series:
    VI. Series: Federation of European Biochemical Societies. FEBS
    symposium; no. 62. [DNLM: 1. Oxygen—Congresses. 2. Superoxide
    dismutase—Congresses. W1 DE997VG v. 11B / QU140 I61 1979b]
QP535.01I53 1979   574.19′2   80-19728
ISBN 0-444-00443-2

Manufactured in the United States of America

v

Contents

Preface

There has been immense interest in the biological production of super-oxide, and its consequences, after the discovery of enzymatic catalysis of superoxide dismutation by McCord and Fridovich in 1968-69. Fridovich and co-workers went on to show the existence of a family of superoxide dismutases in living organisms, with copper, manganese or iron at the active site. Besides being of considerable interest in themselves as metalloproteins, these enzymes raise a fundamental question: Is superoxide toxic and enzymatic catalysis of superoxide dismutation a detoxification mechanism? The evolution and existence of enzymes which scavenge superoxide with nearly perfect catalytic efficiency makes the view that the superoxide radical is in some manner cytotoxic very appealing. This view is supported by certain kinds of evidence in bacteria, blue-green algae and higher organisms relating to such diverse phenomena as oxygen tolerance, photo-oxidative death, radiation damage and inflammation. However, conceptual problems exist regarding the toxicity of superoxide and the significance of the superoxide dismutases. Monovalent reduction of oxygen, with superoxide as the first product, is not disputed for biological systems, but it is difficult to reconcile what Fee has seen fit to call the chemical docility of aqueous superoxide with a direct role of the radical in cellular degradative processes. Granted the monovalent reduction of oxygen, one can shift the des-tructive potential in metathetical fashion from superoxide to the hydroxyl radical. This raises the question of the biological production of hydroxyl radical. Fridovich and collaborators were the first to draw attention to this possibility. Current ideas favour some form of metal-catalysed Haber-Weiss reaction or super-oxide-driven Fenton chemistry, but not without problems regarding the control of transition metal ions, such as iron, in biological media.

The discovery of superoxide dismutase opened a new field in the fundamental endeavour of biologists to see and to show the impact of oxygen on living organisms. There is no sign after a decade of research that the field has been exhausted. Current studies exemplified by the articles in this book show the continuing potential of superoxide dismutase as a conceptual and experimental tool to un-cover what Fridovich has aptly called the benign and malignant faces of oxygen, and its potential as a therapeutic agent against radiation damage and in certain forms of autoimmune and inflammatory disease. Since its discovery the enzyme has clearly pervaded all work and thinking on oxygen-centred radicals, or active oxy-gen species, in biology. The abundantly thriving nature of current work and interest in superoxide, as can be seen in this book and its companion volume, augurs well for future research and understanding and application of the enzyme.

Oxford - March 1980   W.H. Bannister - J.V. Bannister

Acknowledgments

The papers published in this book were presented at an International Symposium on Superoxide and Superoxide Dismutases held in Malta between 1st and 5th October, 1979. The meeting was sponsored by the Federation of European Biochemical Societies. The help received from Professors G. Bernardi (Paris), P.N. Campbell (London), S.P. Datta (London) and T.W. Goodwin, FRS (Liverpool) in getting this meeting funded and organised is gratefully acknowledged.

American participation at the meeting and also support for the meeting was made possible through sponsorship from the Office of Naval Research, National Cancer Institute, National Heart Lung and Blood Institute, National Institute of Environmental Health Sciences and Fogarty International Center.

Further support was obtained from Diagnostic Data, Inc. of Mountain View, California, USA and from Lederle Laboratories, A Division of the American Cyanamid Company, Pearl River, New York, USA. Thanks are also due to Professors A.M. Michelson, J. McCord and I. Fridovich for generously donating royalties towards the meeting.

Professor W.H. Bannister and Dr.J.V. Bannister gratefully acknowledge the Wellcome Trust, London, England for Research Fellowships. This publication would not have been possible without the support of the Wellcome Trust.

THE NATURE OF INTERMEDIATES DURING BIOLOGICAL OXYGEN ACTIVATION

WOLF BORS,[+] MANFRED SARAN[+] AND GIDEON CZAPSKI[++]
[+]Abteilung für Strahlenbiologie, Insitute für Biologie Gesellschaft für Strahlen- and Umweltforschung, 8042 Neuherberg, W. Germany; [++]Department of Physical Chemistry, Hebrew University, Jerusalem, Israel

INTRODUCTION

During the last years quite a few reviews have appeared on the chemistry of the oxygen radical species, the hydroxyl radical ($\cdot$OH) and, in particular, the superoxide anion (O_2^-) (Refs. 42, 43, 93, 121, 174, 196, 343). The data on O_2^- as well as the growing number of investigations on the reactivity of this species with a variety of chemicals[9,33,70,71,109,158,167,290] - taken at face value, should lead to a reappraisal of the relevance of this species in biological systems. While in a few cases doubts have already been voiced,[94,95,121] O_2^- still is one of the favorite candidates for the causative agent in oxygen toxicity.[104,106,225,226] Likewise a reassessment is necessary for the function of the hydroxyl radical in biology, which gradually is taking over the role of O_2^- due to the growing realization of the latter's poor reactivity.[42,68]

PHYSIO-CHEMICAL ASPECTS OF SUPEROXIDE AND HYDROXYL RADICALS

(a) _Physical properties_. There is no need to reiterate the physical properties of the oxygen radicals because of the exceptional review by Fee and Valentine.[93] The only ambiguity concerned the spectral data of $HO_2^{\cdot}/O_2^-$, which were recently re-evaluated by Bielski,[37] together with kinetics and pK value, and are now definite. A noteworthy increase is observed for the absorption of O_2^- in organic solvents, going from 2350 (Ref. 37) to 2686 $M^{-1}cm^{-1}$ (Ref. 175), accompanied by a much higher stability.[3] The latter value corroborates earlier data by Fee and Hildenbrand,[92] while the value given by Ozawa et al.[258] is certainly too low.

(b) _Chemical properties_. Most of the pertinent properties of these radicals were obtained after generation by physico-chemical methods, such as ionizing radiation,[33,35,77] in particular pulse radiolysis,[2,300] photolysis,[129,152,160] electrochemistry,[3,4,72,185,258] rapid-mixing techniques combined with EPR,[76,218] or stopped-flow experiments using KO_2,[71,109,167,212a,213a] dissolved in organic solvents with the help of crown ether.[325] Only these sources yield defined qualities and quantities of the radicals in question, whereas all biological sources, while being more convenient, are far less specific.

As background to the later discussion, the reactions of the pertinent radicals with each other are compiled Table 1. The rate constants were obtained from the most recent compendium[90] and two papers by Bielski and coworkers.[37,337] While the high rate constants of all reactions involving ˙OH radicals are self-evident (except for the reaction with undissociated H_2O_2), the slow reactions of O_2^- with itself respectively H_2O_2 or HO_2^- should be pointed out. Even lower is the uncatalyzed dismutation rate of H_2O_2, with its maximum at the pK-value.[84] Nevertheless this non-radical raction is included, as it has been proposed - notwithstanding the low rate constant - to generate, in nonaqueous solutions, either singlet oxygen[312] for O_2^- and ˙OH radicals[281] via

$$H_2O_2 + HO_2^- \rightarrow O_2^- + \text{˙OH} + H_2O \tag{1}$$

For some of the reactions thermodynamic data were compiled by Singh and Koppenol.[309]

Concerning the reactivity of O_2^- with other substances, in aprotic solvents it reacts preferentially as a nucleophile. It is capable to reduce or oxidize suitable compounds but - contrary to its acid $HO_2^˙$ - only rarely acts as a radical species.[107,293] Even though nucleophilic reactions are less likely in aqueous solutions, they have already been suggested for biological systems.[241] The non-radical behaviour of O_2^- has recently been demonstrated in aqueous solutions as the water-soluble carotenoid crocin is bleached by radiolytically generated $HO_2^˙$ but not by O_2^-.[50]

The very high reactivity of ˙OH radicals, being one of the strongest oxidizing agents and attacking most compounds, renders the site of formation tantamount to the site of reaction. It is therefore inconceivable that _free_ ˙OH radicals are important metabolic intermediates. Generally the rate constants are approaching diffusion-controlled limits[83,90] yet may be influenced by the reactivity of hydrogen bonds in hydrogen abstraction reactions.[302,303]

(c) _Generation processes._ There is no need to emphasize the growing number of biochemical oxygenating systems which are thought to involve oxygen radicals. Reviewing the biological sources, however, is outside the scope of this article. We shall therefore only highlight those reactions which

(a) serve as standard generating systems (Table 2), and

(b) have been proposed to generate ˙OH radicals (Table 3)

All of the listed reactions are steady-state processes and generate far less oxygen radicals per time unit than the most efficient physico-chemical sources. Xanthine oxidase with either purines or acetaldehyde as substrate, the most commonly used biochemical source, is everything but specific. It yields a number of radical species, with H_2O_2 being the usual reduction product of oxygen[255] and O_2^- formed only under special conditions.[103,151] It has also been suggested to generate ˙OH radicals,[24,150] as was the case for autoxidation of dihydroxyfumaric acid,[119] and another hydroxylating agent, NADH in the presence of phenazine methosulfate.[120] However, in all three cases, the results of kinetic investigations with p-nitrosodimethylaniline argued against the presence of ˙OH radicals

TABLE 1

CHEMICAL REACTIVITIES OF OXYGEN RADICALS: INTERSPECIES REACTIONS

Reactions			$k(M^{-1}s^{-1})$	References
1. $HO_2^{\cdot} + HO_2^{\cdot}$	$H_2O_2 + O_2$		8.6×10^5	37
$HO_2^{\cdot} + O_2^-$	$HO_2^- + O_2$		1.02×10^8	"
$O_2^- + O_2^-$	H^+	$HO_2^- + O_2$	0.35	"
2. $^{\cdot}OH + ^{\cdot}OH$	H_2O_2		5.2×10^9	90
$^{\cdot}OH + O^-$	HO_2^-		2.6×10^{10}	"
$O^- + O^-$	$H+$	HO_2^-	$0.9-1 \times 10^9$	"
3. $HO_2^{\cdot} + ^{\cdot}OH$	$H_2O + O_2$		1.5×10^{10}	90
$O_2^- + ^{\cdot}OH$	$H+$	$H_2O + O_2$	1.01×10^{10}	"
4. $HO_2^{\cdot} + H_2O_2$	$^{\cdot}OH + H_2O + O_2$		0.50	337
$O_2^- + H_2O_2$	$^{\cdot}OH + OH^- + O_2$		0.13	"
5. $^{\cdot}HO + H_2O_2$	$HO_2^{\cdot} + H_2O$		$1.2-6.5 \times 10^7$	90
$^{\cdot}OH + HO_2^-$	$O_2^- + H_2O$		8×10^9	"
$O^- + HO_2$	$O_2^- + OH^-$		$0.7-1 \times 10^9$	"
6. $H_2O_2 + H_2O_2$	$2 H_2O + O_2$		$-$	
$H_2O_2 + HO_2^-$	$H_2O + OH^- + {}^1O_2$		2.7×10^{-4}	84
$HO_2^- + HO_2^-$	$2 OH^- + O_2$		$-$	

themselves.[48] At present, photochemical or enzymatic reduction of autoxidizable
compounds - the reactions with flavins were reviewed by Massey et al.[210] - are
the only specific biochemical sources of O_2^- and none exist for $^{\cdot}OH$ radicals.

We now turn to the biological systems in which the formation of $^{\cdot}OH$ radicals
has been postulated. They are listed in Table 3. The general implications of
the formation of $"^{\cdot}OH"$ radicals have been reviewed by Cohen.[74] Only in a few
cases have authors refrained from postulating freely diffusable $^{\cdot}OH$ radicals.[45,64,142]
Also the so-called Haber-Weiss reaction

$$H_2O_2 + O_2^- \rightarrow {}^{\cdot}OH + OH^- + O_2 \qquad (2)$$

nominally discredited for at least two years,[79,96,144,180] and now definitely
established as being too slow,[221,337] is still used as an explanation for the
formation of $^{\cdot}OH$ radicals.

TABLE 2

STANDARD BIOCHEMICAL SOURCES OF OXYGEN RADICALS

Source	Substrate	Remarks
xanthine oxidase	purines, acetaldehydes	most commonly used source of oxygen radicals, however, it generates a number of activated species (Refs. 24, 103, 150, 151); no free ˙OH is formed (Ref. 48)
peroxidase	dihydroxyfumaric acid	hydroxylating system; inhibitable by ˙OH scavengers (Ref. 120), but not by DTPA (Ref. 123); no free ˙OH is formed during autoxidation of DHF (Ref. 48)
NADPH-diaphorase	methyl viologen	generates only O_2^- (Elstner, in preparation)
photosensitization (EDTA, methionine)	flavins (FMN, riboflavin)	generates O_2^- (Refs. 25, 229, 277), preferentially in alkaline solutions (210)
reduction of phenazine methosulfate	NADH	hydroxylating system (Ref. 186), formation of ˙OH radicals assumed (Ref. 120) but not detectable (Ref. 48)

(d) Assay methods. Aside from the uncertainty concerning the biochemical sources of the oxygen radicals, further doubts result from the unspecificity of most assays. A recent review on this subject[45] and the inclusion of more novel data can be summarized as follows:

(i) Assays, based on the reduction by O_2^- should be limited to the use of derivatized cytochrome c,[13,188] as tetrazolium salts are too unspecific[5,12] (the proposed reduction of cytochrom c by ˙OH radicals[197,308] can probably be neglected during routine assay conditions).

(ii) The enzymatic assay of O_2^-, based on the lactate dehydrogenase-catalysed oxidation of NADH, has been extensively examined[32,34] and is very specific as uncatalyzed oxidation of NADH by O_2^- is quite slow.[32]

(iii) Oxidation by O_2^- may be specific but often leads to the same compounds as are formed by ˙OH radicals. Even the standard procedure of control experiments with superoxide dismutase would not differentiate between reactions by O_2^- or ˙OH if O_2^- is merely a precursor of the more potent ˙OH radical.

TABLE 3

BIOCHEMICAL SYSTEMS GENERATING HYDROXYL RADICALS

Substances/Systems	Evidence for 'OH radicals
1. Drugs	
Hydroxylated compounds:	
6-hydroxydopamine 6-aminodopamine 6.7-dihydroxytryptamine dialuric acid	ethylene formation from methional, its inhibition by 'OH scavengers and stimulation by H_2O_2 (Ref. 73)
alloxan	inhibition of induced diabetes by various 'OH scavengers (Ref. 141)
antineoplastic agents with quinoid structure:	
bleomycin	spin trapping with dimethyl-pyrroline-N-oxide (DMPO: Ref. 251) and phenyl butyl nitrone (PBN: Ref. 315) spin trapping with PBN (Ref. 204)
mitomycin B, C and	
streptonigrin	
2. Enzymes:	
xanthine oxidase	ethylene formation from ethional, inhibited by catalase + SOD and 'OH scavengers, stimulated by H_2O_2 (Ref. 24)
prostaglandin synthase	inhibition of self-deactivation by phenol or methional (Ref. 87) and formation of ethylene from the latter compound (Ref. 262)
guanyl cyclase	activation by SOD, blocked with catalase, Mn^{2+} and 'OH scavengers (Ref. 234)
3. Complex systems:	
microsomal electron transfer system	ethylene formation from methional and keto-mercapto-butyric acid as well as methane formation from DMSO, inhibited by 'OH scavengers (Ref. 75)
microsomal ethanol oxidizing system	diminishing of oxidation rate by 'OH scavengers (Ref. 253), in particular if catalase is blocked by azide (Ref. 64)
Phagocytosis	formation of ethylene from methional of keto-mercapto butyric acid both with monocytes (Ref. 338) and granulocytes (neutrophils; Ref. 339)
Inflammation	inhibitory effects of 'OH scavengers and catalase + SOD, stimulation by H_2O_2 (Ref. 285)

(iv) Generation of ethylene from methione and derivatives is initiated by radiolytically produced 'OH radicals,[291] yet the most commonly used precursor of ethylene, methional, is even highly unspecific for 'OH radicals.[273]

(v)   The method of spin trapping has come of age.[101,161,163,164,165,169] Of particular interest is the trapping of ·OH radicals[59,129,132,191,204,315] which is established for phenyl tert-butyl nitrones.[162,163] Doubts remain for other nitrones such as dimethyl-pyrroline-N-oxide (DMPO),[163] which has mainly been used for the trapping of O_2^- respectively its conjugate acid $HO_2^{\cdot}$.[59,130,131] Spin trapping of ·OH radicals may be extremely sensitive due to the accumulation in form of more stable spin adducts, thus giving an integral amount over longer period.

(e)   Conclusions.   Our scepticism on the actual function of the oxygen radicals O_2^- and ·OH in biology is justified, as the chemical facts outlined above are at variance with the high efficiency and high selectivity of most biological oxidation reactions involving oxygen radicals.   Willson[341] also cautioned against the impression, that all radicals behave "promiscuously".   Thus we have to assume that the biological fate of the oxygen radicals per se is different than previously thought.   What, then, is the nature of biochemically generated oxygen radicals? Or - to phrase the question differently - what species are formed in biology which mimic O_2^- and ·OH?   Alternative hypotheses as answers to this query have to
- circumvent irreconcilable discrepancies on known reactivities of the oxygen radicals per se with the characteristics of a number of reactions;
- delineate the actual probability of the oxygen radicals reacting themselves;
- come up with more probable candidates supplementing O_2^- and ·OH as agents of oxygen toxicity.

What has to be kept in mind is the fact, that the whole hypothetic structure is only valid if the basic properties of the radicals do not change if going from model reactions to biological systems - except for the influences by complex formation or hydrophobic environments - an assumption which should be correct if one considers only main reactions and not an accidental side reaction.

BIOLOGICAL ASPECTS OF OXYGEN RADICALS

To arrive at a realistic picture of the role of the oxygen radicals in biology, we best start with a catalogue of the problem areas, in which activated oxygen species are encountered (Table 4).   Under each of the headings examples are listed, for which we shall propose superior explanations, which take into account the known properties of the individual intermediates.

Reactivity of superoxide and function of superoxide dismutase.   The ubiquity of superoxide dismutases in practically all aerobic and aerotolerant organisms has led to the presently accepted function of this enzyme as a protective protein,[104,105,122] which is inducible by O_2^-.[104,137,138]   This view, however, is considered untenable by Fee[95] on account of
- the low reactivity of O_2^-

TABLE 4

PROBLEM AREAS OF BIOLOGICAL OXYGEN ACTIVITATION

---

1. Reactivity of superoxide anions:
   inhibitory effects of superoxide dismutase
   induction of superoxide dismutase by superoxide
   stoichiometric enzymatic generation
   obligatory enzymatic cofactor
   effect of quinoid biocides

---

2. Probability of formation of free hydroxyl radicals:
   radiolysis
   Fenton-type reactions
   inhibition studies with ·OH-scavengers
   spin trapping results

---

3. Nature of ·OH-mimicking species:
   higher specificity/selectivity than ·OH
   lower reactivity with certain ·OH-assays
   propagation of chain reactions; delayed effects

---

4. Nature of intermediate in biological hydroxylation:
   Cytochrome P-450-dependent oxygen activation
   flaving-dependent oxygen activation

- the propensity of a number of metallo-proteins or metal complexes to act as SOD-analogs
- the occurrence of SOD in anaerobic organisms and the high amount of this enzyme in some species

A nevertheless observable protective effect of SOD would then be indirect with O_2^- acting as the precursor of more deleterious species. Yet considering the accelerated supply of H_2O_2 for Fenton-type reactions (see below), it may merely serve as a regulatory enzyme involved in oxygen metabolism.

Winterbourn and colleagues[344, 345] recently proposed still another function of SOD, namely to shift the equilibrium of the redox reactions of semiquinones and quinones to the right:

$$Sq^{\cdot} + O_2 \xrightarrow{\text{SOD}} Q + O_2^- \qquad (3)$$

In their opinion O_2^- would be scavenged by SOD, and the amount of the semiquinone (of hemolytic drug menadione) to react with methemoglobin would be diminished:

$$Hb^{3+} + Sq^{\cdot} \longrightarrow Hb^{2+} + Q \tag{4}$$

Viewed differently, SOD <u>stimulates</u> an O_2^--generating reaction which we could also establish by kinetic methods in the case of adrenaline.[46] We did, however, base this effect on the complex formation of the semiquinone of adrenaline with the prosthetic copper of SOD[230,277,340] - being a general feature of cupric ions.[57,189,278] It is therefore conceivable that the effect observed by Winterbourn et al.[344,345] may similarly be caused by such complex formation.

(a) <u>Enzymatic production and toxicological aspects</u>. To argue for O_2^- to be a toxic species, one has to reconcile the poor reactivity of this species with

- the presence of highly efficient biological sources, producing O_2^- almost stoichiometrically (e.g. NADPH-oxidase in phagocytes,[14] mitochondrial respiration[81,99] or in illuminated chloroplasts;[10]

- the requirement of O_2^- as co-substrate in a new class of enzymes, represented by indolamine-2.3-dioxygenase[148] or nitropropane dioxygenase;[313] and finally

- the seemingly unavoidable production of O_2^- during autoxidation of enzymatically reduced biocides with quinoid structure (e.g. paraquat[106,139] or some antibiotics, with examples listed in Table 3).

(b) <u>Function as precursor/transmitter</u>. As noted briefly, the purpose of SOD as protective enzyme in view of the sluggish nature of O_2^- must be an indirect one, with O_2^- involved in the formation of the active species. This mediating role can be further amplified by two features, directly related to the low reactivity of this species. One is further stabilization by complex formation, while improved diffusion can be an outcome of this complex-formation or the result of the small number of readily attackable target molecules.

Stabilization by complex formation: Formation <u>and</u> stabilization of O_2^- through reaction with transition metal complexes is an important facet of oxygen activation/toxicity involving this radical. There is such an amount of literature on complex formation of the oxygen molecule with transition metals that only reviews can be cited.[168,193,198,209,242a,324,328] Less is known on complexes with other metals,[56,327] as they are less relevant for biological systems and are only interesting from a chemist's viewpoint. Basically one has to differentiate between reversible complex formation as exemplified by hemoglobin[279] and irreversible complex formation involving electron transfer.[41,134] There is ample evidence that O_2^- also participates in such complexes;[219,294,305] in some cases only the addition of O_2^- resulted in their formation.[326,327] The importance of such complexes for the understanding of metal proteins cannot be over-emphasized and was already the subject of several reviews.[23,28,94]

Diffusion probabilities: It goes without saying that a longer lifetime is synonymous to lower reactivity. To what extent the lack of the hydration of O_2^- (Ref. 95) in non-aqueous solvents is responsible for the enhanced stability (e.g. in dimethylsulfoxide,[3] is unknown. While both O_2^- (Ref. 206) and H_2O_2 (Ref.236) have been proposed to penetrate erythrocyte membranes, all transport processes, at the moment, are merely speculative and may just represent the effect of chain sequences involving organic oxygen radicals.[26,27,282] Since secondary formation of O_2^- from various peroxy radicals has been demonstrated,[1,52,155,227] it would be well-nigh impossible to differentiate whether an O_2^- molecule passing through a membrane is the same entity throughout or is reformed at another site.

Membrane structure plays an essential part in all transport-related phenomena and structural integrity at the same time protects membranes against autoxidation or _vice versa_ structural deterioration entails an increased liability of membrane lipids to autoxidize.[80,181,275] Thus one important function of vitamin E (α-tocopherol), besides being an $^1O_2$ scavenger,[54,212] is to stabilize lipid membranes.[8,117,231,237] Another antioxidant, silymarin, which is even more potent than vitamin E may function in the same way.[38]

(c) List of likely target molecules. The above statements were not intended to strike an entirely fatal blow concerning the toxicity of O_2^- itself. There may actually exist a few target molecules in the cell which do react with O_2^- with appreciable reaction rates – nucleic acid bases[62a] and amino acids,[37a] however, are not among those substances. An albeit incomplete list is given in Ref. 90, which certainly is to be expanded in the near future. Foremost among the biological targets are sulfhydryl compounds[9] and thus also SH-containing enzymes.[200,201] Rate constants of enzymatic cofactors with O_2^- are known only for ascorbic acid and tocopherol model compounds[242] and do not exceed the spontaneous dismutation rate. No rate constants are known for the reaction of leucoflavins with O_2^-, while NADH reacts much better if complexed to lactate dehydrogenase.[34]

Quite fast electron-transfer reactions from and to O_2^- have been determined pulse-radiolytically for a number of quinones and hydroquinones,[276] or catechols.[47,49] In the latter case, however, the reaction of O_2^- was completely reversible in less than 20 msec (Ref. 49). Contrary to semiquinones, bipyridylium radical cations, which are produced easily by radiolytic or enzymatic reduction, have been shown by pulse radiolysis to react with O_2^- in an almost diffusion-controlled manner.[91,265] This rapid reaction with O_2^- and the equally fast formation of O_2^- from attack of oxygen at the organic radical are thought to form the principle of action of the bipyridylium herbicides,[91,265] this view being supported by enzymatic studies.[106] Certain carcinogens are also quite reactive with O_2^- (Ref. 112a). A frequently proposed reaction is the reduction of ferric iron or other oxidized metal ions,

$$Fe^{3+} + O_2^- \longrightarrow Fe^{2+} + O_2 \tag{5}$$

which in combination with the Fenton reaction (see below) should account for the cumulative inhibitory effects of SOD and catalase.

Caution, however, should be exercised if one wants to extrapolate the reaction of O_2^- with hydroperoxides:

$$ROOH + O_2^- \longrightarrow RO^{\cdot} + OH^- + O_2 \qquad (6)$$

to more complex systems. The formation of alkoxy radicals has thus far been postulated only for tert-butyl hydroperoxide[267,274] and linoleic acid hydroperoxide,[274,319] yet was already adopted by Tauber et al.[318] to obviate the need to postulate free $\cdot$OH radicals in phagocytosis. Using the most specific radical generating source of pulse radiolysis, we could demonstrate that O_2^- does not react with the above-mentioned hydroperoxides[51] - to the extent that the rate constants are even lower than reported[274,319] and thus irrelevant. Our results corroborate those of Gibian and Ungermann,[110] who negated such a reaction for aprotic solvents, thereby contradicting an earlier proposal by Peters and Foote.[267] Yet there remains the option that more active hydroperoxides, e.g. as derived from linolenic acid or arachidonic acid, do react with O_2^- with appreciable rates - a hypothesis which needs to be tested.

(d) Conclusions. As shown, the number of substances reacting with O_2^- in an irreversible and kinetically appreciable manner, is exceedingly low. It is therefore fully justifiable to negate any deleterious effect for O_2^- itself. Assuming that SOD does function as a protective enzyme, a view which is seriously contested,[95] this could only mean that O_2^- serves as a precursor of the active species. Due to its low reactivity it is actually quite well suited for this purpose, being able to store and transmit radical properties over wider distances.

Reactivity of "free" hydroxyl radicals or OH-mimicking species. Since we excluded O_2^- from the list of compounds responsible for the toxic effect of oxygen, we have to search for an alternative candidate. As noted in the beginning, the hydroxyl radical is presently the favorite species. Because of the high reactivity of this species as determined pulse-radiolytically or by competition with effective scavengers, its generation would almost certainly lead to immediate reaction with its closest neighbor. While this poses a severe handicap against the postulate of free $\cdot$OH radicals in biological systems, there are some instances where attack by $\cdot$OH radicals themselves must be taken into consideration, such as ionizing radiation or Fenton-type reactions.

(a) Radiolysis. Generation of $\cdot$OH radicals (as well as of hydrated electrons, hydrogen atoms - or O_2^- in oxygenated solutions) by ionizing radiation of water or aqueous solutions has only recently come to the attention of biochemists. While the method is one of the most selective and versatile generation processes, the radiolytic deposition of various types of radicals in cell tissues or organisms and their subsequent reactions have still not been fully elucidated.

More to our point of interest in activated oxygen species is the aspect of the potentiation of radiation damage in the presence of oxygen, expressed as the so-called oxygen enhancement ratio (OER; Ref. 299). There is conflicting evidence on the participation of the respective oxygen radicals in the cause of the OER effect. The implications of O_2^- are based on the protection afforded by SOD after ionizing radiation of mice,[270] bacteria,[232,244,250,268] cell cultures,[243,270] chromosomes[248,249] and model membranes.[209] This is juxtaposed to the postulated exclusive attack of $\cdot$OH radicals[22,288,289] with subsequent fixation of the damage by formation of peroxy radicals. The latter reaction, which we prefer, involves membrane-derived organic oxygen radicals (see below) and O_2^- only as a chain propagator.[136]

(b) <u>Fenton-type reactions</u>. Fenton-type reactions, i.e. the reaction of reduced metal ions with H_2O_2

$$Me^{n+} + H_2O_2 \longrightarrow Me^{(n+1)+} + \cdot OH + OH^- \qquad (7)$$

are presently the most popular concept - and at the same time the only alternative to radiation - for the generation of free $\cdot$OH radicals in biological systems.[42,95] Indirect evidence for such reactions in biochemical systems was first obtained from cumulative inhibition by SOD and catalase, and elimination of an initial lag-phase after addition of H_2O_2.[24] They are the most comprehensively reviewed chemical sources of $\cdot$OH radicals[19,79,333] and the products are generally identical to those formed by radiolytically generated $\cdot$OH radicals.

Kinetic aspects: There is a dearth of kinetic data for this reaction. The rate constant of the Fenton reaction itself involving ferrous iron is rather low, reportedly ranging from 23 (Ref. 118) to 76.5 (Ref. 133), and averaging around 50 $M^{-1}s^{-1}$ after including the values given by Barb et al.[20,21] This is in contrast to the very rapid reaction of cuprous ions with H_2O_2, also producing $\cdot$OH radicals (2.3×10^9 $M^{-1}s^{-1}$; Ref. 182).

To allow the Fenton reaction to proceed continuously, a reducing compound has to be present. With the mediating role of O_2^- in mind, reaction (5) would be the obvious reaction to fulfil this function. Table 5 summarizes the presently known rate constants of O_2^- or $HO_2^\cdot$ with a number of ferrous or ferric iron compounds.

Aside from the obvious scarcity of the data there are several important features - and some gaping omissions, such as the crucial reaction of uncomplexed ferric iron with O_2^-! It is evident that the reduction by O_2^- is stimulated when ferric iron is complex-bound or an integral part of a protein. $HO_2^\cdot$, in contrast to O_2^-, can be reduced to H_2O_2 by ferrous iron, the reaction being less influenced by the complex state. This is the result of the considerable increase in the oxidation potential of the $HO_2^\cdot/O_2^-$ - H_2O_2 couple at low pH (see Figure 6 in Ref. 93).

Arguments against "free" $\cdot$OH radicals: There are several discrepancies concerning the generation of <u>free</u> $\cdot$OH radicals by the Fenton reaction. First, Janzen et al.[163] were unable to detect $\cdot$OH adducts by spin trapping in Fe^{2+}/H_2O_2 solutions - in contrast to other reports[190,251] - yet they did observe this species after addition of H_2O_2 to the Fe^{3+}/ADP complex. This can best be explained by an

TABLE 5

REACTION RATES OF $HO_2^{\cdot}/O_2^{-}$ WITH IRON SALTS, COMPLEXES AND PROTEINS

Valence state		Rate constants ($M^{-1}s^{-1}$)		References
ferrous	ferric	$HO_2^{\cdot}$	O_2^{-}	
Fe^{2+}(aq)		1.2×10^6	−	166
	Fe^{3+}(aq)	2.1×10^{-2}[a]	−	"
$Fe(CN)_6^{4-}$		3.0×10^4	−	348
	$Fe(CN)_6^{3-}$	−	2.7×10^2	"
$Fe(EDTA)^{2+}$		−	6×10^3	213
			$2-4 \times 10^6$	156
	$Fe(EDTA)^{3+}$	−	1×10^7	213
			0.52×10^6	156
	Fe^{3+}(ferredoxin)	−	1×10^4	88
Fe^{2+}(cytochrome c)		$0.5-5 \times 10^6$	0	61
	Fe^{3+} (cyt. c)	0	1.4×10^6	"
	Fe^{3+} (peroxidase)	1.5×10^5		254
	$Fe^{3+} \cdot H_2O_2$	2.2×10^8	1.6×10^6	31
	(POD-compound I)			
	Fe-SOD	5.5×10^{8}[b]		195

[a] The value was obtained from $\dfrac{k(HO_2^{\cdot} + Fe^{3+})}{k(HO_2^{\cdot} + Fe^{2+})} \times K(HO_2^{\cdot}) =$ 3.6×10^{-3} M (Ref. 304), with $K(HO_2^{\cdot}) = 2.05 \times 10^{-5}$ M (Ref. 37) and $k(Fe^{2+} + HO_2^{\cdot})$ from the entry in the first line.

[b] Rate constant for the overall reaction

unfortunate choice of the respective concentrations of the reactants (too high amounts of both Fe^{2+} and H_2O_2 would successfully compete with the spin trap for the escaping $\cdot$OH radicals). Free $\cdot$OH radicals have actually been disclaimed for $Fe^{2+}/Fe^{3+}-H_2O_2$ in acetonitrile[113,346] and in the case of iron/EDTA complexes the intermediacy of a preferryl compound (see below) was proposed.[40,316]

It was the scepticism on the actual presence of free $\cdot$OH radicals which led us to propose "Fenton-type oxidants" for the ferredoxin-catalyzed ethylene formation in isolated chloroplasts.[88] We are now pursuing the approach initiated by Buettner et al.[59] and continued by Halliwell,[123,124] i.e. the use of several chelators to inhibit such reactions, using selective scavengers of ferrous and ferric iron. Both the previously used diethylene triamine pentaacetic acid and bathophenanthroline sulfonate do not discriminate sufficiently between these ions (according to their complex stability constants).[208]

   (c) Reactions of "oxygen complexes". Stimulation of Fenton reactions: We have emphasized before that Fenton-type reactions have to be promoted by complex formation, otherwise their reactivity would be too low. There exist already a number of reports that both H_2O_2 (Refs. 97, 183, 256), $\cdot OH$ radicals (Refs. 18, 78) and/or O_2^- (Refs. 18, 286, 287) can be complex-bound in Fenton-type reactions. Also known are complexes of O_2^- (Refs. 219, 294, 305) or $\cdot OH$ (Refs. 36, 317, 334) with other transition metals – and as special case protein-bound O^- during oxygen reduction to water by laccase.[7,55]

Particularly well-known examples of iron complexes involved in oxygen activation represent ferrous/ferric iron chelated with EDTA or similar ligands. They have been studied in flow systems,[213,263,332] by pulse radiolysis,[156] as tertiary complexes together with H_2O_2,[257,330] and have been shown to promote lipid peroxidation[172,205,246,316] and degradation of hyaluronic acid.[124] In addition they may function as analogs of SOD.[213,263] As complex-type Fenton reagent, they are mainly thought to generate $\cdot OH$ radicals,[59,123,124,171,215] in which case the stimulation may be due to the reported decrease of the reduction potential of ferric iron in the complex.[298,309,349]

Aside from the Fenton reaction proper, there exist other metal complexes, which in the presence of oxygen or H_2O_2 show strongly oxidizing properties. They are discussed by Fee (see Table 4 in Ref. 95). Two of them, the Udenfriend[322] and the Hamilton[125,127] system, consist of Fe^{2+}/EDTA/ascorbic acid/O_2 and Fe^{3+}/catechol/H_2O_2, respectively. Both systems have been applied extensively, yet no conclusive evidence concerning the active intermediates is available. Based on hydroxylation reactions, the formation of oxygen radicals was disclaimed for the Udenfriend system.[314] Contrarily, a re-investigation of the Hamilton system now suggests that here oxygen radicals are involved in the chain reactions.[349]

Metal-catalyzed autoxidation as chemical analogy of complex-mediated oxygen activation may be just a general name for a number of reactions of complexes with electron-transfer properties, whose detailed mechanism is hypothetical.[39,209,224,245,335] As demonstrated, even minute amounts of contaminating metal ions can initiate such reactions.[3,63] It is therefore likely that such unrecognized contamination is the reason for the reported $\cdot OH$ formation by the Haber-Weiss reaction itself.[215,259] Whether transition metal charge-transfer complexes containing oxygen generate O_2^- or H_2O_2 seems to depend on the ligand structure;[114,350] also whether the species remain complex-bound and thus stabilized.[203] In addition, it has been shown that both an organic copper complex consisting of methoxylated $CuCl_2$ in pyridine[283,320] and metal-catalyzed peracetic acid oxidation[261] cleave catechols similar to O_2^- (Ref. 239). The former complex has even been shown not to involve a genuine oxygenation step.[283,320]

Reductive activation of hydrogen peroxide: There is still another explanation for the delayed effect of some radical processes, which we shall call reductive activation of hydrogen peroxide. The fact that H_2O_2 precedes ·OH in the reduction sequence of oxygen to water is the underpinning of this hypothesis. There are several reactions for which an inhibitory effect of catalase, but none of superoxide dismutase was reported.[17,120,228,235,238] As H_2O_2 itself has only occasionally been implicated as a toxic agent,[6,236,280] its activation to a more potent species is required - most likely via Fenton-type reactions. Instead of yielding ·OH radicals directly, the reaction in our opinion involves an intermediary charge-transfer complex which is the actual reactive species. Still speculative is the proposal that O_2^- and H_2O_2 are present in the same complex with O_2^- ultimately supplying the reducing equivalent for H_2O_2. Such a reaction is identical to a "catalyzed Haber-Weiss reaction" not involving two discriminate steps and could proceed much faster than the uncatalyzed one.

Complexes of H_2O_2 with prosthetic iron of hemoproteins as catalase[297] or peroxidases[347] may be the biological equivalent of reductive activation of H_2O_2. This was proposed as early as 1952 by George.[108] However, catalase as well as the hypohalide reaction with H_2O_2 do not reduce but disproportionate H_2O_2. The latter reaction is an efficient source of singlet oxygen - another activated oxygen species:[143]

$$OCl^- + H_2O_2 \longrightarrow Cl^- + H_2O + {}^1O_2 \qquad (8)$$

Yet it has been questioned whether halogenating systems of peroxidases, H_2O_2 and hypohalide also produce ${}^1O_2$ under in vivo-conditions.[135]

Not quite in line with the reductive activation of H_2O_2, yet of certain relevance, is the postulated increased reactivity of dissociated hydrogen peroxide, HO_2^- as compared to H_2O_2 itself.[58,184,295] This may correspond to the enhanced nucleophilicity of organic peroxy anions.[216] Formation of HO_2^- must not be limited to pH-regions close to the pK value of 11.6, rather it could be the initial product of reduction or dismutation of O_2^- (Ref. 179).

General implications of complex formation: Several general aspects of oxygen activation through complex formation should not be overlooked:
- Such complexes stabilize the radical and enable it to be transported to more remote sites.
- Re-dissociation of the complex can occur reversibly to the parent compounds, thus representing only a more stable means of transport.
- Alternatively, re-dissociation as activated oxygen species may only be triggered

by suitable reactants, probably interacting in a ternary complex similar to
enzyme mechanisms involving $(E-S-O_2)$-complexes.

- Reacting in this latter mode at the same time enhances the specificity/selecti-
vity, thereby taking care of one of the strongest implications against free
oxygen radicals.

(d) <u>Chain reactions via alkoxy and peroxy radicals</u>. The previous section
detailed the limited chances of $\cdot OH$ radicals themselves to be present in biological
systems. In addition there is sufficient evidence that other species besides $\cdot OH$
radicals are capable to mimic this species, at the same time expressing a higher
specificity by lowering the reactivity towards other molecules.[46] Furthermore,
the high reactivity of $\cdot OH$ cannot explain some effects which are still manifest
at later times after the radical-generating process occurred. Such delayed effects
can both be caused by chain propagation involving organic oxygen radicals or transport phenomena. The interplay of the various radical species in chain reactions
is depicted in the following scheme.

<u>Interrelationship of Oxygen Radicals during Chain Reactions</u>

(1) $RH_2 + \cdot OH \longrightarrow \cdot RH + H_2O$

(2) $RH_2 + RHO \cdot \longrightarrow \cdot RH + RHOH$

(3) $\cdot RH + O_2 \longrightarrow RHOO \cdot$

(4) $RH_2 + RHOO \cdot \longrightarrow RHOOH + \cdot RH$

(5) $RHOO \cdot + RHOO \cdot \longrightarrow 2 RHO \cdot + O_2$

(6) $RHOO \cdot + O_2^- \xrightarrow{+H^+} RHOOH + O_2$

(7) $RHOO \cdot \xrightarrow{-H^+} R + O_2^-$

(8) $O_2^- + Fe^{3+} \longrightarrow Fe^{2+} + O_2$

(9) $Fe^{2+} + H_2O_2 \longrightarrow \cdot OH + OH^- + Fe^{3+}$

(10) $Fe^{2+} + RHOOH \longrightarrow RHO \cdot + OH^- + Fe^{3+}$

Considering the wide occurrence of metal ion redox centers as well as of H_2O_2, the reaction can also account for the re-emergence of $\cdot OH$ radicals at sites remote from their original formation. This can be envisaged by a sequence of events, expostulated by reactions (1), (3), (7) (8) and (9) in the above scheme. It is, however, unlikely that two species derived from the same reduction chain (i.e. O_2^- and H_2O_2) do participate in this sequence of reactions in a statistically relevant time period.

Formation and reactivity of organic oxygen radicals: We first turn to the more likely intermediacy of alkoxy and peroxy radicals in chain reactions as resembling $\cdot OH$-derived reactions. Evidence for the reaction of $\cdot OH$ and oxygen with lipid components of membranes leading to alkoxy and peroxy radicals ($RO\cdot$ or $ROO\cdot$) has been accumulating. In the case of $\cdot OH$ attack, binding of oxygen to an organic radical, formed by the initial H-abstraction, leads directly to peroxy radicals.[102,190] Alkoxy radicals, on the other hand, are mainly produced in a Fenton-analog reaction of ferrous iron with organic hydroperoxides instead of H_2O_2:

$$Fe^{2+} + ROOH \longrightarrow Fe^{3+} + RO\cdot + OH^- \qquad (9)$$

This type of reaction has already been reviewed[176] and was recently demonstrated by EPR (Ref. 111). It may be termed an "organic Fenton reaction". Formation via a Russell-type reaction (Ref. 284; reaction (5) in scheme above) is favored only in highly autoxidized material with the increased chance of the encounter of two peroxy radicals. A similar electron transfer from O_2^- (reaction (6) in the text) is to be discounted (see above), while the formation in radiolytic systems via

$$e_{ag}^- + ROOH \longrightarrow RO\cdot + OH^- \qquad (10)$$

represents a highly effective generation method[51,292] – aside from homolytic dissociation of dialkylperoxides by photolysis or thermolysis.[153] A complete list of all possible generation processes for alkoxy radicals is given by Kochi (see Table 6 in Ref. 177). This author also summarizes all reactions of alkoxy and peroxy radicals known from organic chemistry, such as hydrogen abstraction, fragmentation, rearrangement or addition to double bonds.

$RO\cdot$ or $ROO\cdot$ radicals have repeatedly been investigated in radiation chemistry,[136,307] and have been proposed as intermediates in lipoxygenase-catalyzed fatty acid oxidation[329] and lipid autooxidation.[192,217,272,274] Their physical and chemical properties[69,136,146,306,311] have mainly been determined after investigation in organic chemistry[111,146,187,202,271] and were already the subject of several reviews.[29,98,147,153,177]

Reactivity of semiquinones: A special form of alkoxy radicals are represented by semiquinones. Their formation by reaction of O_2^- with a number of catechols or quinones has been discussed earlier. Depending on the respective redox potentials, they are capable to transfer one electron to oxygen, thereby generating O_2^-.[276]

A reaction of semiquinones with O_2^- itself has to be discounted (see also Ref. 42) due to the low probability of two radicals encountering each other.

Bielski and Begicki[30] were the first to attempt a classification of the types of reaction of catechols, semiquinones and quinones with each other, O_2, O_2^- and H_2O_2. Recently we added another type of reaction — the reversible formation of semiquinones from catechols by O_2^- and H_2O_2 (Ref. 49). Yet this classification may be an oversimplification and may be true only for reactions with radiolytically generated oxygen radicals. The whole complexity of redox reactions of catechol derivatives with O_2^- becomes apparent if one considers electron-transfer equilibria,[220,264] the controversy on base-catalyzed hydroxylation[266] versus dimerization[11] and ring-opening in organic solvents.[239] A very detailed review on all oxygenation reactions of hydroxylated aromatic compounds — among them catechols — was recently presented by Matsuura.[211]

Semiquinones are capable of forming complexes with transition metals,[57,189,278] among them the prosthetic copper of the cytosolic SOD.[230,277,340] In this form they are stabilized in a similar way as discussed before for O_2^- itself. The combination of redox cycling of such species with the concomitant formation of O_2^- as well as complex stabilization is therefore an important way of storing and/or transmitting radical properties.

Catechols and quinones are fundamental hormonal and metabolic intermediates. Thus one may argue, on the one hand, that the metabolism of catecholamines via O-methylation and N-oxidation[301] could well have evolved to avoid the well-documented participation of O_2^- in the oxidation of the catechol moiety.[44,85,296,323] On the other hand, O_2^--formation by such compounds as ubiquinone[53,62] or quinoid drugs[15,82,112] may be an initial step for the generation of required activated oxygen species.

Analytical aspects: Problems are encountered in the search for suitable assay methods which allow to differentiate between the oxygen radicals proper and such organic "oxygen-centered radicals". Even spin trapping has the disadvantage, that those nitrone compounds which give adducts with ˙OH and probably O_2^- like phenyl tert-butyl nitrone (PBN) do not trap RO˙ at all and the ROO˙ adduct is stable only at low temperatures.[154,163,223] The spin trap t-butyl nitrone, heavily used in radiation studies (see Abstracts 6th Int. Congr. Radiat. Res, Tokyo), does trap RO˙ and ROO˙,[154,157,260] but has not been investigated with ˙OH and HO_2^-/O_2^- themselves.

Generally available spectroscopic methods are not known, as the use of p-nitrosodimethylaniline is limited to simple systems due to the reactive nitroso group,[100] otherwise kinetic investigations do allow such a differentiation if the rate constants are widely dissimilar.[48] We are presently investigating other spectroscopic assay methods for this differentiation, one of these being the subject of the contribution by Saran et al.[292]

On the whole we think that the implication of such substituted ˙OH and HO_2^-

radicals is a very "hot" topic and is worth purusing. The reason is that the hypothesis takes into consideration both the complexity of biological systems, the probable formation of oxygen radicals versus their divergent reactivity, and the generally observed specificity of the reactions involving "oxygen radicals".

(e) _Biological hydroxylation reactions._ Implication of intermediates: As final problem area of biological oxygen activation we turn to the question of the intermediates in certain enzymatic oxidation reactions. Both during cytochrome P-450 (Refs. 89, 310) and flavin-dependent hydroxylations[145] or bacterial luminescence (Refs. 140, 195), two electrons are formally transferred to molecular oxygen, reducing one oxygen atom to water while the other atom, without net charge, is inserted into an existing C-H or N-H bond:

$$RH + O_2 + 2\ e^- + 2\ H^+ \longrightarrow ROH + H_2O \tag{11}$$

Oxygen radicals themselves - both O_2^- and $\cdot OH$ - have been proposed to be generated by these systems;[173,207,252] however, their participation during hydroxylation is unlikely from mechanistic and product pattern standpoints. Based on theoretical considerations,[128] results with _in vitro_ enzymatic systems containing organic hydroperoxides[115,247,310] or iodosobenzene,[66,116,199] as well as ferrous/ferric iron complexes of EDTA,[40,316] two mechanisms are presently discussed.

Oxenoid mechanism: The more fundamental one, the so-called "oxenoid mechanism", was first promulgated by Hamilton[126] and is now favored for flavin-dependent reactions.[145] It derives from the fact that the oxygen atom being inserted formally contains only six electrons thus having a similar structure as the hypothetical "carbene" (Ref. 126). Since such a neutral oxygen atom could not exist as a free species, oxygen transfer necessarily has to take place in the encounter complex. This mechanism was first proposed for the model system enediol/H_2O_2, the so-called "Hamilton system".[125,127] As discussed earlier, the non-radical nature of this hydroxylation reaction has now been question.[349]

Perferryl compounds: The second alternative is the hypothesis of the "perferryl intermediate" - $Fe(IV)O^{2+}$ - with Fenton-type oxidants probably representing a border-line case. While such a species has been disclaimed for the genuine Fenton reaction on the basis of EPR experiments,[18] it is still being discussed for ferrous/ferric iron complexes of EDTA.[40,316] Mainly, however, the theory was invoked for iron-catalyzed enzymatic hydroxylations, in which organic hydroperoxides serve as oxidants.[115,247,310]

That the oxenoid and perferryl hypotheses are actually synonymous becomes clear if one views the results of cytochrome P-450-catalyzed hydroxylations, further stimulated by iodosobenzene.[66,116,199] There is agreement that this organic compound serves as the donor of one oxygen atom which forms an intermediary perferryl compound as oxygen-inserting species. With perferryl intermediates limited to iron catalysis, the more general term "oxenoid mechanism" includes the oxygen insertion via flavin hydroperoxides.[140,145]

(f) Conclusions. Site-specific reactions and scavenger problems: We repeatedly pointed out the minimal chance of free ·OH radicals to escape from their site of generation. The realization of this fact led to the hypothesis of "site-specific free radicals", the term being coined by Bachur et al.[15] This hypothesis of the generation of the radicals at the site of the reaction was earlier used to explain the inactivation of SOD by its product H_2O_2 (Ref. 149) and again to account for the site-specific oxidation of heme methine groups.[159] The hypothesis in a straight-forward way alleviates the most common inconsistencies of the reactions involving the oxygen radicals O_2^- and, particularly, ·OH. On the one hand, it allows for the formation of free ·OH, and taking into account its very high reactivity, it presumes its immediate reaction with a closely located chemical moiety. Even more to the point of our arguments is the fact that in two of these examples, ·OH radicals are thought to be formed in Fenton-type reactions.[15,149]

While this reaction alternative may be quite attractive, it is at the same time very difficult to verify. To ensure the site-specific reaction, a compartimentiza-tion[178,233,342] or crevice theory[159] has to be invoked, which necessarily does not allow for scavenging of these radicals. The only feasible approach - as has been done with iron-chelated antibiotics such as bleomycin[16,251,315] - is investigation in model reactions and extrapolation to biological systems.

In the preceding example we mentioned the inability of scavengers to block site-specific radical reactions. However, the fallacy of obtaining accurate information from inhibition/scavenging studies is wide-spread enough to warrant a more detailed repudiation. The efficacy of a certain scavenger is determined by the following factors:

- reaction probabilities of the radicals with compounds in the surrounding medium respectively with scavenger molecules (depending on the reaction rates and the respective concentrations);

- diffusion characteristics of the radicals from their site of generation (influ-enced by the potential presence of diffusion barriers - compartimentization - and the various reaction probabilities);

- distribution of the scavenger in living tissue (which cannot be assumed to be homogenuous)

When these parameters are known, one may calculate the scavenger concentrations which are required to inhibit a given reaction in the bulk solution by 50%. General-ly one arrives at values which are clearly in excess of anything reasonable. Examples for such deliberations are given by Chapman and Reuvers.[67]

A further complication arises, if one considers the case that, say, only 1 in 1000 ·OH radicals, being deposited close to the target molecule, causes a manifest damage, while all the others are scavenged indiscriminately in the bulk of the organism without leaving permanent traces. Considering the size and sensitivity towards radiation damage of the single DNA molecule in a cell (see Ref.336, p.85). such an event may very well be envisaged. At the same time, according to the fact that this is merely a minuscule side reaction, it would be all but practicable to detect by chemical analysis or inhibition studies.

Importance of complex formation: We were faced before with the unarguable facts, that the presently favored reactions for the generation of ·OH radicals – reactions (5) and (7), alternatively denoted "superoxide-promoted Fenton reactions"[95] or "catalyzed Haber-Weiss reaction"[215] – show only a poor kinetic probability to occur with uncomplexed iron. To avoid the same fate as befell the predecessor of these reactions, the uncatalyzed Haber-Weiss reaction, it is of utmost importance to determine whether complex formation of iron indeed stimulates the overall reaction. There are arguments in favor of this assumption. First, the <u>oxidation</u> of H_2O_2 to $HO_2^·/O_2^-$ by ferric iron is accelerated by three orders of magnitude in the presence of a macrocyclic chelator (1.5×10^{-3}, Ref. 331 versus $1.41 \ M^{-1}2^{-1}$, Ref. 222). Second, the Fe^{3+}/ADP complex, in combination with a NADPH-generating source is known as an efficient lipid peroxidation initiator,[212,246,316,321] with H_2O_2 and O_2^- probably supplied by NADPH-cytochrome <u>c</u> (P-450) reductases.[170,191]

Outlook: In the course of this presentation, we offered several alternatives to the involvement of the free oxygen radicals ·OH and O_2^-, based on the fact that the reactivities of these species, generated by physico-chemical methods, do not correlate with their almost regular proposal in biological oxidation reactions. They are summarized in Table 6 as answers to the four problem areas defined in Table 4.

To restate our theme we propose:

(i)   The mediating role of O_2^- can only be achieved by complex-stimulated electron transfer; metal complexes with O_2^- or semequinones can furthermore serve as transport or storage vessels of radical properties.

(ii)   Organic oxygen radicals, in particular alkoxy radicals produced in chain reactions, are superior to ·OH radicals themselves in explaining higher selectivity <u>and</u> delayed reactions after the initial radical-generating process.

(iii)   Enzymatic hydroxlation reactions do not involve either one of the organic or inorganic oxygen radicals. Rather the "oxenoid" entity is inserted into an existing bond in a concerted reaction.

What remains to be established, are more methods to differentiate between the individual proposed radicals, more detailed investigations into the reactivity of organic oxygen radicals, elucidation of the change in properties (e.g. redox potential) during complex formation, lifetime respectively reactivity of oxygen radicals in micellar and membrane systems, etc. As long as such data are not available it is best to assume that physico-chemically determined properties are basically the same in biological systems as they are in aqueous or non-aqueous

TABLE 6

PROBABLE EXPLANATIONS OF BIOLOGICAL OXYGEN ACTIVATION

---

1. Toxicity of superoxide anions:
   few direct target molecules (e.g. SH-compounds, catechols
       quinones, ferric iron, but not hydroperoxides)
   most likely operating as precursor/transmitter
   further stabilization by complex formation enhances chance
       of transmitting damaging properties
   decreased reactivity tantamount to longer lifetime (e.g.
       for diffusion through lipophilic matrix)

---

2. Probability of free ˙OH radicals:
   low probability to exist for appreciable times
   effective only if generated/deposited at site of attack
   most probable generation by Fenton-type reactions or
       radiation
   misconceptions of effective scavenger concentrations

---

3. Nature of ˙OH-mimicking species:
   organic oxygen radicals, in particular alkoxy radicals
       (RO˙)
   reactivity of semiquinones (special type of RO˙)
   reductive activation of hydrogen peroxide
   potential electron transfer in complex

---

4. Intermediates in biological hydroxylation reactions:
   oxenoid mechanisms
   perferryl compounds
   model reactions with hydroperoxides or iodosobenzene
       (insertion of one oxygen atom)

---

solutions. This negates both O_2^- or diffusible free ˙OH radicals as major contributing factors of oxygen toxicity.

ACKNOWLEDGEMENTS

The stimulating discussions with Drs. E.F. Elstner, P. Hemmerich and A. Wessiak
are gratefully acknowledged. We are also very much obliged for the opportunity
to quote from two review articles (Refs. 95, 145), made available to us prior to
publication. Finally we appreciate greatly the continuous and valuable collaboration of Mrs. Christa Micel and Mr. Michael Erben-Russ.

REFERENCES

1. Abramovitch, S. and Rabani, J. (1976) J. Phys. Chem. 80, 1562-1565.
2. Adams, G.E. and Wardman, P. (1977) in Free Radicals in Biology, Vol. III, Pryor, W.A. ed., Academic Press, New York, pp. 53-95.
3. Afanas'ev, I.B., Prigoda, S.V. and Samokhalov, G.I. (1977) J. Gen. Chem. USSR, 47, 2291-2293.
4. Airey, P.L. and Sutton, H.C. (1976) J. Chem. Soc. Faraday Trans. I, 72, 2452-2461.
5. Amano, D., Kagosaki, Y., Usui, T., Yamamoto, S. and Hayaishi, O. (1975) Biochem. Biophys. Res. Commun. 66, 272-279.
6. Ananthaswamy, H.N. and Eisenstark, A. (1976) Photochem. Photbiol. 24, 439-442.
7. Andreasson, L.-E. and Reinhammar, B. (1979) Biochim. Biophys. Acta, 568, 145-156.
8. Arkhipenko, Yu.V., Lobrina, S.K., Kagan, V.E., Kozlov, Yu.P., Nadirov, N.K., Pisarev, VA., Ritov, B.B. and Khafizov, R.K. (1977) Biochemistry Biokhimiya, 42, 1194-1198.
9. Asada, K. and Kanematsu, S. (1976) Agr. Biol. Chem. Tokyo, 40, 1891-1892.
10. Asada, K., Kanematsu, S., Takahashi, M.-A. and Kono, Y. (1976) in Iron and Copper Proteins, Yasunobu, K.T., Mower, H.F. and Hayaishi, O. eds., Plenum Press, New York, pp. 551-564.
11. Ashworth, P. and Dixon, W.T. (1973) J.Chem. Soc., Perkin Trans. II, 2128-2132.
12. Auclair, C., Torres, M. and Hakim, J. (1978) FEBS Lett. 89, 26-28
13. Azzi, A., Montecucco, C. and Richter, C. (1975) Biochem. Biophys. Res. Commun. 65, 597-603.
14. Babior, B.M., Curnutte, J.T. and McMurrich, B.J. (1976) J. Clin. Invest. 58, 989-996.
15. Bachur, N.R., Gordon, S.L. and Gee, M.V. (1978) Cancer Res. 38, 1745-1750.
16. Bachur, N.R., Gordon, S.L., Gee, M.V. and Kon, H. (1979) Proc. Nat. Acad. Sci. USA, 76, 954-957.
17. Baehner, R.L., Boxer, L.A., Allen, J.M. and Davis, J. (1977) Blood, 50, 327-335.
18. Bains, M.S. (1976) J. Indian Chem. Soc. 53, 83-88.
19. Barb, W.G., Baxendale, J.H., George, P. and Hargrave, K.R. (1949) Nature, 163, 692-694.
20. Barb, W.G., Baxendale, J.H., George, P. and Hargrave, K.R. (1951) Trans. Faraday Soc. 47, 462-500.
21. Barb, W.G., Baxendale, J.H., George, P. and Hargrave, (1951) Trans. Faraday Soc. 47, 591-616.
22. Bartosz, G., Leyko, W., Kedziora, J. and Jeske, J. (1977) Int. J. Radiat. Biol. 31, 197-200.
23. Bayer, E., Krauss, P., Röder, A. and Schretzmann, P. (1973) in Oxidase and Related Redox Systems, Vol. I, King, E.T., Mason, H.S. and Morrison, M., eds., University Park Press, Baltimore, pp. 227-263.
24. Beauchamp, C.O. and Fridovich, I. (1970) J. Biol. Chem. 245, 4641-4646.
25. Beauchamp, C.O. and Fridovich, I. (1971) Anal. Biochem. 44, 276-287.
26. Benedetti, A., Casini, A.F. and Ferrali, M. (1977) Res. Commun. Chem. Pathol. Pharm. 17, 519-527.
27. Benedetti, A., Casini, A.F., Ferrali, M. and Comporti, M. (1979) Biochem. J. 180, 303-312.
28. Bennett, L.E. (1973) in Current Research Topics in Bioinorganic Chemistry (Progr. Inorg. Chem., Vol. 18, Lippard, S.J. ed., Wiley-Interscience, New York), pp. 1-176.
29. Benson, S.W. and Shaw, R. (1970) in Organic Peroxides, Vol. I, Swern, D., ed., Wiley-Interscience, New York, pp. 105-139.
30. Bielski, B.H.J. and Gebicki, J.M. (1970) in Advances in Radiation Chemistry, Vol. II, Burton, M. and Magee, J.L., eds., Wiley Interscience, New York, pp. 177-279.
31. Bielski, B.H.J. and Gebicki, J.M. (1974) Biochim. Biophys. Acta, 364 233-235.

32. Bielski, B.H.J. and Chan, P. (1976) J. Biol. Chem. 251, 3841-3844.
33. Bielski, B.H.J. and Richter, H.W. (1977) J. Amer. Chem. Soc. 99, 3019-3023.
34. Bielski, B.H.J. and Chan, P.C. (1977) in Superoxide and Superoxide Dismutases, Michelson, A.M., McCord, J.M. and Fridovich, I. eds., Academic Press, New York, pp. 409-416.
35. Bielski, B.H.J. and Gebicki, J.M. (1977) in Free Radicals in Biology, Vol. III, Pryor, W.A., ed., Academic Press, New York, pp. 1-51.
36. Bielski, B.H.J. and Chan, P.C. (1978) J. Amer. Chem. Soc. 100, 1920-1921.
37. Bielski, B.H.J. (1978) Photochem. Photobiol. 28, 645-649.
37a. Bielski, B.H.J. and Shine, G.C. (1979) in Oxygen Free Radicals and Tissue Damage, Ciba Foundation Symposium, New Series, Vol. 65, Excerpta Medica, Amsterdam, pp. 43-56.
38. Bindoli, A., Cavallini, L. and Siliprandi, N. (1977) Biochem. Pharmcol. 26, 2405-2409.
39. Black, J.F. (1978) J. Amer. Chem. Soc. 100, 527-535.
40. Blair, J.A. and Pearson, A.J. (1975) J. Chem. Soc., Perkin Trans. II, 245-249.
41. Bodini, M.E. and Sawyer, D.T. (1976) J. Amer. Chem. Soc. 98, 8366-8371.
42. Borg, D.C., Schaich, K.M., Elmore, J.J. and Bell, J.A. (1978) Photochem. Photobiol. 28, 887-907.
43. Bors, W., Saran, M., Lengfelder, E., Spöttl, R. and Michel, C. (1974) Curr. Top. Radiat. Res. 9, 247-309.
44. Bors, W., Michel, C., Saran, M. and Lengfelder, E. (1978) Biochim. Biophys. Acta, 540, 162-172.
45. Bors, W., Saran, M., Lengfelder, E., Michel, C., Fuchs, C. and Frenzel, C. (1978) Photochem. Photobiol. 28, 629-638.
46. Bors, W., Michel, C., Saran, M. and Lengfelder, E. (1978) Z. Naturforsch. 33c, 891-896.
47. Bors, W., Saran, M. and Michel, C. (1979) Biochim. Biophys. Acta, 582, 537-542.
48. Bors, W., Michel, C. and Saran, M. (1979) Eur. J. Biochem. 95, 621-627.
49. Bors, W., Saran, M. and Michel, C. (1979) J. Phys. Chem., 83, 2447-2452.
50. Bors, W., Saran, M., Michel, C., Elstner, E.F., Grosch, W. and Laskawy, G. (1979), submitted for publication.
51. Bors, W., Michel, C. and Saran, M. (1979) FEBS Lett., in press.
52. Bothe, E., Schuchmann, M.N., Schulte-Fröhlinde, D. and von Sonntag, C. (1978) Photochem. Photobiol. 28, 639-644.
53. Boveris, A., Cadenas, A. and Stoppani, A.O.M. (1976) Biochem. J. 156, 435-444.
54. Brabham, D.E. and Lee, J. (1976) J. Phys. Chem. 80, 2292-2296.
55. Brahden, R. and Deinum, J. (1978) Biochim. Biophys. Acta, 524, 297-304.
56. Bray, R.C. Mautner, G.M., Fielden, E.M. and Carle, C.I. (1977) in Superoxide and Superoxide Dismutase, Michelson, A.M., McCord, J.M. and Fridovich, I., eds., Academic Press, New York, pp. 62-76.
57. Brown, D.G. and Johnson, W.L. (1979) Z. Naturforsch. 34b, 712-715.
58. Brown, S.B., Hatzikonstantinou, H. and Herries, D.G. (1978) Biochem. J. 174, 901-907.
59. Buettner, G.R., Oberley, L.W. and Chan Leuthauser, S.W.H. (1978) Photochem. Photobiol. 28, 693-695.
60. Buettner, G.R. and Oberley, L.W. (1979) FEBS Lett. 98, 18-20.
61. Butler, J., Jayson, G.G. and Swallow, A.J. (1975) Biochim. Biophys. Acta, 408, 215-222.
62. Cadenas, E., Boveris, A., Ragan, C.I. and Stoppani, A.O.M. (1977) Arch. Biochem. Biophys. 180, 248-257.
62a. Cadet, J. and Theoule, R. (1978) Photochem. Photobiol. 28, 661-667.
63. Carlsson, J., Nyberg, G. and Wrethen, J. (1978) Appl. Environ. Microbiol. 36, 223-229.
64. Cederbaum, A.I., Dicker, E. and Cohen, G. (1978) Biochemistry, 17, 3058-3064.
65. Cederbaum, A.I., Dicker, E., Rubin, E. and Cohen, G. (1979) Biochemistry, 18, 1187-1191.
66. Chang, C.K. and Kuo, M.-S. (1979) J. Amer. Chem. Soc. 101, 3413-3415.

67. Chapman, J.D. and Reuvers, A.P. (1977) in Radioprotection – Chemical Compounds, Biological Means, Locker, A. and Flemming, K. eds., Adler Publ. Co., Chicago, Ill, pp. 9-18.
68. Chedekel, M.R., Smith, S.K., Post, P.W., Pokora, A. and Vessell, D.L. (1978) Proc. Nat. Acad. Sci. USA, 75, 5395-5399.
69. Chenier, J.H.B., Tong, S.G. and Howard, J.A. (1979) Can. J. Chem. 56, 3047-3053.
70. Chern, C.-I. and San Filippo, J. (1977) J. Org. Chem. 42, 178-180.
71. Chern, C.-I., DiCosimo, R., de Jesus, R. and San Filippo, J. (1978) J. Amer. Chem. Soc. 100, 7317-7327.
72. Chevalet, J., Rouelle, F., Gierst, L. and Lambert, J.P. (1972) J. Electroanal. Chem. 39, 201-216.
73. Cohen, G., Heikkila, R.E. and MacNamee, D. (1974) J. Biol. Chem. 249, 2447-2452.
74. Cohen, G. (1978) Photchem. Photobiol. 28, 669-675.
75. Cohen, G. and Cederbaum, A.I. (1979) Science, 204, 66-68.
76. Czapski, G. (1971) J. Phys. Chem. 75, 2957-2967.
77. Czapski, G. (1971) Annu. Rev. Phys. Chem. 22, 171-208.
78. Czapski, G., Samuni, A. and Meisel, D. (1971) J. Phys. Chem. 75, 3271-3280.
79. Czapski, G. and Ilan, Y.A. (1978) Photochem. Photobiol. 28, 651-653.
80. Demopoulos, H.B. (1973) Fed. Proc. 32, 1903-1908.
81. Dionisi, O., Galeotti, T., Terranova, T. and Azzi, A. (1975) Biochim. Biophys. Acta, 403, 292-300.
82. DoCampo, R., Cruz, F.S., Boveris, A., Muniz, R.P.A. and Esquivel, D.M.S. (1979) Biochem. Pharmacol. 28, 723-728.
83. Dorfman, L.M. and Adams, G.E. (1973) NSRDS Report No. 46, U.S. Department of Commerce, Nat. Bur. Standards, Washington, D.C.
84. Duke, F.R. and Haas, T.W. (1961) J. Phys. Chem. 65, 304-306.
85. Dybing, E., Nelson, S.D., Mitchell, J.R., Sasame, H.A. and Gillette, J.R. (1976) Mol. Pharmacol. 12, 911-920.
86. Eastland, G.W. and Symons, M.C.R. (1977) J. Phys. Chem. 81, 1502-1504.
87. Egan, R.W., Paxton, J. and Kuehl, F.A. (1976) J.Biol. Chem. 251, 7329-7335.
88. Elstner, E.F., Saran, M., Bors, W. and Lengfelder, E. (1978) Eur. J. Biochem. 89, 61-66.
89. Estabrook, R.W. and Werringloer, J. (1977) in Microsomes and Drug Oxidations, Ullrich, V., Roots, I., Hildebrandt, A., Estabrook, R.W. and Conney, A.H., eds., Pergamon Press, London, pp.748-757.
90. Farhataziz and Ross, A.B. (1977) NSRDS Report No. 59, U.S. Department of Commerce, Nat. Bur. Standards, Washington, D.C. ·
91. Farrington, J.A., Ebert, M. and Land, E.J. (1978) J. Chem. Soc. Faraday Trans. I, 74, 665-675.
92. Fee, J.A. and Hildenbrand, P.G. (1974) FEBS Lett. 39, 79-82.
93. Fee, J.A. and Valentine, J.S. (1977) in Superoxide and Superoxide Dismutases, Michelson, A.M., McCord, J.M. and Fridovich, I., eds., Academic Press, New York, pp. 19-60.
94. Fee, J.A. and McClune, G.J. (1978) in Mechanism of Oxidizing Enzymes, Singer, T.P. and Ondarza, R.N., eds., Elsevier/North-Holland, Amsterdam, pp. 273-284.
95. Fee, J.A. (1979) in Proc. Third. Int. Symposium on Oxidases and Related Oxidation-Reduction Systems, Albany, New York, in press.
96. Ferrandini, C., Foos, J., Houee, C. and Pucheault, J. (1978) Photochem. Photobiol. 28, 697-700.
97. Fischer, H. (1967) Ber. Bunsenges. Physik. Chem. 71, 685-690.
98. Fish, A. (1970) in Organic Peroxides, Vol. I, Swern, D., ed., Wiley-Interscience, New York, pp. 141-198.
99. Flohe, L., Azzi, A., Loschen, G. and Richter, C. (1977) in Superoxide and Superoxide Dismutases, Michelson, A.M., McCord, J.M. and Fridovich, I., eds. Academic Press, New York, pp. 323-334.
100. Floyd, R.A., Soong, L.M., Stuart, M.A. and Reigh, D.L. (1978) Arch. Biochem. Biophys. 185, 450-457.
101. Floyd, R.A., Soong, L.M., Stuart, M.A. and Reigh, D.L. (1978) Photochem. Photobiol. 28, 857-862.

102. Fong, K.-L., McCay, P.B., Poyer, J.L., Misra, H.P. and Keele, B.B. (1976) Chem.-Biol. Inter. 15, 77-89.
103. Fridovich, I. (1970) J. Biol. Chem. 245, 4053-4057.
104. Fridovich, I. (1976) in Free Radicals in Biology, Vol. I, Pryor, W.A., ed., Academic Press, New York, pp. 239-277.
105. Fridovich, I. (1977) in Biochemical and Medical Aspects of Active Oxygen, Hayaishi, O. and Asada, K., eds., University of Tokyo Press, Tokyo, pp. 171-181.
106. Fridovich, I. and Hassan, H.M. (1979) Trends Biochem. Sci. 4, 113-115.
107. Frimer, A.A. and Rosenthal, I. (1978) Photochem. Photobiol. 28, 711-719.
108. George, P. (1952) Adv. Catalysis, 4, 367-428.
109. Gibian, M.J., Sawyer, D.T., Ungermann, T., Tangpoonpholvivat, R. and Morrison, M.M. (1979) J. Amer. Chem. Soc. 101, 640-644.
110. Gibian, M.J. and Ungermann, T. (1979) J. Amer. Chem. Soc. 101, 1291-1293.
111. Gilbert, B.C., Holmes, R.G.G., Laue, H.A.H. and Norman, R.O.C. (1976) J. Chem. Soc. Perkin Trans. II, 1047-1052.
112. Golberg, B. and Stern, A. (1976) Biochim. Biophys. Acta, 437, 628-632.
112a. Greenstock, C.L. and Ruddock, G.W. (1978) Photochem. Photobiol. 28, 877-880.
113. Groves, J.T. and McCluskey, G.A. (1976) J. Amer. Chem. Soc. 98, 859-861.
114. Güntensperger, M. and Zuberbühler, A.D. (1977) Helv. Chim. Acta, 60, 2584-2594.
115. Gustafsson, J.-A., Hrycay, E.G. and Ernster, L. (1976) Arch. Biochem. Biophys. 174, 440-453.
116. Gustafsson, J.-A., Rondahl, L. and Bergman, J. (1979) Biochemistry, 18, 865-870.
117. Gutteridge, J.M.C. (1978) Res. Commun. Chem. Pathol. Pharm. 22, 563-569.
118. Haber, F. and Weiss, J. (1934) Proc. Roy. Soc., Ser. A, 147, 332-351.
119. Halliwell, B. (1977) Biochem. J. 163, 441-448.
120. Halliwell, B. (1977) Biochem. J. 167, 317-320.
121. Halliwell, B. (1977) in Superoxide and Superoxide Dismutases, Michelson, A.M., McCord, J.M. and Fridovich, I., eds., Academic Press, New York, pp. 335-349.
122. Halliwell, B. (1978) Cell Biol. Int. Rep. 2, 113-128.
123. Halliwell, B. (1978) FEBS Lett. 92, 321-326.
124. Halliwell, B. (1978) FEBS Lett. 96, 238-242.
125. Hamilton, G.A., Workman, R.J. and Woo, L. (1964) J. Amer. Chem. Soc. 86, 3390-3391.
126. Hamilton, G.A. (1964) J. Amer. Chem. Soc. 86, 3391-3392.
127. Hamilton, G.A., Friedman, J.P. and Campbell, P.M. (1966) J. Amer. Chem. Soc. 88, 5266-5268.
128. Hamilton, G.A. (1974) in Molecular Mechanism of Oxygen Activation, Hayaishi, O., ed., Academic Press, New York, pp. 405-451.
129. Harbour, J.R. Chow, V. and Bolton, J.R. (1974) Can. J. Chem. 52. 3549-3556.
130. Harbour, J.R. and Bolton, J.R. (1975) Biochem. Biophys. Res. Commun. 64, 803-809.
131. Harbour, J.R. and Hair, M.L. (1978) J. Phys. Chem. 82, 1397-1399.
132. Harbour, J.R. and Bolton, J.R. (1978) Photochem. Photobiol. 28, 231-234.
133. Hardwick, T.J. (1957) Can. J. Chem. 35, 428-436.
134. Harris, W.R., Bess, R.C., Martell, A.E. and Ridgway, T.H. (1977) J. Amer. Chem. Soc. 99, 2958-2961.
135. Harrison, J.E., Watson, B.D. and Schultz, J. (1978) FEBS Lett. 92, 327-332.
136. Hasegawa, K. and Patterson, L.K. (1978) Photochem. Photobiol. 28, 817-823.
137. Hassan, H.M. and Fridovich, I. (1977) J. Bacteriol. 129, 1574-1583.
138. Hassan, H.M. and Fridovich, I. (1977) J. Bacteriol. 132, 505-510.
139. Hassan, H.M. and Fridovich, I. (1978) J. Biol. Chem. 253, 8143-8148.
140. Hastings, J.W., Eberhard, A., Baldwin, T.O., Nicoli, M.Z., Cline, T.W. and Nealson, K.N. (1973) in Chemiluminescence and Bioluminescence, Cormier, M.J., Hercules, D.M. and Lee, J., eds., Plenum Press, New York, pp. 369-380.
141. Heikkila, R.E., Winston, B., Cohen, G. and Barden, H. (1976) Biochem, Pharmacol. 25, 1085-1092.
142. Heikkila, R.E. and Cabbat, F.S. (1977) Res. Commun. Chem. Pathol. Pharm. 17, 649-662.
143. Held, A.M., Halko, D.J. and Hurst, J.K. (1978) J. Amer. Chem. Soc. 100, 5732-5740.

144. van Hemmen, J.J. and Meuling, W.J.A. (1977) Arch. Biochem. Biophys. 182, 743-748.
145. Hemmerich, P. and Wessiak, A. (1979) in Oxygen - Biochemical and Clinical Aspects, Caughey, W.S., ed., Academic Press, New York, in press.
146. Hendry, D.G. and Schuetzle, D. (1976) J. Org. Chem. 41, 3179-3182.
147. Hiatt, R. (1971) in Organic Peroxides, Vol. II, Swern, D., ed., Wiley-Interscience, New York, pp. 1-56.
148. Hirata, F. and Hayaishi, O. (1977) in Superoxide and Superoxide Dismutases, Michelson, A.M., McCord, J.M. and Fridovich, I., eds., Academic Press, New York, pp. 395-406.
149. Hodgson, E.K. and Fridovich, I. (1975) Biochemistry, 14, 5294-5299.
150. Hodgson, E.K. and Fridovich, I. (1976) Arch. Biochem. Biophys. 172, 202-205.
151. Hodgson, E.K. and Fridovich, I. (1976) Biochim. Biophys. Acta, 430, 182-188.
152. Holroyd, R.A. and Bielski, B.H.J. (1978) J. Amer. Chem. Soc. 100, 5796-5800.
153. Howard, J.A. (1973) in Free Radicals, Vol. II, Kochi, J.K., ed., Wiley-Interscience, New York, pp. 5-10.
154. Howard, J.A. and Tait, J.C. (1978) Can. J. Chem. 56, 176-183.
155. Ilan, Y., Rabani, J. and Henglein, A. (1976) J. Phys. Chem. 80, 1558-1562.
156. Ilan, Y.A. and Czapski, G. (1977) Biochim. Biophys. Acta, 498, 386-394.
157. Ingall, A., Lott, K.A.K., Slater, T.F., Finch, S. and Stier, A. (1978) Biochem. Soc. Trans. 6, 962-964.
158. Isbell, H.S. and Frush, H.L. (1977) Carbohyd. Res. 59, C25-C31.
159. Itano, H.A. Hirota, K. and Vedvick, T.S. (1977) Proc. Nat. Acad. Sci. USA, 74, 2556-2560.
160. Jacob, N., Balakrishnan, I. and Reddy, M.P. (1977) J. Phys. Chem. 81, 17-22.
161. Janzen, E.G. (1971) Account. Chem. Res. 4, 31-40.
162. Janzen, E.G., Wang, Y.Y. and Shetty, R.V. (1978) J. Amer. Chem. Soc. 100, 2923-2925.
163. Janzen, E.G., Nutter, D.E., Davis, E.R., Blackburn, B.J., Poyer, J.L. and McCay, P.B. (1978) Can. J. Chem. 56, 2237-2242.
164. Janzen, E.G., Evans, C.A. and Davis, E.R. (1978) ACS Symp. Ser. 69, 433-445.
165. Janzen, E.G., Dudley, R.L. and Shetty, R.V. (1979) J. Amer. Chem. Soc. 101, 243-245.
166. Jayson, G.G., Parsons, B.J. and Swallow, A.J. (1973) J. Chem. Soc., Faraday Trans. I, 69, 236-242.
167. Johnson, R.A. Nidy, E.G. and Merritt, M.V. (1978) J. Amer. Chem. Soc. 100, 7960-7966.
168. Jones, R.D., Summerville, D.A. and Basolo, F. (1979) Chem. Rev. 79, 139-179.
169. Joshi, A., Rustgi, S., Moss, H. and Riesz, P. (1978) Int. J. Radiat. Biol. 33, 205-229.
170. Kameda, K., Ono, T. and Imai, Y. (1979) Biochim. Biophys. Acta, 572, 77-82.
171. Kaneko, H., Nozaki, K. and Ozawa, T. (1978) J. Electroanal. Chem. 87, 149-153.
172. Kappus, H., Kieczka, H., Scheulen, M. and Remmer, H. (1977) Arch. Pharmacol. 300, 179-187.
173. Kemal, C., Chan, T.W. and Bruice, T.C. (1977) J. Amer. Chem. Soc. 99, 7272-7286.
174. Khan, A.U. (1978) Photochem. Photobiol. 28, 615-627.
175. Kim, S., DiCosimo, R. and San Fillippo, J. (1979) Anal. Chem. 51, 679-681.
176. Kochi, J.K. (1973) in Free Radicals, Vol. I, Kochi, J.K., ed., Wiley-Intersciences, New York, pp. 628-642.
177. Kochi, J.K. (1973) in Free Radicals, Vol. II, Kochi, J.K., ed., Wiley-Intersciences, New York, pp. 665-710.
178. Koch-Schmidt, A.-C., Mattiasson, B. and Mosbach, K. (1977) Bur. J. Biochem. 81, 71-78.
179. Koppenol, W.H. and Butler, J. (1977) FEBS Lett. 83, 1-6.
180. Koppenol, W.H., Butler, J. and van Leeuwen, J.W. (1978) Photochem. Photobiol. 28, 655-660.
181. Kotelevtseva, N.V., Kagan, V.E., Lankin, V.Z. and Kozlov, Yu.P. (1976) Vopr. Med. Khim. 22, 395-400.

182. Kozlov, Yu.P. and Berdnikov, V.M. (1973) J. Phys. Chem. USSR, 47, 338-340.
183. Kremer, M.L. and Stein, G. (1959) Trans. Faraday Soc. 55, 959-973.
184. Kremer, M.L. (1963) Trans. Faraday Soc. 59, 2535-2542.
185. Kudo, S. and Iwase, A. (1979) Bull. Chem. Soc. Japan, 52, 908-910.
186. Kumar, A.A., Rao, B.S.S.R., Vaidyanathan, C.S. and Rao, N.A. (1975) Ind. J. Biochem. Biophys. 12, 163-167.
187. Kurz, M.E. and Pryor, W.A. (1978) J. Amer. Chem. Soc. 100, 7953-7959.
188. Kuthan, H., Tsuji, H., Graf, H., Ullrich, V., Werringloer, J. and Estabrook, R.W. (1978) FEBS Lett. 91, 343-345.
189. Kuz'min, V.A., Khudyakov, I.V., Popkov, A.V. and Koroli, L.L. (1975) Bull. Acad. Sci. USSR, 2319-2322.
190. Lai, C.-S. and Piette, L.H. (1977) Biochem. Biophys. Res. Commun. 78, 51-59.
191. Lai, C.-S., Grover, T.A. and Piette, L.H. (1979) Arch. Biochem. Biophys. 193, 373-378.
192. Lai, E.K., Fong, K.-L. and McCay, P.B. (1978) Biochim. Biophys. Acta, 528, 497-506.
193. Lapidot, A. and Irving, C.S. (1974) in Molecular Oxygen in Biology, Hayaishi, O., ed., North-Holland Publishing Co., Amsterdam, pp. 33-80.
194. Lavelle, F., McAdam, M.E., Fielden, E.M., Roberts, P.B., Puget, K. and Michelson, A.M. (1977) Biochem. J. 161, 3-11.
195. Lee, J. and Murphy, C.L. (1975) Biochemistry, 14, 2259-2265.
196. Lee-Ruff, E. (1977) Chem. Soc. Rev. 6, 195-214.
197. van Leeuwen, J.W., Tromp, J. and Nauta, H. (1979) Biochim. Biophys. Acta, 577, 394-399.
198. Lever, A.B.P. and Gray, H.B. (1978) Accounts. Chem. Res. 11, 348-355.
199. Lichtenberger, F., Nastainczyk, W. and Ullrich, V. (1976) Biochem. Biophys. Res. Commun. 70, 939-946.
200. Lin, W.S., Lal, M., Gaucher, G.M. and Armstrong, D.A. (1977) Faraday Disc. 63, 226-236.
201. Lin, W.S., Armstrong, D.A. and Lal, M. (1978) Int. J. Radiat. Biol. 33, 231-243.
202. Lindsay, D., Howard, J.A., Horswill, E.C., Iton, L., Ingold, K.U., Cobbley, T. and Ll, A. (1973) Can. J. Chem. 51, 870-880.
203. Lindsay Smith, J.R., Shaw, B.A.J., Foulkes, D.M., Jeffrey, A.M. and Jerina, D.M. (1977) J. Chem. Soc. Perkin Trans. II, 1583-1589.
204. Lown, J.W., Sim, S.-K. and Chen, H.-H. (1978) Can. J. Biochem. 56, 1042-1047.
205. Lyakhovich, V.V., Pospelova, L.N., Mishin, V.M. and Pokrovsky, A.G. (1976) FEBS Lett. 71, 303-305.
206. Lynch, R.E. and Fridovich, I. (1978) J. Biol. Chem. 253, 1838-1845.
207. Mager, H.I.X. (1976) in Flavins and Flavoproteins, Singer, T.P., ed., Elsevier, Amsterdam, pp. 23-37.
208. Martell, A.E. (1971) in Stability Constants of Metal-Ion Complexes, Suppl. I, Part II, Special Publ. No. 25, The Chemistry Society, London.
209. Martell, A.E. and Taqui Khan, M.M. (1973) in Inorganic Biochemistry, Vol. II, Eichhorn, G.L., ed., Elsevier, Amsterdam, pp. 654-688.
210. Massey, V., Palmer, G. and Ballou, D. (1973) in Oxidases and Related Redox Systems, Vol. I, King, T.E., Mason, H.S. and Morrison, M.M., eds., University Park Press, Baltimore, pp. 25-49.
211. Matsuura, T. (1977) Tetrahedron, 33, 2869-2905.
212. McCay, P.B., Fong, K.-L., Lai, E.K. and King, M.M. (1978) in Tocopherol, Oxygen and Biomembranes, de Duve, E. and Hayaishi, O. eds., Elsevier/North-Holland, Amsterdam, pp. 41-57.
212a. McClune, G.J. and Fee, J.A. (1976) FEBS Lett. 67, 294-298.
213. McClune, G.J., Fee, J.A., McClusky, G.A. and Groves, J.T. (1977) J. Amer. Chem. Soc. 99, 5220-5222.
213a. McClune, G.J. and Fee, J.A. (1978) Biophys. J. 24, 65-69.
214. McCord, J.M., Keele, B.B. and Fridovich, I. (1971) Proc. Nat. Acad. Sci. USA, 68, 1024-1027.
215. McCord, J.M. and Day, E.D. (1978) FEBS Lett. 86, 139-142.
216. McIsaac, J.E., Subbaraman, L.R., Subbaraman, J., Mulhausen, H.A. and Behrman, E.J. (1972) J. Org. Chem. 37, 1037-1039.

28

217. Mead, J.F. (1976) in Free Radicals in Biology, Vol. I, Pryor, W.A., ed., Academic Press, New York, pp. 51-68.
218. Meisel, D., Czapski, G. and Samuni, A. (1973) J. Chem. Soc., Perkin Trans. II, 1702-1708.
219. Meisel, D., Ilan, Y.A. and Czapski, G. (1974) J. Phys. Chem. 78, 2330-2334.
220. Meisel, D. and Fessenden, R.W. (1976) J. Amer. Chem. Soc. 98, 7505-7506.
221. Melhuish, W.H. and Sutton, H.C. (1978) J. Chem. Soc. Chem. Commun., 970-971.
222. Melnyk, A.C., Kildahl, N.K., Rendina, A.R. and Busch, D.H. (1979) J. Amer. Chem. Soc. 101, 3232-3240.
223. Merritt, M.V. and Johnson, R.A. (1977) J. Amer. Chem. Soc. 99, 3713-3719.
224. Michelson, A.M. (1973) Biochimie, 55, 925-942.
225. Michelson, A.M. (1977) in Superoxide and Superoxide Dismutases, Michelson, A.M., McCord, J.M. and Fridovich, I., eds., Academic Press, New York, pp. 245-256.
226. Michelson, A.M. (1977) in Biochemical and Medical Aspects of Active Oxygen, Hayaishi, O. and Asada, K., eds., University of Tokyo Press, Tokyo, pp. 155-170.
227. Micic, O.I. and Nenadovic, M.T. (1976) J. Phys. Chem. 80, 940-944.
228. Mieyal, J.J., Ackerman, R.S., Blumer, J.L. and Freeman, L.S. (1976) J. Biol. Chem. 251, 3436-3441.
229. Miller, R.W. (1970) Can. J. Biochem. 48, 935-939.
230. Miller, R.W. and Rapp, U. (1973) J. Biol. Chem. 248, 6084-6090.
231. Mino, M. and Sugita, K. (1978) in Tocopherol, Oxygen and Biomembranes, de Duve, C. and Hayaishi, O., eds., Elsevier/North-Holland, Amsterdam, pp. 71-81.
232. Misra, H.P. and Fridovich, I. (1976) Arch. Biochem. Biophys. 176, 577-581.
233. Mitchell, P. (1979) Eur. J. Biochem. 95, 1-20.
234. Mittal, C.K. and Murad, F. (1977) Proc. Nat. Acad. Sci. USA, 74, 4360-4364.
235. Miura, T., Ogawa, N. and Ogiso, T. (1978) Chem. Pharm. Bull. 26, 1261-1266.
236. Miura, T. and Ogiso, T. (1978) Chem. Pharm. Bull. 26, 3540-3545.
237. Molenaar, I., Vos, J. and Hommes, F.A. (1972) in Vitamins and Hormones, Vol. 30, Harrs, R.S., ed., Academic Press, New York, pp. 45-81.
238 Morgan, A.R., Cone, R.L. and Elgert, T.M. (1976) Nucleic Acid Res. 3, 1139-1149.
239. Moro-oka, Y. and Foote, C.S. (1976) J. Amer. Chem. Soc. 98, 1510-1514.
240. Nangia, P.S. and Benson, S.W. (1979) J. Phys. Chem. 83, 1138-1142.
241. Niehaus, W.G. (1978) Bioorganic Chem. 7, 77-84.
242. Nishikimi, M. and Yagi, K. (1977) in Biochemical and Medical Aspects of Active Oxygen, Hayaishi, O. and Asada, K., eds., University of Tokyo Press, Tokyo, pp. 79-87.
242a. Nishinaga, A. (1977) in Biochemical and Medical Aspects of Active Oxygen, Hayaishi, O. and Asada, K., eds., University of Tokyo Press, Tokyo, pp. 13-27.
243. Niwa, T., Yamaguchi, H. and Yano, K. (1977) in Biochemical and Medical Aspects of Active Oxygen, Hayaishi, O. and Asada, K., eds., University of Tokyo Press, Tokyo, pp. 209-225.
244. Niwa, T., Yamaguchi, H. and Yano, K. (1978) Agr. Biol. Chem. Tokyo, 42, 689-695.
245. Nofre, C., Cier, A. and Lefier, A. (1961) Bull. Soc. Chim. France, 530-535.
246. Noguchi, T. and Nakano, M. (1974) Biochim. Biophys. Acta, 368, 446-455.
247. Nordblom, G.D., White, R.E. and Coon, M.J. (1976) Arch. Biochem. Biophys. 175, 524-533.
248. Nordenson, I. (1977) Hereditas, 86, 147-150.
249. Nordenson, I. (1978) Hereditas, 89, 163-167.
250. Oberley, L.W., Lindgren, A.L., Baker, S.A. and Stevens, R.H. (1976) Radiat. Res. 68, 320-328.
251. Oberley, L.W. and Buettner, G.R. (1979) FEBS Lett. 97, 47-49.
252. O'Brien, P.J. (1978) Pharmacol. Ther. A 2, 517-536.
253. Ohnishi, K. and Lieber, C.S. (1978) Arch. Biochem. Biophys. 191, 798-803.
254. Olsen, L.F. (1978) Biochim. Biophys. Acta, 527, 212-220.
255. Olsen, J.S., Ballou, D.P., Palmer, G. and Massey, V. (1974) J. Biol. Chem. 249, 4350-4362.

256. Orhanovic, M. and Wilkins, R.G. (1967) J. Amer. Chem. Soc. 89, 278-282.
257. Orhanovic, M. and Wilkins, R.G. (1967) Croat. Chem. Acta, 39, 149-154.
258. Ozawa, T., Hanaki, A. and Yamamoto, H. (1977) FEBS Lett. 74, 99-102.
259. Ozawa, T. and Hanaki, A. (1978) Chem. Pharm. Bull. Tokyo, 26, 2572-2575.
260. Packer, J.E., Slater, T.F. and Willson, R.L. (1978) Life. Sci. 23, 2617-2620.
261. Pandell, A.J. (1976) J. Org. Chem. 41, 3992-3996.
262. Panganamala, R.V., Gavino, V.C. and Cornwell, D.G. (1979) Prostaglandins, 17, 155-163.
263. Pasternack, R.F. and Halliwell, B. (1979) J. Amer. Chem. Soc. 101, 1026-1031.
264. Patel, K.B. and Willson, R.L. (1973) J. Chem. Soc. Faraday Trans. I, 69, 814-825.
265. Patterson, L.K., Small, R.D. and Scaiano, J.C. (1977) Radiat. Res. 72, 218-225.
266. Pedersen, J.A. (1973) J. Chem., Perkin Trans. II, 424-431.
267. Peters, J.W. and Foote, C.S. (1976) J. Amer. Chem. Soc. 98, 873-875.
268. Petkau, A. and Chelack, W.S. (1974) Int. J. Radiat. Biol. 26, 421-426.
269. Petkau, A. and Chelack, W.S. (1976) Biochim. Biophys. Acta, 433, 445-456.
270. Petkau, A. (1978) Photochem. Photobiol. 28, 765-774.
271. Porter, N.A., Funk, M.O., Gilmore, D., Isaac, R. and Nixon, J. (1976) J. Amer. Chem. Soc. 98, 6000-6005.
272. Pryor, W.A. (1976) in Free Radicals in Biology, Vol. I, Pryor, W.A., ed., Academic Press, New York, pp. 1-49.
273. Pryor, W.A. and Tang, R.H. (1978) Biochem. Biophys. Res. Commun. 81, 498-503.
274. Pryor, W.A. (1978) Photchem. Photobiol. 28, 787-801.
275. Raleigh, J.A. and Kremers, W. (1978) Int. J. Radiat. Biol. 34, 439-447.
276. Rao, P.S. and Hayon, E. (1975) J. Phys. Chem. 79, 397-402.
277. Rapp, U., Adams, W.C. and Miller, R.W. (1973) Can. J. Biochem. 51, 158-171.
278. Razuvaev, G.A., Shal'nova, K.G., Abakumova, L.G. and Abakumov, G.A. (1977) Bull. Acad. Sci. USSR, 26, 1512-1515.
279. Reed, C.A. and Cheung, S.K. (1977) Proc. Nat. Acad. Sci. USA, 74, 1780-1784.
280. Rhaese, H.-J. and Freese, E. (1968) Biochim. Biophys. Acta, 155, 476-490.
281. Roberts, J.L., Morrison, M.M. and Sawyer, D.T. (1978) J. Amer. Chem. Soc. 100, 329-330.
282. Roders, M.K., Glende, E.A. and Recknagel, R.O. (1978) Biochem. Pharmacol. 27, 437-443.
283. Rogic, M.M. and Demmin, T.R. (1978) J. Amer. Chem. Soc. 100, 5472-5487.
284. Russell, G.A. (1957) J. Amer. Chem. Soc. 79, 3871-3877.
285. Salin, M.L. and McCord, J.M. (1977) in Superoxide and Superoxide Dismutases, Michelson, A.M., McCord, J.M. and Fridovich, I., eds., Academic Press, New York, pp. 257-270.
286. Samuni, A. and Czapski, G. (1970) Israel J. Chem. 8, 563-573.
287. Samuni, A. and Czapski, G. (1970) J. Phys. Chem. 74, 4592-4594.
288. Samuni, A., Chevion, M., Halpern, Y.S. Ilan, Y.A. and Czapski, G. (1978) Radiat. Res. 75, 489-496.
289. Samuni, A. and Czapski, G. (1978) Radiat. Res. 76, 624-632.
290. San Filippo, J., Chern, C.-I. and Valentine, J.S. (1976) J. Org. Chem. 41, 1077-1078.
291. Saran, M., Bors, W., Michel, C. and Elstner, E.F. (1979), submitted for publication.
292. Saran, M., Michel, C. and Bors, W., this publication, Vol. I.
293. Sawyer, D.T., Gibian, M.J., Morrison, M.M. and Seo, E.T. (1978) J. Amer. Chem. Soc. 100, 627-628.
294. Schastnev, P.V., Bashurova, V.S., Moskovskaya, T.E. and Luk'kanov, A.E. (1975) J. Struct. Chem. 16, 687-693.
295. Scheller, F., Renneberg, R., Mohr, P., Jänig, G.-R. and Ruckpaul, K. (1976) FEBS Lett. 71, 309-313.
296. Schenkman, J.B., Jansson, I., Powis, G. and Kappus, H. (1979) Mol. Pharmacol. 15, 428-438.
297. Schonbaum, G.R. and Chance, B. (1976) in The Enzymes, Vol. XIII, Boyer, P.D., ed., Academic Press, New York, pp. 363-408.

30

298. Schwarzenbach, G. and Heller, J. (1951) Helv. Chim. Acta, 34, 576-591.
299. Scott, O.C.A. and Revesz, L. (1974) in Molecular Oxygen in Biology, Hayaishi, O., ed., North-Holland Publ. Co., Amsterdam, pp. 137-162.
300. Shafferman, A. and Stein, G. (1975) Biochim. Biophys. Acta, 416, 287-317.
301. Sharman, D.F. (1973) Brit. Med. Bull. 29, 110-115.
302. Shinohara, H., Imamura, A., Masuda, T. and Kondo, M. (1978) Bull. Chem. Soc. Japan, 51, 98-102.
303. Shinohara, H., Imamura, A., Masuda, T. and Kondo, M. (1979) Bull. Chem. Soc. Japan, 52, 1-7.
304. Shubin, V.N. and Dolin, P.I. (1963) Radiat. Res. 19, 345-358.
305. Shuvalov, V.F., Moravskii, A.P. and Lebedev, Y.S. (1977) Proc. Acad. Sci. USSR, 235, 778-781.
306. Simic, M. and Hayon, E. (1973) Biochem. Biophys. Res. Commun. 50, 364-369.
307. Simic, M. (1975) in Fast Processes in Radiation Chemistry and Radiation Biology, Adams, G.E., Fielden, E.M. and Michael, B.B., eds., Inst. Physics and Wiley and Sons, New York, pp. 162-179.
308. Simic, M.G. and Taub, I.A. (1977) Faraday Disc. 63, 270-278.
309. Singh, A. and Koppenol, W.H. (1978) Photochem. Photobiol. 28, 429-433.
310. Sligar, S.G., Shastry, B.S. and Gunsalus, I.C. (1977) in Microsomes and Drug Oxidations, Ullrich, V., Roots, I., Hildebrandt, A., Estabrook, R.W. and Conney, A.H., eds., Pergamon Press, London, pp. 202-209.
311. Small, R.D., Scaiano, J.C. and Patterson, L.K. (1979) Photochem. Photobiol. 29, 49-51.
312. Smith, L.L. and Kulig, M.J. (1976) J. Amer. Chem. Soc. 98, 1027-1029.
313. Soda, K., Kido, T. and Asada, K. (1977) in Biochemical and Medical Aspects of Active Oxygen, Hayaishi, O. and Asada, K., eds., University of Tokyo Press, Tokyo, pp. 119-133.
314. Staudinger, H.-J. and Ullrich, V. (1964) Z. Naturforsch, 19b, 409-413.
315. Sugiura, Y. and Kikuchi, T. (1978) J. Antibiot. 31, 1310-1312.
316. Svingen, B.A., O'Neal, F.O. and Aust. S.D. (1978) Photochem. Photobiol. 28, 803-809.
317. Takakura, K. and Ranby, B. (1968) J. Phys. Chem. 72, 164-168.
318. Tauber, A.I., Gabig, T.G. and Babior, B.M. (1979) Blood, 53, 666-676.
319. Thomas, M.J., Mehl, K.S. and Pryor, W.A. (1978) Biochem. Biophys. Res. Commun. 83, 927-932.
320. Tsuji, J. and Takayanagi, H. (1978) Tetrahedron, 34, 641-644.
321. Tyler, D.D. (1975) FEBS Lett. 51, 180-183.
322. Udenfriend, S., Clark, C.T., Axelrod, J. and Brodie, B.B. (1954) J. Biol. Chem. 208, 731-740.
323. Uemura, T., Chiesara, E. and Cova, D. (1977) Mol. Pharmacol. 13, 196-215.
324. Valentine, J.S. (1973) Chem. Rev. 73, 235-245.
325. Valentine, J.S. and Curtis, A.B. (1975) J. Amer. Chem. Soc. 97, 224-226.
326. Valentine, J.S. and Quinn, A.E. (1976) Inorg. Chem. 15, 1997-1999.
327. Valentine, J.S., Tatsuno, Y. and Nappa, M. (1977) J. Amer. Chem. Soc. 99, 3522-3523.
328. Vaska, L. (1976) Account. Chem. Res. 9, 175-183.
329. Veldink, G.A., Vliegenthart, J.F.G. and Boldingh, J. (1977) Progr. Chem. Fats Lipids 15, 131-166.
330. Walling, C., Kurz, M. and Schugar, H.J. (1970) Inorg. Chem. 9, 931-937.
331. Walling, C., Goosen, A. (1973) J. Amer. Chem. Soc. 95, 2987-2991.
332. Walling, C., Partch, R.E. and Weil, T. (1975) Proc. Nat. Acad. Sci. USA, 72, 140-142.
333. Walling, C. (1975) Account. Chem. Res. 8, 125-131.
334. Waltz, W.L., Woods, R.J. and Whitburn, K.D. (1978) Photochem. Photobiol. 28, 681-635.
335. Waters, W.A. (1971) J. Amer. Oil Chem. Soc. 48, 427-433.
336. Watson, J.D. (1965), Molecular Biology of the Gene, W.A. Benjamin, Inc., New York.
337. Weinstein, J. and Bielski, B.H.J. (1979) J. Amer. Chem. Soc. 101, 58-62.
338. Weiss, S.J., King, G.W. and LoBuglio, A.F. (1977) J. Clin. Invest. 60, 370-373.

339. Weiss, S.J., Rustagi, P.K. and LoBuglio, A.F. (1978) J. Exp. Med. 147, 316-323.
340. Weser, U. and Schubotz, G. (1979) Bioinorg. Chem. 9, 505-512.
341. Willson, R.L. (1977) Chem. Industry, 183-193.
342. Willson, R.L. (1977) New Scientist 76, 558-560.
343. Wilshire, J. and Sawyer, D.T. (1979) Account. Chem. Res. 12, 105-110.
344. Winterbourn, C.C., French, J.K. and Claridge, R.F.C. (1978) FEBS Lett. 94, 269-272.
345. Winterbourn, C.C., French, J.K. and Clardige, R.F.C. (1979) Biochem. J. 179, 665-673.
346. Yamamoto, H., Takei, H., Yamamoto, T. and Kimura, M. (1979) Chem. Pharm. Bull. 27, 789-792.
347. Yamazki, I. (1974) in Molecular Mechanism of Oxygen Activation, Hayaishi, O., ed., Academic Press, New York, pp. 535-558.
348. Zehavi, D. and Rabani, J. (1972) J. Phys. Chem. 76, 3703-3709.
349. Zhuravleva, O.S. and Berdnikov, V.M. (1976) J. Phys. Che. USSR, 50, 1369-1372.
350. Zuberbühler, A.D. (1976) Helv. Chim. Acta, 59, 1448-1459.

THE BIOLOGICAL SIGNIFICANCE OF THE HABER-WEISS REACTION

BARRY HALLIWELL, RAMSAY RICHMOND, SUSAN F. WONG AND JOHN M.C. GUTTERIDGE
Department of Biochemistry, University of London King's College, London WC2 2LS, England

The superoxide radical, O_2^-, is formed in virtually all aerobic organisms. In aqueous solution O_2^- can act as a powerful reducing agent, giving up its extra electron, or as a weak oxidising agent, becoming reduced to H_2O_2. For example, it reduces cytochrome c but oxidises ascorbic acid, adrenalin and thiols, although the rate of this last oxidation is slow.[1,2] It also reacts with the metal ion Mn^{2+} to produce a more powerful oxidant. For example, O_2^- itself only slowly oxidises NADH but there is a rapid oxidation if Mn^{2+} is present.[3] The product of reaction of Mn^{2+} with O_2^- is probably the species MnO_2^+.[4] The protonated form of O_2^-, written as $HO_2^{\bullet}$, is a much more powerful oxidant than in O_2^- (Ref.5), although since pK_a for the dissociation reaction

$$HO_2^{\bullet} \rightleftharpoons H^+ + O_2^- \tag{1}$$

is 4.88 there is little $HO_2^{\bullet}$ present at physiological pH values. Superoxide in aqueous solution undergoes dismutation. Although the overall process may be represented by the equation

$$2O_2^- + 2H^+ \longrightarrow H_2O_2 + O_2 \tag{2}$$

the direct reaction of two O_2^- radicals is in fact a very slow process, probably because of electrostatic repulsion. Dismutation actually occurs by the formation of $HO_2^{\bullet}$ followed by the reactions shown below (Ref.1):

$$HO_2^{\bullet} + O_2^- + H^+ \longrightarrow H_2O_2 + O_2 \quad k = 8 \times 10^7 M^{-1}s^{-1} \tag{3}$$

$$HO_2^{\bullet} + HO_2^{\bullet} \longrightarrow H_2O_2 + O_2 \quad k = 8 \times 10^5 M^{-1}s^{-1} \tag{4}$$

Dismutation is thus most rapid at the acidic pH values needed to protonate O_2^- and is much slower at physiological pH. As a result, O_2^- formed in vivo will have a significant lifetime, which allows it to diffuse away from its site of formation and to reach other structures within the cell.

In comparison with other oxygen-derived species such as hydroxyl radicals or singlet oxygen, O_2^- is not a highly reactive entity. In particular it is incapable of reacting directly with membrane lipids to produce lipid peroxides.[6] Yet systems generating O_2^- have been observed to kill bacteria, inactivate viruses, damage enzymes and membranes, induce lipid peroxidation and destroy animal cells in culture.[1] Because of the low reactivity of O_2^-, such deleterious effects must be caused by some more reactive species derived from O_2^-. There are two principal candidates: hydroxyl radicals (OH$\cdot$) and singlet oxygen $^1\Delta$g. The latter species has been suggested to be produced either during dismutation of O_2^- (eqns. 3 and 4) or during an iron-catalysed Haber-Weiss reaction (see below). Results are very confused, since most studies have employed "singlet O_2 scavengers" that are not specific for reaction with singlet O_2, such as 2,5-diphenylfuran, histidine, diphenylisobenzofuran or 1,4-diazabicyclooctane. Superoxide itself at high concentrations acts as a singlet O_2 scavenger. Before singlet oxygen generation can be accepted as a mechanism to explain the damaging effects of O_2^-, there must be a clear demonstration of its production by a O_2^--generating system. Since to the best of my knowledge this has not been achieved, I do not intend to discuss singlet O_2 further here.

Generation of the hydroxyl radical from superoxide. Many of the deleterious effects of O_2^--generating systems are decreased by scavengers of the hydroxyl radical (e.g. mannitol, formate, thiourea, alcohols) to extents that can be correlated with the rate constants for reaction of the scavengers with OH$\cdot$. Damage can also often be prevented by superoxide dismutase (SOD) as well as by catalase. Assuming that the catalase samples are free of SOD, which must be checked,[7] it may be concluded that H_2O_2, OH$\cdot$ and O_2^- are all involved in the damage with OH$\cdot$ being the most likely deleterious entity because of its high and indiscriminate reactivity. It was originally suggested[1,2] that O_2 and H_2O_2 could react together to generate OH$\cdot$ as shown below

$$O_2^- + H_2O_2 \longrightarrow OH^- + OH\cdot + O_2 \qquad (5)$$

Reaction (5) is often referred to as the Haber-Weiss reaction. Unfortunately, the rate constant for this reaction is very small and it certainly could not occur at the low concentrations of O_2^- and H_2O_2 likely to be present in vivo.[8-12] The Haber-Weiss reaction is thermodynamically possible, however,[13] and it is therefore amenable to catalysis. Since transition metal ions, especially iron, are present in significant amounts in living systems, several workers have proposed that such ions could catalyse reaction (5) (Refs.14-17). For example, iron complexes react rapidly with O_2^- (Ref.18) and one can envisage a mechanism such as the following generation of OH$\cdot$ radicals

$$\text{Fe}^{3+}\text{-complex} + O_2^- \longrightarrow \text{Fe}^{2+}\text{-complex} + O_2 \quad (O_2^- \text{ acting as a reductant}) \qquad (6)$$

$$\text{Fe}^{2+}\text{-complex} + H_2O_2 \longrightarrow \text{Fe}^{3+}\text{-complex} + OH\cdot + OH^- \quad \text{(Fenton-type reaction)} \qquad (7)$$

Net: $\qquad\qquad\qquad O_2^- + H_2O_2 \longrightarrow O_2 + OH\cdot + OH^- \qquad (5)$

McCord and Day[19] investigated the feasibility of reactions (6) and (7) by measuring UV-absorbance changes due to the reaction of OH$\cdot$ with tryptophan, and they were able to demonstrate formation of OH$\cdot$ by a O_2^--generating system to which iron chelates had been added. However, the absorbance changes observed were small and the authors' attempts to develop this system further were hampered by the UV-absorbance changes of the iron chelates themselves as they react with O_2^- (Ref.20). Spin-trapping methods were also successful in detecting reactions (6) and (7) (Ref.21) but this technique is very prone to artefacts.[22]

Hydroxyl radicals can be detected more sensitively by their ability to hydroxylate aromatic compounds and, in an attempt to confirm the results of McCord and Day,[19] we studied the hydroxylation of such compounds by a O_2^--generating system.[23] When 2-hydroxybenzoate (salicylate) was incubated with xanthine and xanthine oxidase at pH 7.4, no formation of bihydroxylated products was detected. However, addition of small amounts of Fe^{2+} or Fe^{3+} salts caused significant hydroxylation. Hydroxylation of salicylate in the Fe^{2+}/xanthine/ xanthine oxidase system was stimulated by addition of EDTA, since iron-EDTA chelates react with O_2^- and with H_2O_2 more rapidly than do the unchelated salts.[18] However it was inhibited by SOD or by catalase (Table 1). It was also inhibited by scavengers of $OH^{\cdot}$ (Table 2) and the degree of inhibition could be correlated with the published rate constants for reaction of $OH^{\cdot}$ with the scavengers. Urea and KCl, which react poorly with $OH^{\cdot}$ radicals, did not inhibit.

TABLE 1

EFFECTS OF SUPEROXIDE DISMUTASE AND CATALSE ON HYDROXYLATION BY XANTHINE/XANTHINE OXIDASE IN THE PRESENCE OF IRON-SALTS[a]

Iron salt added	Inhibitor added	nmol diphenol produced/h	% Inhibition
$FeSO_4$	None	109	0
	Superoxide dismutase		
	10 units	63	42
	20 units	51	53
	50 units	44	60
	100 units	11	90
	Heated superoxide dismutase		
	100 units[b]	108	1
	Catalase		
	18 units	55	50
	36 units	32	71
	93 units	11	90
	150 units	0	100
	Heated catalase		
	150 units[b]	110	0
$FeCl_3$	None	110	0
	Superoxide dismutase		
	100 units	17	85
	Catalse		
	150 units	0	100

[a]Reaction mixtures were described in Ref.23; units of SOD were as defined (Ref.24); units of catalase as in (Ref.23).

[b]Activity before heating the enzyme.

At lower (and more physiological) concentrations of Fe^{2+}, $OH^.$ can still be generated by reaction (9), but to obtain significant amounts of this radical it is necessary to recycle the Fe^{3+} so produced back to Fe^{2+}. This can be achieved by O_2^- (reaction (6): reversal of reaction (8) at low $[Fe^{2+}]$), so producing an "iron-catalysed Haber-Weiss reaction" (eqn. 5) which is inhibited by SOD. Other physiological reductants such as ascorbate[29] might also recycle Fe^{3+} in this way, although ascorbate is apowerful scavenger of $OH^.$ (Ref.25) and its net effect in vivo is therefore difficult to predict.

Our results confirm and extend the work of others on different systems[9,14,21] and the overall conclusion is that an iron-catalysed Haber-Weiss reaction is a feasible way of making $OH^.$ in biological systems, given their significant content of iron compounds. The extent to which SOD inhibits the $OH^.$ generation will depend on the iron concentration in the system, which perhaps explains some conflicting results in the literature obtained during studies on in vitro systems, although SOD would almost certainly be inhibitory at the low concentrations of free iron salts likely to be found in vivo.

To assess the biological role of the iron-catalysed Haber-Weiss reaction we have sought for chelators of iron which, unlike EDTA, actually prevent $OH^.$ generation. Three such chelators have so far been discovered, viz. diethylene-triaminepentaacetic acid[21,23] bathophenanthroline sulphonate[27] and desferri-oxamine.[30] The last of these three is by far the most effective (Table 3 shows some typical results). It could be argued that the chelators are acting merely as scavengers of $OH^.$, but bathophenanthroline sulphonate (BPS) and especially desferrioxamine are effective at lower concentrations than is thiourea, which reacts with $OH^.$ at essentially a diffusion-controlled rate.[25] Diethylenetri-aminepentaacetic acid (DETAPAC) at concentrations $\geqslant 1$-3 mM inhibits hydroxyla-tion of aromatic compounds neither by the NADH/phenazine methosulphate system, which is known to be due to the generation of $OH^.$ from H_2O_2 (Ref.31), nor by the peroxidase-dihydroxyfumarate system (see below) and so it cannot be a $OH^.$ scavenger at these concentrations. In any case it is difficult to see how DETAPAC could be such a powerful $OH^.$ scavenger whereas EDTA, which has a similar structure, actually stimulates $OH^.$ generation.[27]

As well as being the most effective inhibitor of the iron-catalysed Haber-Weiss reaction, desferrioxamine is a specific chelator of iron, whereas BPS also chelates copper ions and DETAPAC forms complexes with a wide range of metal ions.[32] The above chelators, and especially desferredoxins, can be used to test the biochemical significance of the iron-catalysed Haber-Weiss reaction.

Role of the iron-catalysed Haber-Weiss reaction in biochemical systems

Dihydroxyfumaric acid. A mixture of peroxidase and dihydroxyfumaric acid hydroxylates added aromatic compounds and hydroxylation is inhibited by SOD, by catalase or by scavengers of the $OH^.$ radical.[33,34] We naturally proposed that

TABLE 2

EFFECTS OF HYDROXYL RADICAL SCAVENGERS ON HYDROXYLATION BY THE XANTHINE/XANTHINE OXIDASE SYSTEM IN THE PRESENCE OF IRON SALTS[a]

Scavenger added	Final conc.(mM) in reaction mixture	nmol diphenol produced/h	% Inhibition of hydroxylation	Rate constant for scavenger reaction with $OH^{\cdot} (M^{-1}s^{-1})$
None	–	102	0	–
Mannitol	5	64	37	1.0×10^9
	10	50	51	(estimated
	25	36	65	from that of
	47	15	85	similar
Sodium				sugars)
formate	5	49	52	2.7×10^9
	10	41	60	
	25	34	67	
	47	7	93	
Thiourea	1	57	44	4.7×10^9
	2	44	57	
	5	24	76	
KCl	50	104	0	$< 10^3$
Urea	5	102	0	$< 7 \times 10^5$

[a]Reaction mixtures were as described in ref.(23). Similar results were obtained when $FeCl_3$ replaced $FeSO_4$. None of the above scavengers affected the assay for diphenol production. Neither formate nor mannitol affected xanthine oxidase activity. Thiourea inhibited xanthine oxidase slightly as assayed by O_2 uptake (final conc. 5 mM inhibited by 15%, 8 mM by 24% and 16.7 mM by 40%). These inhibitions are clearly too small to account for more than a small part of the striking effect of thiourea on hydroxylation. Rate constants were taken from the compilation in Ref.(25).

During the course of these experiments we observed that concentrations of Fe^{2+} or Fe^{2+}-EDTA greater than 100 μM produced significant hydroxylation even in the absence of xanthine or xanthine oxidase, whereas omission of either of these completely abolished hydroxylation under the conditions used for the experiments shown in Tables 1 and 2. Indeed, it is possible to hydroxylate a wide range of aromatic compounds merely by treating them with a high concentration of $FeSO_4$ in phosphate buffer under aerobic conditions at pH 7.2 (Refs.26,27). Such hydroxylation is inhibited by catalase or by scavengers of $OH^{\cdot}$, but not by SOD (Ref.27).

Our results can be explained quite easily. Iron(II) salts in solution autoxidise, especially in phosphate buffer at alkaline pH, and O_2^- is generated (Ref.28):

$$Fe^{2+} + O_2 \rightleftharpoons Fe^{2+} - O_2 \rightleftharpoons Fe^{3+} + O_2^- \qquad (8)$$

Superoxide can dismute to H_2O_2 (eqn. 2) which can undergo a Fenton reaction with remaining Fe^{2+} (eqn. 9).

$$Fe^{2+} + H_2O_2 \longrightarrow Fe^{3+} + OH^{\cdot} + OH^- \qquad (9)$$

If Fe^{2+} is present in large amounts, much $OH^{\cdot}$ can be produced (Refs.27,28) and hydroxylation will be inhibited by removal of H_2O_2 or of $OH^{\cdot}$. Superoxide dismutase should not inhibit, however, since it merely increases the supply of H_2O_2 needed for reaction (9), by catalysing the dismutation of O_2^-.

TABLE 3

EFFECT OF DESFERRIOXAMINE ON OH$^{\cdot}$ GENERATION BY THE XANTHINE/XANTHINE OXIDASE Fe^{2+} SYSTEM[a]

Final conc. of desferrioxamine present in reaction mixture (mM)	Diphenol produced (nmol/h)	Inhibition of hydroxylation (%)
0	42	0
0.005	38	9
0.05	18	57
0.2	9	79

[a]Assays of hydroxylation and xanthine oxidase activity were carried out as described in Ref.(2). Desferrioxamine had no effect on the assay used to measure hydroxylated product formation, nor on xanthine oxidase activity. For comparison, 1 mM bathophenanthroline sulphonate (BPS) produces a 67% inhibition of hydroxyl radical generation and 1 mM diethylene-triamine pentaacetic acid (DETAPAC) a 53% inhibition.

hydroxylation was due to formation of OH$^{\cdot}$ by a Haber-Weiss reaction.[33] Unfortunately this cannot be the case because it is not inhibited by DETAPAC, tested at 1-3 mM concentrations.[23] Either the OH$^{\cdot}$ radicals generated from O_2^- and H_2O_2 in this system are produced by some other mechanism or else the hydroxylating species is not OH$^{\cdot}$, although we would then have to postulate a species that reacts with scavengers of OH$^{\cdot}$ with rate constants comparable to those for OH$^{\cdot}$ itself, which seems unlikely. Perhaps the peroxidase itself can catalyse a Haber-Weiss reaction, although we found no evidence for this in studies with the xanthine/xanthine oxidase system.[23] However, lipid peroxidation induced by dihydroxyfumarate in the absence of peroxidase requires traces of metal ions[30] and DETAPAC protects isolated lung cells against the toxic effects of autoxidising dihydroxyfumarate.[35]

Alloxan. This compound causes degeneration of the β-cells of the pancreas when injected into animals, resulting in diabetes. Its damaging effects on isolated β-cells seem to be due to its reduction by cellular enzymes to dialuric acid, which then autoxidises with production of O_2^- and of OH$^{\cdot}$ (Ref.36). Addition of SOD, catalase, OH$^{\cdot}$ scavengers or DETAPAC to the β-cells provided significant protection[36,38] which is consistent with the occurrence of an iron-catalysed Haber-Weiss reaction, although the effects of desferrioxamine need to be studied to prove this. Hydroxyl radical is probably the major damaging species in vivo, since scavengers of it protect against alloxan-induced diabetes when fed to whole animals.[37]

6-Hydroxydopamine. The destructive effects of 6-hydroxydopamine on catecholamine nerve terminals may well be due to generation of oxygen radicals, since autoxidation of this compound in vitro produces O_2^- and OH$^{\cdot}$ (Ref.37). Formation of OH$^{\cdot}$ is inhibited by DETAPAC and by desferrioxamine,[39] strongly implicating an iron-catalysed Haber-Weiss reaction. It would be of great interest to see if these chelators affect the action of 6-hydroxydopamine in vivo.

Hyaluronic acid. In inflammatory joint diseases the hyaluronic acid present in synovial fluid has an abnormally low intrinsic viscosity. During acute inflammation there is usually extensive infiltration of polymorphonuclear leucocytes into the joint space, and it is known that these cells release O_2^- by the action of an enzyme located in their plasma membranes.[40] Exposure of synovial fluid or of solutions of hyaluronic acid itself to a O_2^--generating system causes depolymerisation, which can be prevented by adding scavengers of OH·, catalase or SOD (Ref.41). Using commercial samples of hyaluronic acid we have confirmed and extended these results[27,30] and have shown that depolymerisation is inhibited by DETAPAC, BPS and desferrioxamine. Hence the degradation is due to reaction of the hyaluronic acid with OH· generated by an iron-catalysed Haber-Weiss reaction. Atomic absorption analysis has shown the presence of sufficient iron in "purified" hyaluronic acid to catalyse such a reaction, and even more iron is present in synovial fluid in vivo.[42] Since synovial fluid has little SOD or catalase activities[41] the Haber-Weiss reaction may well be responsible for loss of viscosity during inflammation.

If hyaluronic acid samples low in contaminating iron are exposed to O_2^- generating systems, no depolymerisation occurs. If high concentrations of Fe^{2+} (≥100 µM) are added to such hyaluronic acid solutions depolymerisation occurs without it being necessary to provide O_2^-; Fe^{3+} is ineffective. Loss of viscosity can be prevented by catalase and by OH· scavengers, but not by SOD. In this system it seems likely that reactions (8) and (9) can provide sufficient OH· to cuase depolymerisation, but at more physiological concentrations of iron salts it is necessary to recycle Fe^{3+} by means of a reduction by O_2^-, as discussed previously.

Caeruloplasmin: a physiological inhibitor of the iron-catalysed Haber-Weiss reaction? The chelators DETAPAC, BPS and desferrioxamine inhibit the iron-catalysed Haber-Weiss reaction presumably because, unlike EDTA, they complex the iron in a form which cannot undergo the redox cycles shown in reactions (6) and (7). Free iron salts are damaging not only because of OH· generation but also because they can initiate lipid peroxidation.[30] The extracellular fluids of the body, which usually contain only low activities of SOD and catalase, contain high concentrations of the copper protein caeruloplasmin. This protein is able to catalyse the oxidation of Fe^{2+} to Fe^{3+} and the amount of it is serum rises in inflammatory conditions such as rheumatoid arthritis.[43] Like most copper proteins and complexes, caeruloplasmin can react with O_2^-, but its rate constant is only 7×10^5 $M^{-1}s^{-1}$ as compared with 2×10^9 $M^{-1}s^{-1}$ for the copper-zinc SOD enzyme.[44] We have found that caeruloplasmin is a powerful inhibitor of the Fe^{2+}-induced peroxidation of purified phospholipid liposomes in vitro.

Catalase and SOD also inhibited partially, but the effect of caeruloplasmin was far too great to be explained by its low "SOD-like" activity. Iron(II) salts in vitro also cuase strand scission of closed circular viral DNA, and again damage can be decreased by caeruloplasmin as well as by SOD or by catalase. Our results suggest that caeruloplasmin may be a natural protective mechanism against the deleterious effects of free iron salts, since its ferroxidase action would serve to keep them in the Fe^{3+} state and so prevent them from undergoing the redox cycles necessary to generate OH$^{\cdot}$. Further details of these experiments will be published shortly.

ACKNOWLEDGEMENTS

We thank the Wellcome Trust and the Central Reserach Fund of the University of London for financial support. R.R. thanks the Science Research Council for a research studentship and S.F.W. thanks the Medical Research Council for an intercalated B.Sc. award.

REFERENCES

1. Fridovich, I. (1979) Adv. Inorg. Biochem. in press.
2. Halliwell, B. (1978) Cell Biol. Int. Rep. 2, 113-128.
3. Halliwell, B. (1978) Planta, 140, 81-88.
4. Bielski, B.H.J. and Chan, P.C. (1978) J. Amer. Chem. Soc. 100, 1920-1921.
5. Sawyer, D.T. and Gibian, M.T. (1979) Tetrahedron, 35, 1471-1481.
6. Fee, J.A. (1979) in Proceedings of the Third International Symposium on Oxidases and Related Redox Systems, King, T.E., Mason, H.S. and Morrison, M. eds., in press.
7. Halliwell, B. (1973) Biochem. J. 135, 379-381.
8. McClune, G.J. and Fee, J.A. (1976) FEBS Lett. 67, 294-298.
9. Halliwell, B. (1976) FEBS Lett. 72, 8-10.
10. Rigo, A., Stevanato, R., Finnazzi-Agro, A. and Rotilio, G. (1977) FEBS Lett. 80, 130-132.
11. Weinstein, J. and Bielski, B.H.J. (1979) J. Amer. Chem. Soc. 101, 58-62.
12. Ferradini, C., Foos, J., Houee, C. and Pucheault, J. (1978) Photochem. Photobiol. 28, 697-700.
13. Koppenol, W.H. and Butler, J. (1977) FEBS Lett. 83, 1-6.
14. Fong, K.L., McCay, P.B., Poyer, J.L., Misra, H.P. and Keele, B.B. (1976) Chem. Biol. Inter. 15, 77-89.
15. Van Hemmen, J.J. and Meuling, W.J.A. (1977) Arch. Biochem. Biophys. 182, 743-748.
16. Halliwell, B. and De Rycker, J. (1978) Photochem. Photobiol. 28, 757-763.
17. Cohen, G. (1977) in Superoxide and Superoxide Dismutases, Michelson, A.M., McCord, J.M. and Fridovich, I., eds., Academic Press, New York, pp.317-321.
18. Halliwell, B. (1975) FEBS Lett. 56, 34-38.
19. McCord, J.M. and Day, E.D. (1978) FEBS Lett. 86, 139-142.
20. McClune, G.J., Fee, J.A., McCluskey, G.A. and Groves, J.T. (1977) J. Amer. Chem. Soc. 99, 5220-5222.
21. Buetner, G.R., Oberley, L.W. and Leuthauser, S.W.H.C. (1978) Photochem. Photobiol. 21, 693-695.
22. Finkelstein, E., Rosen, G.M. and Rauckman, E.J. (1979) Mol. Pharmacol. 16, 676-685.
23. Halliwell, B. (1978) FEBS Lett. 92, 321-326.
24. McCord, J.M. and Fridovich, I. (1969) J. Biol. Chem. 244, 6049-6055.
25. Anbar, M. and Neta, P. (1967) Int. J. Appl. Radiat. Isot. 18, 495-523.

26. Nofre, C., Cier, A. and Lefier, A. (1961) Bull. Soc. Chim. France, 530-535.
27. Halliwell, B. (1978) FEBS Lett. 96, 238-242.
28. Michelson, A.M. (1977) in Superoxide and Superoxide Dismutases,Michelson, A.M., McCord, J.M. and Fridovich, I. eds., Academic Press, New York, pp.77-86.
29. Winterbourn, C.C. (1979) Biochem. J. 182, 625-628.
30. Gutteridge, J.M.C., Richmond, and Halliwell, B. (1979) Biochem. J. 184, 469-472
31. Halliwell, B. (1977) Biochem. J. 167, 317-320.
32. Chabarek, S., Frost, A.E., Doran, M.A. and Bicknell, N.J. (1959) J. Inorg. Nucl. Chem. 11, 184-196.
33. Halliwell, B. and Ahluwalia, S. (1976) Biochem. J. 153, 513-518.
34. Halliwell, B. (1977) Biochem. J. 163, 441-448.
35. Autor, A.P. and Fox, A.W. (1979) in Molecular Basis of Environmental Toxicity, Bhatnagar, R.S., ed., Ann Arbor Science Publishers, in press.
36. Grankvist, K., Marklund, S., Sehlin, J. and Taljedal, I.B. (1979) Biochem. J. 17-25.
37. Cohen, G. (1978) Photochem. Photobiol. 28, 669-675.
38. Grankvist, K., Markland, S. and Taljedal, I.B. (1979) FEBS Lett. 105, 15-18.
39. Floyd, R.A. and Wiseman, B.B. (1979) Biochem. Biophys. Acta, 586, 196-207.
40. Babior, B.M. (1978) N. Engl. J. Med. 298, 659-668.
41. McCord, J.M. (1974(Science, 185, 529-531.
42. Sorenson, J.H. (1978) Inorg. Perspect. Biol. Med. 2, 1-26.
43. Gutteridge, J.M.C. (1978) Ann. Clin. Biochem. 15, 293-296.
44. Goldstein, I.M., Kaplan, H.B., Edelson, H.S. and Weissman, G. (1979) J. Biol. Chem. 254, 4040-4045.

IS SUPEROXIDE TOXIC?

JAMES A. FEE
Biophysics Research Division, University of Michigan, Ann Arbor,
Michigan 48109, USA

One of the motivating factors in the area of research involving superoxide dismutases has been the concept that superoxide ion or some product derived from it other than by dismutation is toxic to cellular systems.[1] While there is still no direct evidence for in vivo O_2^- toxicity, there have been many indications of in vitro O_2^- toxicity, and a significant amount of research has gone into ascertaining the molecular basis of this putative toxicity. Examination of the literature reveals an interesting dichotomy of conclusion on the question titling this article. On the one hand careful experiments designed to observe a direct, stoichiometric reaction with various substances have revealed a very limited chemical reactivity in O_2^- in aqueous solution. Thus, as pointed out in several reviews[2,3,4] O_2^- is a facile one-electron reductant but reacts most rapidly with metal ions. While thermodynamic considerations suggest O_2^- to be a strong oxidant a kinetic barrier precludes this type of reaction except when the free dianion, O_2^{2-} is avoided. Thus, certain complexes of lower valent metal ions such as Cu(I) and Fe(II) have been shown to rapidly reduce O_2^- to the level of peroxide, presumably via formation of a metal-peroxide species. The general unreactivity of O_2^- toward substances of biological significance, as assessed in these carefully controlled experiments, is shown in Table 1.

By contrast to the stoichiometric chemical studies examination of complex mixtures has revealed superoxide dependence of a variety of degradative processes such as lipid peroxidation and DNA degradation, summarized in Table 2, which are clearly deleterious to a cell.

What is the origin of the apparent dilemma raised by the observations cited in Tables 1 and 2? At the 1976 EMBO Workshop on Superoxide and Superoxide Dismutase (Banyuls, France) one of the points of rather heated debate was the role of the Haber-Weiss reaction[1] in various manifestations of superoxide toxicity[2,18,19]

$$O_2^- + H_2O_2 + H^+ \rightarrow OH^\cdot + O_2 + H_2O \tag{1}$$

It was clear already at that time that <u>both</u> superoxide and peroxide were somehow involved in the various degradative phenomena under consideration and the Haber-Weiss reaction was a convenient vehicle to explain the results. By this reaction O_2^- could react with H_2O_2 to form the hydroxyl radical which is well known to react with essential cellular components thereby decreasing the ability to perform their cellular function. There was some question, however, as to whether this reaction could proceed directly. Examination of the early literature revealed that other chemists had also studied this reaction finding that it did not occur to a detectable extent.[20] Since 1976, reaction (1) has been tested by a variety of different experimental approaches and found not to occur (Table 1). The objection was skirted[19], however, by the facile inclusion of catalytic amounts of trace metal ions which would catalyze this reaction. Thus, using iron as an example,

$$O_2^- + Fe^{3+} \rightarrow O_2 + Fe^{2+} \qquad (2)$$

$$Fe^{2+} + H_2O_2 \rightarrow OH^. + H_2O + Fe^{3+} \qquad (3)$$

the hydroxyl radical can be formed by reactions (2) and (3) which comprise a minor variation of Fenton chemistry which has been carefully scrutinized for many years.[21-23]

This sequence of reactions can account for essentially all the superoxide dependent processes mentioned in Table 2. The evidence for this statement comes from two lines of observation. First, these processes have been studied for many years prior to the discovery of superoxide dismutase and the publication of the theory of superoxide mediated oxygen toxicity[1] under a variety of "Fenton conditions," some of which are listed in Table 3. The components required for "degradation" are trace elements such as iron or copper, peroxide and a reducing agent. The essential chemistry which occurs is outlined in reactions (4) to (7), using iron as an example:

$$Fe^{3+} + A^- \rightarrow Fe^{2+} + A \qquad (4)$$

$$Fe^{2+} + H_2O_2 \rightarrow "OH^." + Fe^{3+} \qquad (5)$$

$$"OH^." + RH \rightarrow R^. + H_2O \qquad (6)$$

$$R^. \longrightarrow \text{degradation products} \qquad (7)$$

Note the similarity of reactions (4) and (5) to (2) and (3).

The second form of evidence supporting the idea that reactions (2), (3), (6) and (7) are responsible for the evident involvement of O_2^- in the degradative processes of Table 2 comes from an as yet limited number of experiments in which the contribution of O_2^- to a "degradative" system can be overcome by including a small amount of reducing agent, such as ascorbic acid, and completely overriding the contribution of O_2^- to the degradative process. These examples are summarized in Table 4. It is evident that a form of superoxide-mediated Fenton chemistry or a metal-ion catalyzed Haber-Weiss reaction is probably responsible for the <u>in vitro</u> manifestations of superoxide toxicity.

TABLE 1

SUBSTANCES WITH WHICH AQUEOUS SUPEROXIDE HAS BEEN SHOWN NOT TO REACT[a]

Substance	Condition or Type of Experiment	References
$\phi\chi$-174 DNA	Radiolysis centrifugation	5
Nuclear bases	Gamma irradiation/production analysis	6
Poliovirus	KO_2 dissolution/virulence	7
Cholesterol	Product analysis	8
Formate, fumarate, α-keto-gluterate, pyruvate, oxalate, imidazole, Tris, EDTA	Combined pulse radiolysis and stopped-flow/H_2O_2 and O_2 analyses	9[b]
H_2O_2	KO_2 dissolution/O_2 analysis	10
	Stopped-flow kinetics	11
	H_2O_2 inhibition of NBT reduction by O_2^-	12
	Pulsed radiolysis	13
	Search for aromatic hydroxylation by O_2^- and H_2O_2	14,15
Linolenic acid (dispersion)	Stopped-flow kinetics	(Fee and McClune, un-published)
ROOR[c]		16
ROOH[c]		17
Hyaluronic acid	Xanthine/xanthine oxidase/O_2	84

[a]Representative only

[b]Several other related compounds were also found not to react with O_2^-.

[c]In acetonitrile

TABLE 2

IN VITRO DEGRADATIVE PROCESSES IN WHICH SUPEROXIDE HAS BEEN SHOWN TO PARTICIPATE[a]

Process	References
DNA degradation/depolymerization	24-27
Lipid peroxidation	28-35
Depolymerization of polysaccharides	36-37[b]
Hydroxylation of aromatic substances	38-43
In vitro cell killing (bacteria and myoblasts)	44-48
Ethylene formation from methional[c]	49-51

[a]Representative. The participation of superoxide has generally been ascertained by an inhibition of the overall process by superoxide dismutase.

[b]Contains no experimental data supporting superoxide involvement.

[c]Including compounds related to methional.

TABLE 3

FENTON CONDITIONS FOR DEGRADATIVE PROCESSES IN WHICH SUPEROXIDE PARTICIPATION
HAS BEEN OBSERVED

System	Conditions and Requirements	References
Ethylene production for "methional" [a]	H_2O_2 + Fe^{2+}	52
	O_2 + ascorbate and microsomes	53
	Sulfite, phenols and Mn^{2+}	54,54
	O_2^- and H_2O_2 + "trace metals"[b]	49
DNA degradation, depolymerization or loss of biological activity	Effect of ascorbic acid, Cu^{2+}, O_2 and H_2O_2 on pneumoccocal transforming substance.[c]	56
	Effect of H_2O_2, KO_2, ascorbic acid, Fe-EDTA and a variety of other substances of similar nature.[d]	57-59
	Quinones/quinols/cysteine, metal ions and H_2O_2	26,60-64
Degradation of polysaccharides such as hyaluronic acid, alginic acids and other polymeric structures	Effects of ascorbic acid, a variety of reducing agents, metal ions, chelators and H_2O_2 on depolymerization.[e] [f]	59,65-73
Peroxidation of lipids in a variety of biological structures under in vitro conditions	Metal ions, reducing agents, H_2O_2 and oxygen.[g]	74-79
Cell killing	Treatment with riboflavin, light and oxygen or various strongly reducing materials in the presence of normal culture components.	44,45,47

[a]Here "methional" indicates similar substances such as methione.

[b]"Trace metals" are apparently necessary for this process and are assumed here to be present even though not explicity considered by the auhtors to be involved.

[c]McCarty (Ref.56) recognized the similarity of the destructive action of ascorbic acid on transforming substance with its effects on bacterial toxins, viruses and enzymes described by previous workers.

[d]The authors of Refs.57-59 are clearly cognizant of the Fenton chemistry.

[e]While the authors of several of these papers do not explicitly involve metal ions in their proposed mechanism of degradation, it is clear such ions are ubiquitous in purified preparations of nucleic acids (Ref.65, and references therein).

[f]Pigman and co-workers (Refs.69-71) have termed the depolymerization of these substances the oxidative-reduction depolymerization (ORD) reactions, and its analogy to Fenton chemistry is obvious.

[g]This topic has been so widely studied that only a few general or leading references are given which emphasize the essentiality of metal ions, reducing agents and O_2 or peroxide.

TABLE 4

ABILITY OF ADDED REDUCING AGENTS TO SUBSTITUTE FOR SUPEROXIDE IN FENTON TYPE
SYSTEMS WITH LOSS OF INHIBITION BY SUPEROXIDE DISMUTASE

System	Observation	Percent Inhibition by Superoxide Dismutase	References
Microsomal lipids/$FeCl_2$/Fe-EDTA/O_2	Malondialdehyde formation[a]		80
+ xanthine/xanthine oxidase		85	
+ 0.2 mM ascorbic acid		2	
Brain lipids/metal ions/O_2	Fluorescent conjugate of peroxidized lipid		81
+ 100 micromolar Fe^{2+} or Fe^{3+}		∿43	
+ 50 micromolar Cu^+ or Cu^{2+}		∿60	
+ 1 mM ascorbate, no added metals		0	
DNA/bleomycin/Fe/O_2[c]	Malondialdehyde formation[b]		82,83
Hyaluronic acid/xanthine oxidase/ xanthine/O_2[d]	Decrease in viscosity		84
Methional	Ethylene production		
+ xanthine oxidase/xanthine O_2		100	85[e]
+ 50 mM ascorbate		slight	

[a] No overt attempt was made to minimize or sequester contaiminating trace metals in this experiment, but it is certain they were present since extensive lipid peroxidation occurred.

[b] Acid soluble radioactive fragments released from radiolabelled DNA are use as indicators of degradation as well as products which react with α-thiobarbituric acid to form colored products.

[c] The work described in Refs.82 and 83 establishes that the essential requirements for bleomycin induced DNA degradation are Fe, O_2 and a reducing agent. A mixture of xanthine, xanthine oxidase and O_2 can support Fe-bleomycin degradation and is inhibited by superoxide dismutase. However, when other reducing agents are added such as thiols there is virtually no inhibition by added superoxide dismutase. The authors propose a scheme where O_2 can serve as a reductant of the Fe^{3+}-bleomycin complex but is reduced by other reducing agents. These papers describe an excellent example of how the Fenton chemistry can be harnessed for a useful biological purpose.

[d] Metal ions are absolutely required for depolymerization and added ascorbate obviates inhibtion by superoxide dismutase (Ref.84 and B. Halliwell, personal communication). The metal ion requirement is consistent with the observations of previous workers.

[e] This communication is the first to directly address the question of whether O_2^- can compete with ascorbate at typical cellular levels.

46

This conclusion raises some questions of considerable relevance to this field of research:

(a) Is oxygen toxicity due to the presence of both peroxide and metal ions at concentrations capable of carrying on a significant Fenton chemistry within the cell?

(b) Is it realistic to consider O_2^- as being able to effectively compete with the massive amounts of reducing agents known to be present in cells?

(c) Does a totally unknown chemical reactvity of O_2^- exist which would allow it to react in a direct, specific, and destructive fashion with cellular constituents thus endowing it with toxic properties?

There are very good reasons to respond negatively to each of these questions:

(a) Peroxide levels in normal tissues are very carefully controlled and a number of systems exist for its rapid conversion to either water and oxygen or for its direct reduction to water. Similarly, intracellular trace elements such as Fe and Cu can be expected to be sequesteres within specific protein complexes none of which appears to catalyze reaction (4) and (5) indiscriminately.

(b) The ratios of ascorbate, glutathione and other reducing agents to O_2^- within cells are very large, probably exceeding 10^6 - 10^7 under normal conditions, and it appears unlikely that reaction (2) will proceed with a velocity comparable to reaction (4) (Ref.85).

(c) The organic functional groups shown to be unreactive toward O_2^- are representative of those present in living systems and it seems likely that more sensitive functional units would also be subject to other types of chemical modification.

By presuming the answers to these questions to be negative, as argued above, the author has presented a logic construct[86] in which O_2^- was argued to be of no importance in oxygen toxicity, superoxide dismutase activity was argued to be a trivial reaction dictated by the chemistry of the metal ion, and it was concluded that the true biological functions of the metalloproteins, presently called superoxide dismutases, were not known. While this may appear as totally unfounded conjecture, unless we have some solid evidence that in vivo superoxide is truly deleterious, it may not be fruitful to continue to assume that it is toxic. Therefore such questions become perfectly valid bases of scientific inquiry.

ACKNOWLEDGEMENT

This work is supported by US PHS GM 21518.

REFERENCES

1. McCord, J.M. Keele, B.B., Jr. and Fridovich, I. (1971) Proc. Natl. Acad. Sci. USA 68, 1024-1027.
2. Fee. J.A. and Valentine, J.S. (1977) in Superoxide and Superoxide Dismutases, Michelson, A.M., McCord, J.M. and Fridovich, I., eds., Academic Press, New York, pp. 19-60.

3. Valentine, J.S. in Oxygen: Biochemical and Clinical Aspects, W.S.Caughey, ed., Academic Press, New York, in press.
4. Lee-Ruff, E. (1977) Chem. Soc. Rev. 6, 195-214.
5. Blok, J. (1967) in Radiation Research, Sillini, G., ed., North-Holland, Amsterdam, pp. 423-437.
6. Cadet, J. and Teoule, R. (1978) Photochem. Photobiol. 28, 661-667.
7. Pelouz, Y., Nofre, C., Cier, A. and Colobert, L. (1962) Inst. Pasteur, Ann. 102, 6-23.
8. Smith, L.L., Kulig, M.J. and Teng, J.I. (1977) Chem. Phys. Lipids 20, 221-227.
9. Bielski, B.H.J. and Richter, H.W., (1977) J. Amer. Chem. Soc. 99, 3091-3023.
10. George, P. (1947) Disc. Faraday Soc. 2, 196-205.
11. McClure, G.J., and Fee, J.A. (1976) FEBS Lett. 67, 294-298.
12. Halliwell, (1976) FEBS Lett. 72, 8-10.
13. Czapski, G. and Ilan, Y.A. (1978) Photochem. Photobiol, 28, 657-753.
14. Fee, J.A. and Hildebrand, P.G. (1974) FEBS Lett. 39, 79-82.
15. Rigo, A., Stevanato, R., Finazzi-Agro, A., and Rotilio, G. (1977) FEBS Lett. 80, 130-132.
16. Peters, J.W., and Foote, C.S., (1976) J. Amer. Chem. Soc. 98, 873-875.
17. Gibian, M.J. and Ungermann, T. (1979) J. Amer. Chem. Soc. 101, 1291-1294.
18. Cohen, G. (1977) in Superoxide and Superoxide Dismutases, Michelson, A.M., McCord, J.M. and Fridovich, I., eds., Academic Press, New York, pp.317-321.
19. Editorial comment in Superoxide and Superoxide Dimsutases, Michelson, A.M., McCord, J.M. and Fridovich, I., eds., Academic Press, New York, p.320.
20. George, P. (1947) Disc. Faraday Soc. 2, 219-220.
21. Fenton, H.J.H. (1896) J. Chem. Soc. 69, 546-562.
22. Marz, J.H. and Water, W.A. (1949) J. Chem. Soc. pp.2427-2433.
23. Walling, C. (1975) Account. Chem. Res. 8, 125-131.
24. Van Hemmen, J.J., and Meuling, W.J.A. (1975) Biochem. Biophys. Acta, 402, 133-141.
25. White, J.R., Vaughn, T.O. and Yeh, W.S. (1971) Fed. Proc., Abst. 537.
26. Lorentzen, R.J., and Tso, P.O.P. (1977) Biochemistry, 16, 1467-1473.
27. Cone, R., Hasan, S.K., Lown, J.W., and Morgan, A.R. (1976) Can. J. Biochem. 54, 219-223.
28. Fee, J.A. and Teitelbaum, H.D. (1972) Biochem. Biophys. Res. Commun. 49, 150-158.
29. Lynch, R.E. and Fridovich, I. (1978) J. Biol. Chem. 252, 1838-1845.
30. Kellog, E.W.,III and Fridovich, I. (1978) J. Biol. Chem. 250, 8812-8817.
31. Tyler, D.D. (1975) Biochim. Biophys. Acta, 396, 335-346.
32. Pederson, T.C. and Aust, S.D. (1972) Biochem. Biophys. Res. Commun. 48, 789-795.
33. Bus, J.S., Aust, S.D., and Gibson, J.E. (1974) Biochem. Biophys. Res Commun. 58, 749-755.
34. Tyler, D.D. (1975) FEBS Lett. 51, 180-183.
35. Noguchi, T. and Nakano,M. (1974) Biochem. Biophys. Acta, 368, 446-455.
36. McCord, J.M. (1974) Science, 185, 529-530.
37. Isbell, H.S. and Frusch, H.L. (1977)Carbohydrate Res. 59, C25-C31.
38. Strickland, S., and Massey, V. (1973) in Oxidases and Related Redox Systems, King, T.E., Mason, H.S. and Morrison, M., eds., University Park Press, Baltimore, pp. 189-194.
39. Goscin, S.A., and Fridovich, I. (1976) Arch. Biochem. Biophys. 153, 778-783.
40. Halliwell, B. (1955) Eur. J. Biochem. 55, 355-360.
41. Halliwell, B. and Ahluwalia, S. (1976) Biochem. J. 153, 513-518.
42. Halliwell, B. (1977) Biochem. J. 163, 441-448.
43. McCord, J.M., and Day, E.D.,Jr. (1978) FEBS Lett. 86, 139-142.
44. Lavelle, F., Michelson, A.M., and Dimitrijevic, L. (1973) Biochem. Biophys. Res. Commun. 55, 350-357.
45. Michelson, A.M. and Buckingham, M.E. (1974) Biochem. Biophys. Res. Commun. 58, 1079-1086.
46. Gregory, E.M., Yost, F.J. and Fridovich, I. (1973) J. Bacteriol. 115, 987-981.
47. DeChatelet, L.R., Shirley, P.S., Gordson, P.R. and McCall, C.E. (1975) Antimicrob. Agents and Chemother. 8, 146-153.

48. Van Hemmen, J.J. and Meuling, W.J.A. (1977) Arch. Biochem. Biophys. 182, 743-748.
49. Beauchamp, C. and Fridovich, I., J. Biol. Chem. 245, 4641-4646.
50. Weiss, S.J., Rustagi, P.K. and LoBuglio, A.F. (1978) J. Exp. Med. 147, 316-323.
51. Weiss, S.J., Turk, J. and Needleman, P. Blood, in press.
52. Lieberman, M., Kunishi, A.T., Mapson, L.W. and Wardale, D.A. (1965) Biochem. J. 97, 449-459.
53. Lieberman, M. and Hochstein, P. (1966) Science, 152, 213.
54. Yang, S.F. (1967) Arch. Biochem. Biophys. 122, 481-487.
55. Yang, S.F. (1969) J. Biol. Chem. 244, 4360-4365.
56. McCarty, M. (1945) J. Exp. Med. 81, 501-514.
57. Limperos, G. and Mosher, W.A. (1950) Amer. J. Roentgenol. Radiat. Ther. 63, 581-690.
58. Conway, B.E. and Butler, J.A.V. (1952) J. Chem. Soc., pp.834-838.
59. Gilbert, D.L., Gerschman, R., Cohen, J., and Sherwood, W. (1957) J. Amer. Chem. Soc. 79, 5677-5680.
60. Morita, J., Morita, S. and Komano, T. (1971) Biochem. Biophys. Acta, 475, 403-487.
61. Rosenkranz, H.S. and Rosenkranz, S. (1971) Arch. Biochem. Biophys. 146, 483-487.
62. Massie, H.R., Samis, H.V. and Baird, M.B. (1972) Biochem. Biophys. Acta, 272, 539-548.
63. Cone, R., Hasan, S.K., Lown, J.W. and Morgan, A.R. (1976) Can. J. Biochem. 54, 219-223.
64. White, H.L. and White, J.R. (1966) Biochem. Biophys. Acta, 123, 648-651.
65. Srinska, V.A., Messineo, L., Towns, R.L.R. and Pearson, K.H. (1978) Int. J. Biochem. 9, 637-638.
66. Robertson, W.B., Ropes, M.W. amd Bauer, W. (1941) Biochem. J. 35, 903-908.
67. Ogston, A.G. and Sherman, T.F. (1959) Biochem. J. 72, 301-305.
68. Alexander, P. and Fox, M. (1958) J.Polymer Sci. 33, 493-496.
69. Pigman, W. and Rzivi, S. (1959) Biochem. Biophys. Res. Commun. 1, 39-43.
70. Pigman, W., Rizvi, S., and Holley, H.L. (1961) Arthritis Rheum. 4, 240-252.
71. Matsumara, G. and Pigman, W. (1965) Arch. Biochem. Biophys. 110, 526-533.
72. Neidermeier, W., Dobson, C., Laney, R.P. and Dobson, C. (1967) Biochem. Biophys. Acta, 141, 366-373.
73. Neidermeier, W., Laney, P. and Dobson, C. (1967) Biochem. Biophys. Acta, 148, 400-405.
74. Ingold, K.U. (1962) in Lipids and their Oxidation, Schultz, H.W., Day, E.A. and Sinnhuber, R.O., eds., Avi Publishing Co., Westport, Connecticut, pp. 93-121.
75. Wills, E.D. (1965) Biochem. Biophys. Acta, 98, 238-251.
76. Wills, E.D. (1969) Biochem. J. 113, 325-332.
77. Hunter, F.E., Jr., Gebicki, G.M., Haffstein, P.E., Weinstein, J. and Scott, A. (1963) J. Biol. Chem. 238, 828-835.
78. McKnight, R.C., Hunter, F.E. and Oehlert, W.H. (1965) J. Biol. Chem. 240, 3439-3446.
79. Deutsch, H.F., Kline, B.E., and Rusch, H.P. (1947) J. Biol. Chem. 141, 529-538.
80. Pederson, T.C. and Aust, S.D. (1973) Biochem. Biophys. Res. Commun. 52, 1071-1078.
81. Gutteridge, J.M.C. (1977) Biochem. Biophys. Res. Commun. 77, 379-386.
82. Sausville, E.A., Peisach, J. and Horwitz, S.B. (1978) Biochemistry, 16, 2740-2746.
83. Sausville, E.A., Stein, R.W., Peisach, J. and Horwitz, S.B. (1978) Biochemistry, 17, 2746-2754.
84. Halliwell, B. (1978) FEBS Lett. 96, 238-242.
85. Winterbourne, C.C. (1979) Biochem. J. 182, 625-628.
86. Fee, J.A. in Third International Symposium on Oxidases and Related Redox Systems, King, T.E., Mason, H.S.amd Morrison, M., eds., in press.

MYCOPLASMA PNEUMONIAE: A PATHOGEN WHICH MANUFACTURES SUPEROXIDE
BUT LACKS SUPEROXIDE DISMUTASE

R.E. LYNCH[+] AND B.C. COLE[++]
[+]Howard Hughes Medical Institute at the University of Utah, Hematology Division;
and [++]Arthritis Division, Department of Medicine, University of Utah College of
Medicine, Salt Lake City, Utah 84132, USA

INTRODUCTION

The Mycoplasmataceae are simplified prokaryotes. They lack cell walls, require cholesterol for growth, possess a genome only about half the size of the smallest bacterium,[3] and grow only in complex media. One of the mycoplasma, $\underline{M}$. pneumoniae causes a common illness in man, primary a typical pneumonia.

The characteristics of the metabolism of molecular oxygen by $\underline{M}$. pneumoniae suggest that this organism might be well suited to infecting the lung, a tissue bathed in O_2. It is known to produce H_2O_2 by which means it hemolyses erythrocytes in agar.[12,13,26] Its adherance to the surfaces of cells suggests that intimate contact between $\underline{M}$. pneumoniae and its eukaryotic host might serve any of several purposes. Among the possibilities is that $\underline{M}$. pneumoniae exerts its cytopathic effect by the manufacture of toxins which are effective only at very short range. One such toxin might be O_2^- which, by spontaneous dismutation, would give rise to the H_2O_2 which $\underline{M}$. pneumoniae excretes.

If indeed, $\underline{M}$. pneumoniae produced O_2^- as a weapon against the cell it infects, it should be possible to show that it produces and exports O_2^-, that it exerts cytopathic effects as a consequence of the production of O_2^-, and that it is well defended against the O_2^- and H_2O_2 it produces. We have tested the first and third of these predictions and we report the preliminary results of those experiments.

MATERIALS AND METHODS

Reagents. Cytochrome c type III (Sigma), nitro blue tetrazolium (Sigma), xanthine (Sigma), bovine erythrocyte superoxide dismutase (Truett, Diagnostic Data), flavin mononucleotide (Sigma), flavin adenine dinucleotide (Sigma), riboflavin (Nutritional Biochemicals), were obtained from commercial sources. Xanthine oxidase was purified from unpasteurized cream by a procedure which avoids proteolysis.[38]

Assays. Superoxide dismutase was assayed by its ability to inhibit the reduction of cytochrome c which was mediated by O_2^- generated by the enzymatic action of xanthine oxidase.[31] Polyacrylamide gels were stained for superoxide dismutase activity,[4] for catalase activity,[18] and peroxidase activity[18] after electrophoresis by established methods. Catalase was assayed in solution by the rate of change in absorbance at 240 nm of a solution of H_2O_2 20 mM in 16.7 mM potassium phosphate buffer pH 7.0.[5] Peroxidase activity was determined with o-dianisidine as chromogenic substrate.[19] The production of superoxide was estimated from the rates of reduction of cytochrome c (5×10^{-5} M) determined spectrophotometrically at 550 nm of of nitro blue tetrazolium (4.1×10^{-4} M) at 560 nm. The portion attributed to reduction by O_2^- was the part which superoxide dismutase inhibited. Bands of NAD(P)H - NBT oxidoreductase activity in samples electrophoresed in polyacrylamide gels were located by soaking the gels first in a solution of nitro blue tetrazolium 200 mg/dl for 45 min at room temperature in the dark followed by soaking the gel in a solution of either NADH or NADPH (1 mM) in 50 mM potassium phosphate pH 7.8, 10^{-4} M EDTA in the dark until good contract is obtained. When inhibition of the reduction of NBT by O_2^- was attempted superoxide dismutase (10 µg/ml) was included in the solution of NBT. NADH and NADPH oxidase activities were measured spectrophotometrically at 340 nm at 25°C in solutions of 33 uM substrate in 0.15 M NaCl, 10 mM potassium phosphate pH 7.65. Protein concentrations were determined by the method of Lowry.[27] Polyacrylamide gel electrophoresis was performed by a modification[29] of the methods of Davis[16] and Maurer.[30]

Mycoplasma strains and culture procedures. The sources of the mycoplasmas used in this study are as follows: M. fermentans (M713), M. gallisepticum (M722), M. hominis (M711) and A. laidlawii type A (M728), M.F. Barile, Bureau of Biologies, NIH, Bethesda, MD; M. pneumoniae strains 1428 and TW 48-5, J.G. Tully, Laboratory of Infectious Diseases, NIH, Bethesda, MD; M. pneumoniae (Jose), N.L. Somerson, Ohio State University, Columbus, Ohio; M. pulmonis, JB, D. Taylor-Robinson, Harrow, England; M. arthritidis (158 P10P9), B.C. Cole, University of Utah, Salt Lake City, UT; A. laidlawii (1305), R.N. Gourlay, ARC Inst. Res. Anim. Dis., Compton, England.

The M. pneumoniae strains were grown as lawns in 850 cm^2 roller bottles (Falcon, 3027) in Difco PPLO broth supplemented to final concentrations of 20% (vol/vol) fresh yeast extract, 0.01% (wt/vol) NAD and 1000 units/ml penicillin.[10,24] The lawns were harvested by agitation with glass beads. All other mycoplasmas were grown in suspension cultures. Mycoplasma arthritidis and M. hominis were cultured in Difco PPLO broth supplemented to final concentrations of 12 to 15% (vol/vol) inactivated horse serum, 5% (vol/vol) fresh yeast extract, 0.5% (wt/vol) L - arginine HCl and 1000 units/ml penicillin. M. gallisepticum and M. fermentans were grown in Difco PPLO broth supplemented with 20% (vol/vol) heated horse serum, 7.5% fresh yeast extract, 0.5% glucose and 1000 units/ml penicillin. M. pulmonis was similarly cultured but received an additional supplement of 0.01% NAD. Acholeplasma laidlawii was cultured in Tryptose phosphate broth (Difco) supplemented with 2% (vol/vol) Difco PPLO serum fraction and 1000 units/ml penicillin. The identity of each mycoplasma species was confirmed serologically using the growth inhibition test[11] or by epi-immunofluorescence[2] employing refe- rence antisera.

Mycoplasma synoviae, strain WVU 1791, obtained from N.O. Olson (Division of Animal and Veterinary Sciences, W. Virginia University, Morgantown West Virginia) was cultured according to Aldridge.[39]

RESULTS

The production of superoxide in lysates of M. pneumoniae. The ability of extracts of M. pneumoniae to generate O_2^- from NADH was determined in a prepara- tion disrupted by a French pressure cell (Table 1). As a means of locating the site of production of O_2^- three preparations were assayed, the cytosol from which membranes and unlysed cells had been sedimented by centrifugation at 39,000 g, the resulting pellet resuspended in phosphate buffered saline, as well as the whole extract from the French pressure cell. The production of O_2^- was deter- mined from the portion of the reduction of oxidized cytochrome c, measured spectro- photometrically at 550 nm, which superoxide dismutase inhibited. In the prescence of NADH cytochrome c was reduced. The majority of the reduction was attributable to O_2^-. Most of the activity remained in the supernate after centrifugation, indicating that this activity was probably soluble and not membrane-bound. Nearly all the activity in the oxidation of NADH and NADPH was soluble as well, a fea- ture so characteristic of mycoplasma that it is a taxonomic criterion used to distinguish mycoplasma from acholeplasma (Table 2).

The production of superoxide in cell suspensions of M. pneumoniae. The demon- stration that enzymes in the cytosol could produce O_2^- did not bear directly on the ability of these organisms to manufacture O_2^- for export. To examine this possibility the production of O_2^- was measured in suspensions of organisms. In separate experiments both cytochrome c and NBT were used as spectrophotometric indicators of O_2^- which was considered to be present when reduction of either was inhibited by the specific scavenger, superoxide dismutase.

The results of one such experiment with NBT (Table 3) indicate that the forma- tion of O_2^- occurred in suspensions of cells as well as in extracts. In the sus- pensions of cells, as was later shown in extracts, flavins markedly stimulated the formation of O_2^- in the presence of NADH. In all cases a substantial frac- tion of the reduction NBT was attributable to O_2^-. Although some lysis occurred in the suspension of organisms at the end of the experiment 80% of the activity in the reduction of NBT in the presence of NADH and FMN was sedimented at 39,000 g for 30 min whereas in lysates 95% of the NADH oxidase activity was soluble.

The apparent intracellular location of the activity which generated O_2^-, when combined with the demonstration that suspensions of organisms could export O_2^- into the medium surrounding them, suggested that O_2^- made inside the cell might

TABLE 1

SUPEROXIDE PRODUCTION FROM NADH IN LYSATE, CYTOSOL AND MEMBRANE FRACTIONS
OF M. PNEUMONIAE[a]

Fraction	Rates of reduction of cytochrome c		Percent inhibition
	without SOD	with SOD	
A. Lysate (French press)	0.42	0.16	62
B. Cytosol (Supernate from A)	0.44	0.13	71
C. Membranes (Pellet from A)	0.136	0.064	53

[a]A washed suspension of M. pneumoniae in 0.15 M NaCl, 10 mM potassium phosphate, pH 7.65, was lysed by passage through a French pressure cell and divided into two equal parts. One, the lysate, was not processed further. The other was subjected to centrifugation for 30 min at $4^{\circ}C$ at 39,000 g. The resulting supernatant was the cytosol (B). The resulting pellet was resuspended in a volume of 0.15 M NaCl, 10 mM potassium phosphate, pH 7.65 equal to half the original volume of the lysate. Assays were performed in 0.15 M NaCl, 10 mM potassium phosphate, pH 7.65, at $25^{\circ}C$. Concentrations of NADH and SOD were 33 μM and 6.7 μg/ml, respectively. Rate of cytochrome c reduction is ΔA_{550}/min/ml.

TABLE 2

NAD(P)H OXIDASE ACTIVITY IN FRACTIONS OF M. PNEUMONIAE[a]

Fraction	ΔA_{340}/min/ml	
	NADH	NADPH
Homogenate	6.94	0.502
Cytosol (After centrifugation of homogenate)	6.60	0.448
% Activity in cytosol	95	89

[a]The NADH or NADPH oxidase activity was measured by recording the rate of absorbance change at 340 nm in 50 mM potassium phosphate pH 7.8, 10^{-4} M EDTA, with NADH 33 μM or NADPH 333 μM at $25^{\circ}C$.

exit through the membrane. If this were the case, there might be a passage through the membrane for O_2^-. In the membrane of the erythrocyte such a channel for the exchange of stable anions has been extensively studied.[37,20,14,15] Potent inhibitors of the exchange of anions through this channel have been described[6-8] two of which, 4,4 -diisothiocyano-2,2'-disulfonic acid stilbene (DIDS) and 4-acetamido-4'-isothiocyano-2,2'-disulfonic acid stilbene (SITS), appear to inhibit markedly the movement of O_2^- through the membrane of the red cell.[28]

TABLE 3

PRODUCTION OF SUPEROXIDE BY SUSPENSIONS OF M. PNEUMONIAE[a]

| Substrates | | | Rate of NBT reduction | | % Inhibition by SOD |
NADH (100 µM)	FAD (40 µM)	FMN (40 µM)	no SOD	with SOD	
+	+		0.061	0.019	62
+			0.029	0.012	68
	+		0.003	0.002	–
+		+	0.110	0.043	57
+			0.028	0.019	62
		+	0.001	–	–

[a]M. pneumoniae, thrice washed in 0.15 M NaCl, 10 mM potassium phosphate pH 7.65, was incubated at 25°C in a solution containing 4.1×10^{-4} M NBT, 0.15 M NaCl, 10 µM potassium phosphate pH 7.65 with the indicated additions and the absorbance at 560 nm was recorded with time in a Cary 118C recording spectrophotometer with the cuvette positioned immediately in front of the photomultiplier tube to minimize the effect of light scattering by the turbid suspensions. Rate of NBT reduction is observed ΔA_{560}/min/mg protein. SOD concentration was 33 µg/ml.

If O_2^- passed through such a channel in the membrane of M. pneumoniae DIDS might also prevent the reduction of NBT or cytochrome c in the medium by O_2^- formed inside. In accord with this prediction the reduction of extracellular NBT by suspensions of M. pneumoniae in the presence of FAD and FMN was inhibited by 60 and 76%, respectively, values very similar to the amount of inhibition produced by superoxide dismutase in the medium, 62% and 71%. However, it was also found that DIDS inhibits the reduction of NBT in lysates of M. pneumoniae with NADH as substrate, indicating that its effect may not be simply on the transmembrane movement of O_2^-.

The activities of superoxide dismutase and catalase in M. pneumonia. The ability of M. pneumoniae to produce O_2^- suggested that it must, like other cells which reduce O_2, be endowed with superoxide dismutase and catalase. To our surprise no superoxide dismutase was found either in the assay in which its ability to prevent the reduction of cytochrome c by O_2^- generated enzymically in solution[31] is assessed, or in the assay in polyacrylamide gels, in which superoxide dismutase is seen as an achromatic band against a background of dark formazan produced by the reduction of NBT by O_2^- generated photochemically.[4] Catalase and peroxidase were similarly absent, whether assayed in solution or by staining for activity polyacrylamide gels into which the clarified lysate had been electrophoresed. To confirm the findings in the Jose strain of M. pneumoniae two other strains, 1428 and TW 48-5, were also assayed. In neither could

superoxide dismutase or catalase be detected even though both reduced NBT and
superoxide dismutase inhibited the reaction.

Results of assays of other mycoplasma and of Acholeplasma laidlawii for the
presence of superoxide dismutase and catalase. To deter whether M. pneumoniae
differed from its close relatives, other mycoplasmas A. laidlawii were assayed
for superoxide dismutase and catalase (Table 4). As in M. pneumoniae, none of
the mycoplasma tested were shown to possess superoxide dismutase or catalase
either by assays in solution or in polyacrylamide gels, even though in all the
species the reduction of NBT in the presence of NADH was partially inhibitable
by superoxide dismutase. In contract A. laidlawii did contain superoxide dismutase in both types of assay. As with the mycoplasma species, A. laidlawii
lacked catalase.

TABLE 4

RESULTS OF ASSAYS OF VARIOUS MYCOPLASMA SPECIES FOR SUPEROXIDE DISMUTASE
ACTIVITY AND CATALASE ACTIVITY

| Species | Superoxide dismutase | | Catalase | |
	assay[a]	gels	assay[a]	gels
M. hominis	0.04	–	0	–
M. pulmonis	0.11	–	0	–
M. arthritidis	0.02	–	0	–
M. gallisepticum	0.11	–	0	–
M. fermentans	0.21[b]	–	0	–
M. synoviae	0.01	–	0	–
A. laidlawii M728	32.8[c]	+	0	–
1305	32.9			

[a] In units/mg protein.

[b] This was the only value from assays of extracts of the mycoplasma which might
have represented real activity. The percent inhibition of cytochrome c which
the other extracts produced was not significantly different from the uninhibited
rate.

[c] Values obtained on different strains grown and assayed at different times.

DISCUSSION

The abilities of M. pneumoniae to consume O_2, to produce H_2O_2[12,13,26,36] to
reduce certain tetrazolium salts,[1] and to adhere to cells[32,34] have long been
recognized. The data we have presented suggests that these could be related
events. We have considered the possibility that these pathogens infect the
lung by adhering to the surface of cells where they exploit the oxygen-rich
environment to reduce O_2 to O_2^-. By virtue of their close contact with the

infected cell O_2^- can be directed into or onto its target with minimal loss by spontaneous dismutation. The data in support of this suggestion are still fragmentary. They consist of the evidence that in the presence of NADH both cells and lysates of M. pneumoniae make O_2^-. Whether the manufacture of O_2^- confers on M. pneumoniae any advantage remains to be tested.

Our inability to detect superoxide dismutase in mycoplasma is problematical. One possible explanation is that the organisms were grown under conditions of reduced concentrations of oxygen in which the synthesis of superoxide dismutase is not induced.[17,21-23] This seems unlikely since M. pneumoniae were grown in roller bottles in which the "lawns" of organisms are suspended on air on the walls of the flask during most of each revolution of the bottle. In addition A. laidlawii which did produce superoxide dismutase was cultured in still, batch cultures, conditions in which less O_2 would be available. The possibility that there was sufficient superoxide dismutase in the medium to render endogenous synthesis unnecessary was also considered. The superoxide dismutase present in the medium was found to be a constituent of the yeast extract. It could be inactivated by autoclaving, but even when M. pneumoniae were grown in medium containing autoclaved yeast extract no superoxide dismutase was detected.

Still other possibilities have been considered. By virtue of their intimate association with cells endowed with superoxide dismutase M. pneumoniae may avail themselves of the enzyme the host possesses, possibly by establishing intercellular communications.[9] Another possibility is that M. pneumoniae produces O_2^- only for export, removing it so efficiently to the medium that its own vulnerable constituents are spared. Conceivably M. pneumoniae is inherently resistant to the effects of O_2^-. Finally we should acknowledge the possibility that the production of O_2^- occurs only in vitro. Whatever the reason, mycoplasma appear unique among aerobic organisms in their capacity to form O_2^- and H_2O_2 without synthesizing the enzymes which are thought to be indispensable for mitigating the toxic effects of O_2^- and H_2O_2.

ACKNOWLEDGEMENTS

We gratefully acknowledge the technical assistance of Jeanne Dietz and Jacqueline Thomas. This work is supported by Grants AM 20207 and AM 02255 from the National Institutes of Arthritis, Metabolic and Digestive Diseases (NIAMDD), National Institutes of Health. R.E.L. is an Associate Investigator, Howard Hughes Medical Institute.

REFERENCES

1. Aluotto, B.B., Wittler, R.G., Williams, C.O. and Faber, J.E. (1970) Int. J. Systematic Bacteriol. 20, 35-58.
2. Barile, M.F. and Del Giudice, R.A. (1972) Pathogenic Mycoplasmas, A CIBA Foundation Symposium, ASP, Amsterdam, pp. 165-185.

3. Bak, A.L., Black, F.T., Christiansen, C. and Freundt, E.A. (1969) Nature, 224, 1209-1210.
4. Beauchamp, C. and Fridovich, I. (1971) Anal. Biochem. 44, 276-287.
5. Beers, R.F., Jr. and Sizer, I.W. (1952) J. Biol. Chem. 195, 133-140.
6. Cabantchik, Z.I. and Rothstein, A. (1972) J. Membrane Biol. 10, 311-330.
7. Cabantchik, Z.I. and Rothstein, A. (1974) J. Membrane Biol. 15, 227-248.
8. Cabantchik, Z.I. and Rothstein, A. (1974) J. Membrane Biol. 15, 207-226.
9. Cassell, G.H., Davis, J.K., Wilborn, W.H. and Wise, K.S. (1978) in Microbiology, Schlessinger, D. ed., Amer. Soc. Microbiol., Washington, D.C., pp. 399-403.
10. Chanock, R.M., Hayflick, L. and Barile, M.F. (1962) Proc. Nat. Acad. Sci. USA, 48, 41-49.
11. Clyde, W.A. (1964) J. Immunol. 92, 958-965.
12. Cohen, G. and Somerson, N.L. (1969) J. Bacteriol. 98, 547-551.
13. Cole, B.C., Ward, J.R. and Martin, C.H. (1968) J. Bacteriol. 95, 2022-2030.
14. Dalmark, M. and Wieth, J.O. (1972) J. Physiol. 224, 583-610.
15. Dalmark, M. and Wieth, J.O. (1970) Biochim. Biophys. Acta, 219, 525-527.
16. Davis, B.J. (1964) Ann. N.Y. Acad. Sci. 121, 404-427.
17. Gregory, E.M. and Fridovich, I. (1973) J. Bacteriol, 114, 543-548.
18. Gregory, E.M. and Fridovich, I. (1974) Anal. Biochem. 58, 57-62.
19. Guidotti, G., Colombo, J-P. and Foa, P.P. (1961) Anal. Chem. 33, 151-153.
20. Gunn, R.B., Dalmark, M., Tosteson, D.C. and Wieth, J.O. (1973) J. Gen. Physiol. 61, 185-206.
21. Hassan, H.M. and Fridovich, I. (1977) J. Bacteriol. 132, 505-510.
22. Hassan, H.M. and Fridovich, I. (1977) J. Biol. Chem. 252, 7667-7672.
23. Hassan, H.M. and Fridovich, I. (1977) J. Bacteriol. 129, 1574-1583.
24. Hayflick, L. (1975) Tex. Rep. Biol. Med. 23, Suppl. 1, 285-303.
25. Hu, P.C., Collier, A.M. and Baseman, J.B. (1977) J. Exp. Med. 145, 1328-1343.
26. Low, I.E., Eaton, M.D. and Proctor, P. (1968) J. Bacteriol. 95, 1425-1430.
27. Lowry, O.H., Roseborough, N.J., Farr, A.L. and Randall, R.J. (1951) J. Biol. Chem. 193, 265-275.
28. Lynch, R.E. and Fridovich, I. (1978) J. Biol. Chem. 253, 4697-4699.
29. Malinowski, D.P. and Fridovich, I. (1979) Biochemistry, 18, 237-244.
30. Maurer, H.R. (1971) in Disc Electrophoresis and Related Techniques of Polyacrylamide Gel Electrophoresis, Walter de Gruyter, Berlin, pp. 44-45.
31. McCord, J.M. and Fridovich, I. (1969) J. Biol. Chem. 244, 6049-6055.
32. Powell, D.A., Hu, P.C., Wilson, M., Collier, A.M. and Baseman, J.B. (1976) Infec. Immunity, 13, 959-966.
33. Ross, A.H. and McConnell, H.M. (1978) J. Biol. Chem. 253, 4777-4782.
34. Sobeslavsky, O., Prescott, B. and Chanock, R.M. (1968) J. Bacteriol. 96, 695-705.
35. Somerson, N.L. and Morton, H.E. (1953) J. Bacteriol. 65, 245-251.
36. Somerson, N.L. Walls, B.E. and Chanock, R.M. (1965) Science, 150, 226-228.
37. Tosteson, D.C. (1959) Acta Physiol. Scand. 46, 19-41.
38. Waud, W.R., Brady, F.O., Wiley, R.D. and Rajagopalan, K.V. (1975) Arch. Biochem. Biophys. 169, 695-701.
39. Aldridge, K.E. (1975) Infec. Immunity 12, 198-204.

IMPERMEABILITY OF THE ESCHERICHIA COLI CELL ENVELOPE TO SUPEROXIDE RADICAL

H. MOUSTAFA HASSAN AND IRWIN FRIDOVICH
Department of Biochemistry, Duke University Medical Center, Durham,
North Carolina 27710, USA

ABSTRACT

Paraquat mediates a superoxide dismutase-inhibitable reduction of cytochrome
c by suspensions of E. coli B. Paraquat reduction depended upon a NADPH:
paraquat diaphorase, present in the cytosol. Reduced paraquat could diffuse
across the cell envelope and react with dioxygen, in the suspending medium,
thus generating O_2^- in that compartment. Most of the paraquat reduced in the
cell, under the conditions used, reoxidized in situ and most of the O_2^- production was thus intracellular. The partitioning of reduced paraquat between intracellular and extracellular compartments, prior to reaction with dioxygen,
depended upon intracellular pO_2 and any strategy which raised intracellular pO_2
decreased the efflux of reduced paraquat and thus decreased extracellular O_2^-
production. Extracellular O_2^- and H_2O_2 did contribute to cell damage in proportion to the amount produced. Superoxide appeared to be unable to cross the
cell envelope in either direction and the only O_2^- which was effective in raising
the rate of biosynthesis of the manganese-superoxide dismutase, was that generated within the cell.

INTRODUCTION

Paraquat is readily reduced by E. coli, as evidenced by the accumulation of
the univalently reduced blue monocation radical, in anaerobic cultures containing
this substance. The paraquat radical reacts very rapidly with dioxygen to yield
O_2^- ($k_2 = 7.7 \times 10^8 \ M^{-1}sec^{-1}$)(Ref.1). Consequently, the blue paraquat does not
accumulate under aerobic conditions, but rather then serves as a continuing
source of O_2^-, as it undergoes repeated cycles of oxidation and reduction. This
has been demonstrated in diverse biological materials including illuminated
chloroplasts,[2,3] lung microsomes[4] and homogenates of lung, liver and kidney.[5]
Increased production of O_2^- is a major factor in the toxicity of paraquat, which
correlated with oxygen enhancement of this toxicity in plants,[6-8] rats[9] and
in E. coli.[10]

Cyanide-resistant respiration can be used as an approximate measure of the
extent of diversion of the intracellular electron flow by paraquat. In E. coli
1.0 mM paraquat causes a ten-fold increase in cyanide-resistant respiration and

a concomitant marked increase in the rate of synthesis of the manganese-containing superoxide dismutase (MnSOD). Both of these effects were entirely dependent upon the presence of both dioxygen and a source of electrons.[10,11] Furthermore, the increase in MnSOD, occasioned by aerobic paraquat, was an adaptive response which protected the cells against its lethality.[10,11]

It is probable that paraquat must enter E. coli before being reduced to the blue radical and that the O_2^- production, which is significant for cell damage and for induction of the MnSOD is also intracellular. Yet, we have recently reported[12] that paraquat will enable E. coli, suspended in a glucose medium, to reduce exogenous cytochrome c and this was inhibited by superoxide dismutase, added to the medium. It follows that the extracellular flux of O_2^- was increased by paraquat. This raised the possibility that the E. coli envelope might be permeable to O_2^-, as is the erythrocyte stroma.[13] We here report experiments designed to explore this question.

MATERIALS AND METHODS

Escherichia coli B B_{12}^-, ATCC 29682, was used throughout. It was grown at 37°C on a water bath rotary shaker (New Brunswick) at 200 rpm, in minimal medium[10] containing 0.5% glucose. Nutrient broth medium contained 0.8% nutrient broth from the Baltimore Biological Laboratories. Trypticase soy/yeast extract medium (3% trypticase soy broth and 0.5% yeast extract both from BBL) was used during inductions of superoxide dismutase and catalase by paraquat and by pyocyanine.

Reduction of exogenously added cytochrome c by cell suspensions was measured by incubating the cells at 37°C on a water bath rotary shaker at 120-150 rpm in reaction mixtures containing 25 μM cytochrome c, the specified concentrations of paraquat and electron donors, in the presence and absence of 50 μg/ml of the superoxide dismutase from bovine erythrocytes, in 0.05 M potassium phosphate buffer, pH 7.8. The cells were removed by filtration (Millipore filters 0.45 μ) before reading absorbance at 550 nm to determine the extent of cytochrome c reduction. A molar extinction coefficient of 2.1×10^4 (Ref.14) was used to calculate the concentration of reduced cytochrome c.

Experiments performed in the presence of cyanide required some modification of conditions. Since the copper-zinc superoxide dismutase is inhibited by cyanide, whereas the manganese enzyme is not, we then used the manganese enzyme isolated from E. coli. Furthermore, we noted that autoclaved glucose would itself cause the reduction of cytochrome c in the presence of cyanide; whereas non-autoclaved glucose did not. We accordingly used non-autoclaved glucose in measurements of cytochrome c reduction by whole cells in the presence of cyanide. It should be noted that autoclaving did not modify the ability of glucose to support respiration by E. coli, or to support the paraquat-dependent reduction of exogenous cytochrome c.

Oxygen uptake by cell suspensions and cell fractions was measured at 35°C using a Clark type polarographic electrode (Rank Bros., Bottisham, Cambridge, England). Cell fractionation was performed as described previously.[15] Cell-free extracts were prepared and assayed for superoxide dismutase and catalase, as previously described.[16] Protein was estimated by the method of Lowry et al.,[17] using pure bovine serum albumin as standard. Cell density was measured at 600 nm in cuvettes with 1 cm path. Under these conditions one absorbancy unit at 600 nm is equal to 0.28 mg of dry weight cells per ml. Viable cells were enumerated by spreading appropriate dilutions in quadruplicate on TSY solidified with 2% agar and counting colonies after 24 hours of aerobic incubation at 37°C.

RESULTS

Extracellular production of superoxide mediated by paraquat. E. coli, when suspended in fresh glucose-salts medium, caused a very slow reduction of extra-cellular cytochrome c (21 µM per hr per mg dry wt). Paraquat at 6.0 mM increased this rate of cytochrome c reduction by more than 10-fold (230 µM per hr per mg dry wt). SOD added to the medium to 50 µg/ml inhibited cytochrome c reduction 57% in the absence of paraquat and 95% in its presence. These results are shown in Figure 1. The effect of paraquat concentration on the rate of reduction of cytochrome c by E. coli was also explored and is shown in Figure 2. Since E. coli are certainly not freely permeable to cytochrome c, these results suggest that paraquat, added to aerobic E. coli, does increase the level of extracellular O_2^-.

The nature of the proximal electron donor. It seemed likely that either NADH or NADPH was the proximal electron donor to paraquat. NADH and NADPH cannot cross the E. coli envelope and, when tested with whole cells, they were much poorer substrates for extracellular paraquat mediated O_2^- production than was glucose. Indeed it may be that the very small rates seen with NADH and NADPH were due to some cell lysis and release of enzymes into the medium. Purposeful lysis of cells, by sonication for 2 minutes, dramatically increased O_2^- production from 0.5 mM NADPH, in the presence of 6 mM paraquat. NADH was, under identical conditions, less effective and much of the cytochrome c reduction, which was seen with NADH, was insensitive to superoxide dismutase. These results are presented in Table 1. The rate of superoxide dismutase-inhibitable reduction of exogenous cytochrome c by intact E. coli, in the presence of glucose plus paraquat, was less than 5% of the rate seen with lysed cells in the presence of NADPH. The amount of O_2^- appearing in the suspending medium in the presence of paraquat is thus a small fraction of the amount which could be made inside the cells.

Paraquat-mediated respiration. Intact E. coli were unable to utilize NADPH even in the presence of paraquat, but proceeded to do so when the detergent

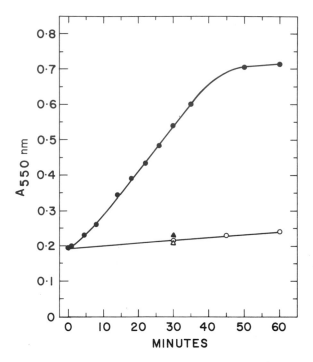

Fig. 1. Kinetics of the paraquat-dependent, superoxide dismutase-inhibitable cytochrome c reduction by suspensions of E. coli. Glucose-grown cells were suspended to A_{600} nm = 0.56, in 50 ml reaction mixtures containing: cytochrome c, 0.025 mM; glucose, 0.5%; ± paraquat, 6 mM; ± CuZnSOD, 50 µg/ml, in a basal-salts medium without vitamin B_{12} at pH 7.0. The flasks were incubated on water bath shaker (150 rpm) at 37°C. At the specified time intervals samples were removed and the cells were separated by passage through 0.45 µ Millipore filter before absorbance at 550 nm was recorded. ●, in presence of paraquat; O, in presence of paraquat and SOD; ▲, in absence of paraquat; Δ, in absence of para-quat but in presence of SOD.

Triton X-100 was added to eliminate the permeability barrier of the cell envelope. This is shown by the upper line in Figure 3 (part A). NADH was not a good sub-strate, even in the presence of the detergent, as shown by the lower line. As expected, glucose was an excellent substrate for whole cells. Cyanide inhibited the glucose-fueled respiration and paraquat reduced the cyanide inhibition and allowed a rate of respiration nearly equal to that seen in the presence of NADPH plus Triton X-100.

Localization of the NADPH:paraquat reductase. Soluble and particulate fractions of E. coli B, disrupted in a French Press at 20,000 lbs/in^2, were isolated as previously described.[15] The particulate fraction, which bears the electron transport assemblies, catalyzed the expected brisk oxidation of NADH which was inhibited by cyanide and not dependent upon the presence of paraquat

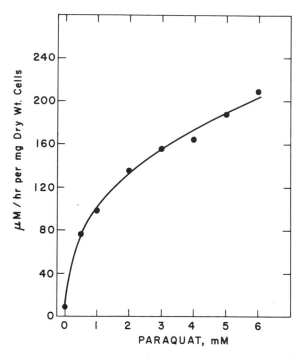

Fig. 2. Effect of paraquat concentration on the rate of superoxide dismutase-inhibitable cytochrome c reduction by whole cells of E. coli B. Experimental conditions were the same as in Figure 1 except that the final concentration of paraquat was varied.

TABLE 1

COMPARISONS OF NADH AND NADPH AS ELECTRON DONORS FOR THE PARAQUAT-MEDIATED, SUPEROXIDE DISMUTASE-INHIBITABLE REDUCTION OF CYTOCHROME c BY WHOLE CELLS AND BY HOMOGENATES[a]

Enzyme Source	Electron Donor	Rate[b] μM/hr/mg dry weight
Whole cells	NADH (0.5 mM)	2.3
Whole cells	NADPH (0.5 mM)	4.5
Whole cells	Glucose (0.5%)	219
Homogenate	NADPH (0.5 mM)	6,120

[a]E. coli B were grown in glucose-minimal medium to the mid logarithmic phase of growth and were washed as described in Table 2. Washed cells were suspended to 168 μg dry weight/ml in 5.0 ml reaction mixtures containing: cytochrome c, 0.025 mM; paraquat, 6 mM; NADH or NADPH, 0.5 mM; $MgSO_4$, 1.0 mM and potassium phosphate, 50 mM. The pH was 7.8 and the temperature was 37°C. After 30 min at 150 rpm on an oscillating water bath in 25 ml Erlenmeyer flasks the suspensions were clarified by passage through a 0.45 μ Millipore filter and absorbance at 550 nm was recorded. Homogenate was prepared by sonication of the cell suspension for 2 min. Rates with homogenates were recorded continuously on the reaction mixtures without filtration. Rates are recorded as the difference between the rates in the absence and in the presence of 50 μg/ml of superoxide dismutase.
[b]All rates were corrected for small endogenous rates seen in the absence of paraquat and for the rates seen in the presence of paraquat, but in the absence of an exogenous electron source.

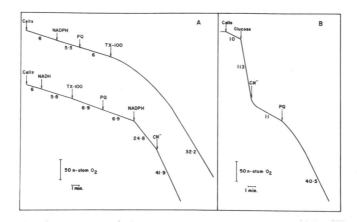

Fig. 3. Paraquat-dependent oxygen uptake by E. coli B. A) Effects of NADH, NADPH, and of Triton X-100. B) Effect of paraquat on the rate of cyanide-resistant respiration. Total volume of the reaction mixture was 2 ml. Reactants were added at the following final concentrations: cells, 168 μg/ml; NADH or NADPH, 0.5 mM; paraquat, 6 mM; Triton X-100, 0.5%; NaCN, 1 mM; glucose, 0.5%; potassium phosphate, 50 mM (pH 7.0). Numbers shown on the tracings indicate the slopes expressed as n-atom O_2/min.

and as shown in Figure 4 (parts A and B), NADH could not be replaced by NADPH. The reverse situation was observed with the soluble fraction. There was then a rapid, paraquat-dependent oxidation of NADPH, but not of NADH (Figure 4, parts C and D). E. coli clearly contains a soluble NADPH:paraquat reductase and the dioxygen consumptions recorded in Figure 4 depend upon the autoxidation of the paraquat radical. Since this autoxidation produces O_2^-, which would dismute to O_2 plus H_2O_2, one might generate H_2O_2 more rapidly than it could be decomposed by endogenous catalase. That this was the case is shown in Figure 4 (part D). Thus, addition of catalase decreased oxygen uptake, because the H_2O_2, which would have accumulated in the absence of exogenous catalase, was decomposed. This effect of catalase was eliminated by cyanide, which inhibited its action. When cyanide itself was added to the soluble fraction, respiring in the presence of NADPH and paraquat, it augmented dioxygen uptake by inhibiting endogenous catalase and thus allowing greater accumulation of H_2O_2. This too is shwon in Figure 4 (part D).

Effects of intracellular pO_2 on extracellular superoxide. Reduced paraquat, generated in the cytosol of E. coli by the action of the soluble NADPH:paraquat reductase, might diffuse from the cell and autoxidize in the suspending medium or alternatively might autoxidize in situ. If the former did not occur then O_2^-, generated within the cells, would have to escape by diffusion in order to account for the paraquat-dependent, superoxide dismutase-inhibitable reduction of extracellular cytochrome c. These two possibilities were distinguished by exploring

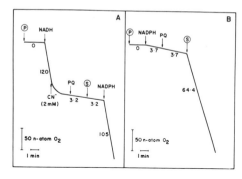

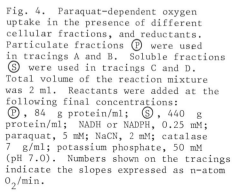

Fig. 4. Paraquat-dependent oxygen uptake in the presence of different cellular fractions, and reductants. Particulate fractions Ⓟ were used in tracings A and B. Soluble fractions Ⓢ were used in tracings C and D. Total volume of the reaction mixture was 2 ml. Reactants were added at the following final concentrations: Ⓟ, 84 g protein/ml; Ⓢ, 440 g protein/ml; NADH or NADPH, 0.25 mM; paraquat, 5 mM; NaCN, 2 mM; catalase 7 g/ml; potassium phosphate, 50 mM (pH 7.0). Numbers shown on the tracings indicate the slopes expressed as n-atom O_2/min.

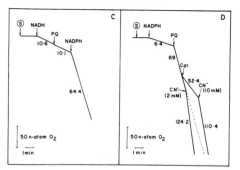

the effects of changes in intracellular pO_2. Intracellular generation of O_2^- should increase with intracellular pO_2. In contrast, extracellular generation of O_2^- should decrease with increasing intracellular pO_2, becuase the lower the intracellular pO_2 the greater the likelihood that the paraquat would escape reaction with dioxygen long enough to allow diffusion from the cell. It would then react with dioxygen and generate O_2^-, in the suspending medium.

Two methods were used to manipulate intracellular pO_2. Rapid dioxygen consumption within the cell depends upon the cytochrome oxidase and inhibiting this enzyme with cyanide should raise the intracellular pO_2 and eliminate the sharp gradient in pO_2 between the cell interior and the suspending medium. As shown in Figure 5, 1.0 mM cyanide markedly decreased the paraquat-dependent, superoxide dismutase-inhibitable reduction of cytochrome c by E. coli suspended in a glucose medium. Previous work[15] has shown that cyanide itself causes an induction of superoxide dismutase in E. coli and augments the induction caused by paraquat. Cyanide therefore increases the intracellular production of O_2^- and augments the intracellular production of O_2^- by paraquat. If O_2^- were able to effuse from the cell, then cyanide should have increased extracellular reduction of cytochrome c. Its opposite effect indicates that O_2^- cannot cross the cell envelope, but that

64

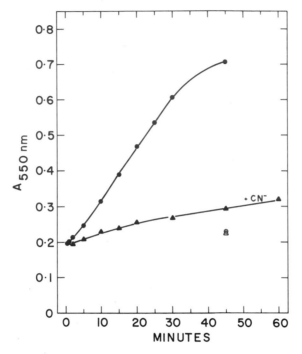

Fig. 5. Effect of cyanide on the rate of paraquat-dependent, superoxide dismutase-inhibitable cytochrome c reduction by whole cells of E. coli. Glucose grown cells were suspended to A_{600} nm = 0.6 (168 μg dry wt cells/ml) in 50 ml reaction mixtures containing: cytochrome c, 0.025 mM; non-sterilized glucose, 0.5%, paraquat, 6 mM; ± MnSOD, 20 μg/ml ± NaCN, 1 mM; in a potassium phosphate buffer (50 mM) pH 7.8. The flasks were incubated on water bath shaker (120 rpm) at 37°C. Cells were separated by Millipore filters before absorbance of the supernatant was measured at 550 nm. ●, no cyanide; ▲, plus 1 mM cyanide; 0, plus SOD, no cyanide; Δ, plus SOD and cyanide.

the paraquat radical can do so and can then generate O_2^- in the medium.

Intracelluar pO_2 can also be manipulated directly without the use of inhibitors, by varying the extracellular pO_2. As shown in Figure 6, elevating pO_2 in the medium decreased the rate of cytochrome c reduction, much as the addition of cyanide had done (Figure 5). The autocatalytic appearance of the rates in Figure 6 was due to progressive depletion of oxygen during the period of observation. These results also lead to the conclusion that O_2^- does not cross the cell envelope.

Manipulation of intracellular superoxide dismutase and catalase. If the O_2^- detected in the medium was derived, by effusion, from the intracellular O_2^-, then elevated levels of superoxide dismutase, within the cell, would decrease the O_2^- in the medium. In contrast, if the O_2^- in the medium was generated there by autoxidation of reduced paraquat, then the level of intracellular superoxide dismutase should not be a relevant factor. The level of intracellular superoxide dismutase

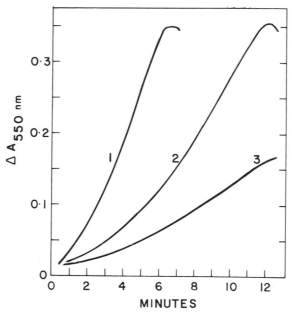

Fig. 6. Effect of oxygenation on the rate of paraquat-depedent, superoxide dismutase-inhibitable cytóchrome c reduction by whole cells of E. coli B. The reaction mixtures contained: glucose grown cells, 168 g/ml; glucose, 0.5%; parquat, 6 mM; potassium phosphate (pH 7.8), 50 mM, in 3 ml total volume. The change in absorbance was recorded at $37^{o}C$ in a double beam spectrophoto-meter; CuZnSOD, 50 μg/ml was added to the reference cuvette. Curve 1, reaction mixture was kept still; curve 2, reaction mixture was partially oxygenated with 100% oxygen; curve 3, reaction mixture was fully oxygenated with 100% oxygen.

in E. coli was modified by growth in aerobic TSY media containing 0.2, 0.5 and 1.0 mM paraquat.[10,11,15] These cells were then tested for their abilities to mediate the superoxide dismutase-inhibitable reduction of cytochrome c in the presence of paraquat. The results are given in Table 2. Growth in 0.2 mM paraquat increased intracellular superoxide dismutase over three-fold. This did not diminish generation of extracellular O_2^-, in the presence of paraquat, but rather increased it. Intracellular O_2^- is clearly not the source of extra-cellular O_2^-. Intracellular superoxide dismutase does protect the cell against damage by O_2^- and increasing the level of this enzyme allowed retention of a greater fraction of normal metabolic function. This actually increased the intracellular electron supply, which increased the rate of reduction of paraquat, which, in turn, increased the extracellular O_2^-.

Cells grown in the presence of higher levels of paraquat (0.5 and 1.0 mM) did show a progressive decrease in extracellular O_2^- and this is explained by the induction of catalase. Thus, 0.5 mM paraquat increased catalase by 30%, while 1.0 mM paraquat increased it by 300% (Table 3). This is not unexpected, since

66

TABLE 2

EFFECTS OF THE INTRACELLULAR LEVELS OF SUPEROXIDE DISMUTASE AND OF CATALASE ON
THE RATE OF THE PARAQUAT-DEPENDENT, SUPEROXIDE DISMUTASE-INHIBITABLE CYTOCHROME c
REDUCTION BY WHOLE CELLS OF E. COLI B

Paraquat-Induced Cells[a]				Pyocyanine-Induced Cells[a]			
Conc.of Inducer	SOD[b]	Catalase[b]	ΔA_{550}/10 min[c]	Conc.of Inducer	SOD[b]	Catalase[b]	ΔA_{550}/10 min[c]
(mM)	(U/mg)	(U/mg)		(mM)	(U/mg)	(U/mg)	
0	9.8	12.1	0.320	0	10.3	13.2	0.237
0.2	32.1	13.6	0.410	0.01	26.6	14.1	0.277
0.5	61.6	16.9	0.317	0.1	48.9	198.1	0.051
1.0	70.3	35.6	0.202	0.2	82.4	397.0	0.047

[a]E. coli B were grown aerobically to late logarithmic phase (16 hr) in a glucose-minimal medium at 37°C. This culture was used to inoculate TSY-media containing the indicated concentrations of paraquat or of pyocyanine, to an intial turbidity ($A_{600 \ nm}$) of 0.15. The test cultures were then incubated for 2½ hours at 37°C in a water-bath shaker at 200 rpm. The cells were then harvested by centrifugation and were washed once with cold TSY medium. Washed cells were resuspended in cold TSY media and kept on ice for 1 hr to rid them of the inducers. Cells were collected by centrifugation and resuspended in a small volume of 0.05 M potassium phosphate (pH 7.8) containing 1 mM $MgSO_4$. The cells were then used immediately to avoid any autolysis.

[b]Portions of the above cell suspensions were disrupted by sonication and used for SOD and catalase assays.

[c]The rate of superoxide dismutase-inhibitable cytochrome c reduction was measured as described in Table 1 but with 0.5% glucose as the electron source. The reaction mixtures were incubated for 10 min at 37°C and 120 rpm.

TABLE 3

EFFECT OF EXOGENOUSLY ADDED SUPEROXIDE DISMUTASE ON THE INDUCTION OF THE ENZYME BY PARAQUAT

Conditions[a]	SOD (U/mg)
TSY	8.7
TSY + 0.2 mM paraquat	26.7
TSY + 0.2 mM paraquat + SOD	24.5[b]
TSY + 0.2 mM paraquat + catalase	26.5
TSY + 0.2 mM paraquat + SOD + catalase	24.1[b]

[a]Growth and induction conditions are the same as in Table 2. CuZnSOD and catalase were added at 50 g/ml and 30 g/ml, respectively.

[b]The superoxide dismutase activity was assayed in the presence of 1 mM NaCN to inhibit any residual CuZnSOD.

paraquat leads to the production of O_2^- and thence of H_2O_2. The action of catalase upon intracellular H_2O_2 increases intracellular pO_2 and we have already shown that raising intracellular pO_2 decreases the efflux of reduced paraquat and thus decreases extracellular production of O_2^-. This proposed explanation was tested by using cells grown in the presence of pyocyanine, which profoundly induced catalase.[15] As shown in Table 2, 0.2 mM pyocyanine increased catalase 30-fold and this markedly decreased the extracellular production of O_2^- in the presence of paraquat. The results in Table 2 thus support the view that O_2^- cannot diffuse from within E. coli and furthermore demonstrate the intracellular action of catalase upon endogenous H_2O_2.

Can exogenous superoxide enter E. coli? Given that O_2^- made within E. coli cannot significantly diffuse into the medium, we are left with the inverse question: Can O_2^- diffuse into E. coli? Table 3 demonstrates that superoxide dismutase and/or catalase, added to the medium, had no effect on the induction of superoxide dismutase by 0.2 mM paraquat. The superoxide dismutase added to the medium was the copper-zinc enzyme from bovine erythrocytes and its activity could be suppressed by cyanide, which does not inhibit the bacterial enzymes. It was thus, possible, in the presence of cyanide, to assay induction of the E. coli superoxide dismutase, without interference from the copper-zinc enzyme. Superoxide,generated in the medium, by autoxidation of the blue paraquat radical, clearly does not contribute to induction of superoxide dismutase seen in the presence of paraquat. Furthermore, as shown in Table 4, an extracellular enzymatic source of O_2^- (the xanthine oxidase reaction) does not cause induction of superoxide dismutase in E. coli. Since O_2^-, generated within the cells does cause this induction, we conclude that O_2^- cannot pass, from the medium, into E. coli.

TABLE 4

EFFECT OF EXOGENOUSLY GENERATED SUPEROXIDE ON THE INDUCTION OF SUPEROXIDE DISMUTASE

Conditions[a]	SOD (U/mg)
TSY	8.7
TSY + 1 mM xanthine	8.4
TSY + 1 mM xanthine + 10^{-8} M xanthine oxidase	7.7
TSY + xanthine + xanthine oxidase + SOD	9.9[b]

[a] Growth conditions were the same as in Table 2. Cells were grown for a total of $2\frac{1}{2}$ hr.

[b] Assayed in the presence of 1 mM NaCN.

Lethality of extracellular superoxide. Fluxes of O_2^- have been seen to inactivate virus,[18] depolymerize hyaluronate,[19] induce lipid peroxidation,[20,21]
lyse erythrocytes and artifical vesicles[22,23] and kill bacteria.[24-27] Paraquat
increases the production of O_2^- by E. coli both intracellularly and extracellularly.
The extracellular production of O_2^-, as judged by the superoxide dismutase-
inhibitable reduction of cytochrome c, is a small fraction (somewhat less than
5%) of the intracellular production of O_2^-, as judged by the extent of cyanide-
resistant respiration and by the NADPH:paraquat reductase activity of soluble
extracts. It nevertheless appeared possible that the extracellular O_2^- and
H_2O_2 might contribute to the lethality of paraquat. This was tested by adding
superoxide dismutase and/or catalase to suspensions of E. coli exposed to
paraquat in the presence of an inhibitor of protein synthesis, to prevent
induction of superoxide dismutase. As shown in Figure 7, exogenous superoxide
dismutase and catalase did provide some protection against lethality of paraquat.
Extracellular production of O_2^- and thence of H_2O_2, by paraquat, did therefore
contribute modestly to its lethality.

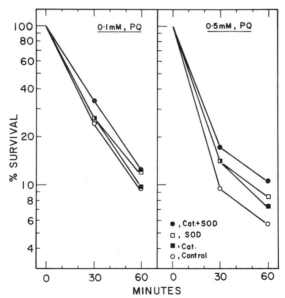

Fig. 7. Effect of exogenously added superoxide dismutase and catalase on the
toxicity of paraquat. E. coli were grown for 17 hr in nutrient broth medium
and were then diluted to 3×10^7 cells/ml into nutrient broth medium containing
0.5% glucose, 0.5 mg/ml of puromycin, and the indicated concentrations of
paraquat (0.1 and 0.5 mM). CuZnSOD and catalase were added at 100 μg/ml of
each. At intervals, viable cells were enumerated by plating appropriate
dilutions on TSY agar.

DISCUSSION

The data presented above can be explained on the basis of the reactions schematically portrayed in Figure 8. NADH, generated within the cell, transmits electrons to oxygen by the electron transport chain. The ultimate enzyme of this pathway, cytochrome oxidase, reduces dioxygen to water without releasing either O_2^- or H_2O_2. NADPH, in contrast, transmits electrons to paraquat, when it is present, and so generates the blue paraquat monocation radical. This paraquat radical can react with dioxygen inside the cell, or it can diffuse from the cell before reacting with dioxygen in the suspending medium. In either case, the product is O_2^-, which does not cross the cell envelope. The reaction of the paraquat radical with dioxygen is a very rapid process[1] and the fraction of paraquat radical which escapes this reaction long enough to diffuse from the cell, depends upon the intracellular pO_2. Inhibition of cytochrome oxidase, with cyanide, or raising the extracellular pO_2, increases intracellular pO_2 and so decreases the efflux of paraquat radical from the cell. Increasing the intracellular catalase likewise increases intracellular pO_2 and also decreases the outflow of reduced paraquat. The intracellular level of catalase was manipulated by prior induction with paraquat or, more effectively, with pyocyanine.

It might be argued that paraquat augmented the rate of superoxide dismutase-inhibitable reduction of exogenous cytochrome c merely by increasing the permeability of the cells, perhaps even by lysing the cells. Several lines of evidence exclude that possibility. Thus the data in Figure 3 and in Table 1 demonstrate that the cells were impermeable to NADH and to NADPH, in the presence of paraquat. The data in Figures 5 and 6 indicate that there was a steep cyanide sensitive gradient of dioxygen between the medium and the cell interior. This was due to rapid intracellular dioxygen consumption by cytochrome oxidase and this gradient could not have been sustained by leaky or lysed cells. Figure 7 demonstrates that superoxide dismutase and catalase added to the medium protected only slightly, in proportion to the calculated production of O_2^- and H_2O_2. Previous studies have shown very large protections by increases in intracellular superoxide dismutase.[10] The data in Table 2 demonstrate that elevations of intracellular catalase increased intracellular pO_2 and such an effect would depend upon intact cell envelopes. The data in Table 3 show that extracellular superoxide dismutase did not diminish intracellular O_2^- in the presence of paraquat, as judged by the extent of induced biosynthesis of superoxide dismutase. Taken in toto the data strongly support the view that the E. coli envelope was intact during these studies with paraquat.

The O_2^- detected in the suspending medium, when paraquat was present, was generated by autoxidation of that small fraction (approximately 5%) of the reduced paraquat which escaped reaction with intracellular dioxygen and diffused from the cell. Extracellular O_2^- and thence H_2O_2 production, was thus much less than intracellular O_2^- and H_2O_2 production but it did measurably

70

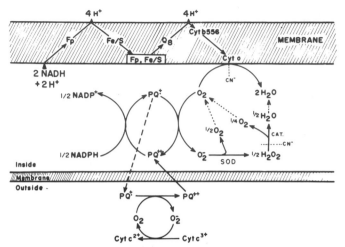

Fig. 8. Schematic representation of pathways of dioxygen reduction in the presence of paraquat (PQ).

contribute to the lethality of paraquat in the presence of puromycin and its effects could be countered by superoxide dismutase and by catalase, added to the suspending medium. Since O_2^- did not measurably cross the cell envelope, only intracellular O_2^- was effective in inducing increased biosynthesis of the manganese superoxide dismutase. Superoxide, generated in the suspending medium, was entirely ineffective in this regard and superoxide dismutase added to the medium did not diminish the inductive effects of intracellular O_2^-.

Cyanide-resistant respiration by whole cells acting on glucose in the presence of 6 mM paraquat was 3,600 µM O_2/hr/mg dry wt cells. Since cyanide would inhibit catalase, this respiration must have been accompanied by the accumulation of H_2O_2. If we assume it was all due to the production of O_2^- by the autoxidation of reduced paraquat, we have 7,200 µM O_2^- produced per hr per mg dry wt cells. This correlates very well with the 6,120 µM cytochrome c reduced per hr per mg dry wt cells when the supernatant fraction was incubated with 6 mM paraquat, 0.025 mM cytochrome c and 0.5 mM NADPH. The soluble NADPH:paraquat reductase can thus generate sufficient O_2^- to account for virtually all of the paraquat-dependent cyanide-resistant respiration by whole cells. The superoxide dismutase-inhibitable, paraquat-dependent reduction of exogenous cytochrome c by whole cells was 219 µM per hr per mg dry wt cells or only 3-4% of the total intracellular process.

ACKNOWLEDGEMENTS

This work was supported by research grants from the National Institutes of Health, Bethesda, Maryland (GM-10287); from the United States Army Research Office, Research Triangle Park, N.C. (DRXRO-PR-P-15319-L); and from Merck,Sharp and Dohme, Rahway, New Jersey.

REFERENCES

1. Farrington, J.A., Ebert, M., Land, E.J. and Fletcher, K. (1973) Biochim. Biophys. Acta, 314, 372-381.
2. Miller, R.W. and MacDowell, F.D.H. (1975) Biochim. Biophys. Acta, 387, 176-187.
3. Harbour, J.R. and Bolton, J.R. (1975) Biochem. Biophys. Res. Commun. 64, 803-807.
4. Montgomery, M.R. (1977) Res. Commun. Chem. Pathol. Pharmacol. 16, 155-158.
5. Baldwin, R.C., Pasi, A., MacGregor, J.T. and Hine, C.H. (1975) Toxicol. Appl. Pharmacol. 32, 298-304.
6. Mees, G.C. (1960) Ann. Appl. Biol. 48, 601-612.
7. Calderbrook, A. (1968) Advan. Pest Control Res. 8, 127-238.
8. Conning, D.M., Fletcher, K. and Swan, A.A.B. (1969) Brit. Med. Bull. 25, 245-249.
9. Fisher, H.K. (1977) in Biochemical Mechanisms of Paraquat Toxicity, Autor, A.P., ed., Academic Press, New York, pp.57-65.
10. Hassan, H.M. and Fridovich, I. (1978) J. Biol. Chem. 253, 8143-8148.
11. Hassan, H.M. and Fridovich, I. (1977) J. Biol. Chem. 252, 7667-7672.
12. Hassan, H.M. and Fridovich, I. (1979) Fed. Proc. 38, Abstr.825.
13. Lynch, R.E. and Fridovich, I. (1978) J. Biol. Chem. 253, 4697-4699.
14. Massey, V. (1959) Biochim. Biophys. ACta, 34, 255-256.
15. Hassan, H.M. and Fridovich, I. (1979) Arch. Biochem. Biophys. in press.
16. Hassan, H.M. and Fridovich, I. (1977) J. Bacteriol. 129, 1574-1583.
17. Lowry, O.H., Rosenbrough, H.J., Farr, A.L. and Randall, R.J. (1951) J. Biol. Chem. 193, 265-275.
18. Lavelle, F., Michelson, A.M. and Dimitrejevic, L. (1973) Biochem. Biophys. Res. Commun. 55, 350-357.
19. McCord, J.M. (1974) Science, 185, 529-531.
20. Kellogg,E.W.,III and Fridovich, I. (1975) J. Biol. Chem. 250, 8812-8817.
21. Gutteridge, J.M.C. (1977) Biochem. Biophys. Res. Commun. 77, 379-386.
22. Kellogg, E.W.,III and Fridovich, I. (1977) J. Biol. Chem. 252, 6721-6728.
23. Goldstein, I.M. and Weissman, G. (1977) Biochem. Biophys. Res. Commun. 75, 604-609.
24. Gregory, E.M. and Fridovich, I. (1974) J. Bacteriol, 117, 166-169.
25. Babior, B., Curnutte, J.T. and Kipnes, R.S. (1975) J. Lab. Clin. Med. 85, 235-244.
26. Klebanoff, S.J. and Rosen, H. (1979) in Oxygen Free Radicals and Tissue Damage, Ciba Foundation Symposium, New Series, Vol. 65, Excerpta Medica, Amsterdam, pp.263-277.
27. Pederson, T.C. and Aust, S.D. (1972) Biochem. Biophys. Res. Commun. 48, 789-795.

THE PRODUCTION AND UTILISATION OF SUPEROXIDE DURING THE CARBOXYLATION
OF GLUTAMYL RESIDUES IN PREPROTHROMBIN

M. PETER ESNOUF,[+] ANNETTE I. GAINEY,[+] MARTIN R. GREEN,[++] H. ALLEN O. HILL,[++]
MONIKA J. OKOLOW-ZUBKOWSKA[++] AND STEPHEN J. WALTER[+]
[+]Nuffield Department of Clinical Biochemistry, Radcliffe Infirmary, Oxford
OX2 6HE, England; [++]Inorganic Chemistry Department, University of Oxford,
Oxford OX1 3QR, England

The synthesis of prothrombin, an essential constituent of the blood clotting
cascade, requires the post ribosomal conversion of specific glutamyl residues in
the molecule to the γ-carboxy analogue.[1,2] The ten glutamyl residues involved
are located in the amino terminal region of the molecule (Figure 1) and their carboxylation is essential for the rapid conversion of prothrombin to thrombin by
physiological activators.

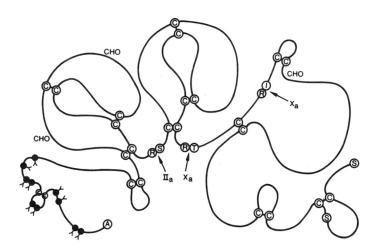

Fig. 1. A schematic representation of the structure of prothrombin, based on
the known sequence, showing the sites of carboxylation (solid circles). The one-
letter notation for amino acids is used. (From Ref. 41).

The carboxylation reaction requires vitamin K_1, NADH, dioxygen, carbon dioxide,
preprothrombin (or synthetic peptide substitutes) and a membrane bound carboxylase

complex present in liver microsomes.[3-5] The microsomes contain a vitamin K reductase[6] which, in the presence of NADH, reduces the vitamin to the semiquinone or
hydroquinone. Carboxylation of preprothrombin and the peptide substitute may also
be achieved by adding the vitamin in a reduced form (vitamin K_1 hydroquinone) and
omitting the NADH.[7]

The reaction of reduced quinones with dioxygen is known to produce both superoxide
and hydrogen peroxide.[8,9] Esnouf et al.[10] have demonstrated that vitamin K_1 will
stimulate superoxide production in rat liver microsomes presumably via the NADH
dependent vitamin K_1 reductase/vitamin K_1 couple required for carboxylase activity.

The carboxylation of prothrombin is accompanied by the formation of vitamin K_1
epoxide.[11] The formation of the epoxide may not require concomitantly the carboxylation of precursor prothrombin since the epoxide can be generated directly from
the reaction of superoxide and hydrogen peroxide with vitamin K_1 (Refs. 12, 13).
Catalase has been found to be ineffective at inhibiting carboxylation in rat liver
microsomes but to be a reasonable inhibitor of epoxide formation under the same
conditions, suggesting that the two reactions can be uncoupled.[10] However, at the
present time the relationship between the epoxidation and carboxylation reactions
remains uncertain.

In experiments to elucidate the mechanism of the carboxylation reaction Larson
and Suttie[14] have shown that glutathione peroxidase inhibits both carboxylation
and epoxidation and have suggested that a hydroperoxide adduct of the vitamin may
be the intermediate in both processes. However, Esnouf et al.[10,15,16] have demonstrated that both carboxylation and epoxidation are inhibited by bovine erythrocyte
superoxide dismutase (SOD), albeit at high concentrations, and have proposed that
superoxide or hydrogen peroxide may react with carbon dioxide to yield various
carbon containing intermediates (e.g. CO_2^-, CO_3^-, CO_4^-, HCO_4^-, $C_2O_6^-$). One or
more of these species may be a substrate for the carboxylase.

The data presented here further implicate the involvement of superoxide in the
carboxylation reaction.

MATERIALS AND METHOD

Microsomes were prepared from the livers of 3-5 day old calves, normal and
vitamin K deficient male Sprague-Dawley rats, and carboxylation assays performed
by the methods previously described.[10] The microsomal suspensions were diluted
differently so that the carboxylase activity of the preparations from each source
was approximately the same. In the case of the vitamin K deficient rats, 1 ml of
the suspension was obtained from 0.5 g liver, while with normal rats the suspension
was twice as concentrated. When calf liver was used 1 ml of the suspension was
derived from 2 g liver. The e.p.r. experiments were carried out using more concentrated suspensions such that 1 ml of microsomes was prepared from 5 g liver. Where
indicated the microsomal preparation was solublised with Triton X-100.[17]

The pentapeptide substrate was synthesised by the active ester procedure.[10]
Superoxide production by rat liver microsomes was assayed by the method of Esnouf
et al.[10]

Bovine erythrocyte SOD was a gift from Dr. J.V. Bannister; catalase, vitamin
K_1, NADH, xanthine oxidase Type III and ferricytochrome c were obtained from Sigma
and $NaH^{14}CO_3$ was from the Radiochemical Center, Amersham.

5,5-dimethyl-1-pyrroline-N-oxide (DMPO) was prepared as described by Bonnet et
al.[18] and further purified by the method of Buettner et al.[19] The concentration of
the stock solution was determined spectrophotometrically (ε_{234}= 7700$M^{-1}cm^{-1}$) in ethanol.

The copper and nickel complexes were prepared according to published methods:
Copper penicillamine ((D-penicillaminato)$_{2n}$(aqua)$_{2n}$(copper(II)$_n$) and copper aspirinate (tetrakis-μ-acetylsalicyato-dicopper(II) as in Ref. 20; copper tyrosine
(bis(L-tyrosinato)-copper(II)) as in Ref. 21; nickel glycinate (diaquobis-(glycinato)-
nickel(II)) as in Ref. 22. Stock solutions (1-8 mM) were made up in water and
the metal content determined by atomic absorption spectroscopy. The superoxide
dismutase activity of the complexes was assayed by the photoreduction method.[23]

Vitamin K reductase was purified as described by Hosoda et al.[24] and assayed
by measuring the rate of reduction of potassium ferricyanide present on the inside
of liposomes which contained vitamin K_1 in the lipid walls.[6] The reduction of
the ferricyanide was followed at 416/470 nm using a double beam spectrometer (DW2,
American Instrument Co. Inc.); $\Delta\varepsilon$ = 960 $M^{-1}cm^{-1}$ under these experimental conditions.
The assay mixture contained 0.4 ml liposomes, 0.1 ml NADH (16 mM), 0.05 ml of sample
and 1.95 ml of 20 mM Tris/HCl buffer, pH 7.5 at 25°C. One unit of vitamin K reductase activity is defined as the amount of enzyme required to reduce 1 μmole of
ferricyanide per minute under the assay conditions.

Electron paramagnetic resonance spectra were recorded using a Varian E-109 spectrometer with 100 KHz field modulation in the first derivative mode. Samples of
the incubation mixtures were drawn rapidly into a Varian flat quartz cell (0.2 ml)
by a simple syringe technique.[25] The incubation mixtures for the e.p.r. experiments
contained 0.1 ml microsomes suspended in 0.15 M KCl/50 mM Tris-HCl buffer, pH 7.5
(buffered KCl), or 0.1 ml of the purified vitamin K reductase preparation in 50
mM sodium phosphate buffer, pH 7.5, DMPO (100 mM), diethylenetriaminepentaacetic
acid (DTPA) (1 mM), NADH (0.6 mM) and 0.25 ml of vitamin K_1 liposomes prepared
without ferricyanide. The volume of the sample was adjusted to 0.5 ml with the
appropriate buffer.

Xanthine oxidase activity was assayed by following the conversion of xanthine
to urate at 295 nm in 10^{-4}M xanthine, 1 mM DTPA, 50 mM sodium phosphate buffer,
pH 7.5 at 25°C. ($\Delta\varepsilon$ = 1.1 x 10^4 $M^{-1}cm^{-1}$) One unit is defined as the amount of
xanthine oxidase required to convert 1 μmole of xanthine to urate per minute. The
activity of the xanthine oxidase was unaffected by including 100 mM purified DMPO
in the assay medium.

RESULTS

The experiment demonstrating the vitamin K_1 dependent production of superoxide from rat liver microsomes[10] has been repeated. A representative set of data are presented in Table 1 together with the results obtained when the microsomes were derived from rats serverely deficient in vitamin K. In these experiments the micro-somes were suspended in buffered KC1; if 0.25 M sucrose was added the vitamin K_1 stimulated production of superoxide was substantially reduced.

TABLE 1

SUPEROXIDE PRODUCTION BY NORMAL AND VITAMIN K DEFICIENT RAT MICROSOMES[a]

Conditions	Superoxide produced (mol/ml/min)
Normal rat microsomes	0
+ NADH	60
+ NADH + vitamin K_1	230
+ NADH + 0.25 M sucrose	79
+ NADH + 0.25 M sucrose + vitamin K_1	117
Vitamin K deficient rat microsomes	0
+ NADH	157
+ NADH + vitamin K_1	545

[a]Incubations contained 2 ml of O_2 saturated 0.15 M KC1/50 mM Tris-HC1 buffer, pH 7.5; 0.1 ml microsomes suspended in buffered KC1; 0.1 ml adrenaline (4mg ml^{-1}); 0.05 ml NADH (10 mg ml^{-1}) and 0.1 ml vitamin K_1 (10 mg ml^{-1}).

Superoxide and hydroxyl radical production in biological systems may also be detected by the technique of spin trapping.[26-28] Figure 2 (spectrum a) shows the resonances obtained on incubating rat liver microsomes in buffered KC1 with the spin trap DMPO (I), DTPA, NADH and liposomes containing vitamin K_1. The spectrum has been assigned (see Refs. 19, 29-31) to a mixture of the products (II and III) formed by the attack of hydroxyl and superoxide radicals on (I), as shown below.

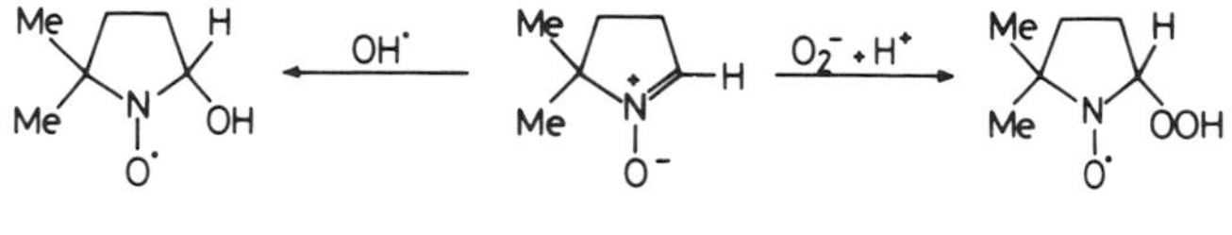

II   I   III

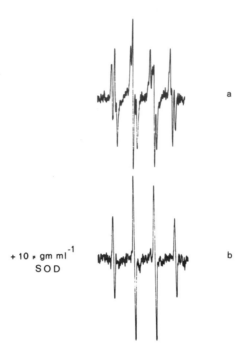

Fig. 2. E.p.r. spectra obtained from incubating vitamin K_1 replete rat liver microsomes suspended in buffered KCl (see MATERIALS AND METHODS) with DMPO: (a) microsomes (0.1 ml), DMPO (100 mM), NADH (0.3 mM), Triton X-100 (2%), 0.25 ml of liposomes containing only vitamin K_1 and sample made up to 0.5 ml with buffered KCl; (b) as in (a) plus SOD (10 µg ml^{-1}). Field 3385 G, frequency 9.50 GHz, power 30 mW, modulation 1 G, time constant 0.5 s, scan rate 0.2 Gs^{-1}, gain 1.25 x 10^5.

The presence of vitamin K_1 in the liposomes stimulates the signal height of (III) approximately 2.5-fold. On addition of superoxide dismutase to the same sample (Figure 2, spectrum b) only (II) is observed. This reflects the relative instability of (III) compared to (II) (see Ref. 19).

The signal intensity of (III) in aqueous media gives a measure of the steady state superoxide concentration and not the cumulative amount of superoxide produced. Control experiments have shown that a similar superoxide steady state concentration (measured by the signal height of (III)) may be generated by the superoxide flux from 0.004 units ml^{-1} xanthine oxidase acting on 10^{-3} M xanthine in the presence of 1 mM DTPA and 50 mM sodium phosphate buffer, pH 7.5. The amount of superoxide produced from this system and hence, the minimum value of the superoxide flux in the microsomal system, is about 1 nmole ml^{-1} min^{-1} (Ref. 32). However, we have found considerable variations in the amount of superoxide generated from separate microsomal preparations using the spin trapping technique. This may reflect the variable amounts of endogenous superoxide dismutase(s) in each preparation.[33,34]

The production of (II) and (III) has also been detected from purified vitamin K reductase isolated from rat microsomes (Figure 3, spectrum a). The signal height due to superoxide adduct (III) was doubled by inclusion of vitamin K_1 in the liposomes and if superoxide dismutase (10 μg ml^{-1}) was present (Figure 3, spectrum b) only a small amount of the hydroxyl radical adduct (II) was detected. A control experiment showed that this residual signal was generated by the liposomes and NADH alone suggesting that most of the hydroxyl radical production in the presence of vitamin K reductase was dependent on the intermediate production of superoxide.

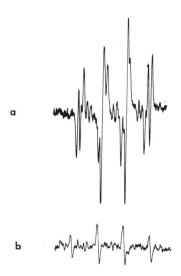

Fig. 3. (a) E.p.r. spectrum obtained by incubating a purified preparation of vitamin K_1 reductase (0.96 units ml^{-1}) isolated from normal rat liver microsomes with NADH (0.65 mM), DMPO (100 mM), DTPA (1 mM) and liposomes containing only vitamin K_1 in 50 mM sodium phosphate buffer, pH 7.5; (b) as (a) plus 25 μg ml^{-1} SOD. E.p.r. conditions as for Figure 2 except time constants 1.0 s and scan rate 0.1 Gs^{-1}.

A signal very similar to that of (III) was also obtained during the reoxidation of vitamin K_1 hydroquinone by dioxygen in an ethanol:water mixture (Figure 4) (A_N = 13.6 G, A_H^β = 10.8, A_H^γ = 1.5 G; g = 2.006). This is further evidence that vitamin K_1 hydroquinone is able to reduce dioxygen to superoxide.

If superoxide is an intermediate in the carboxylation process, the reaction

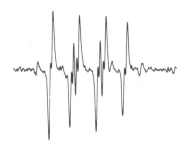

Fig. 4. E.p.r. spectrum obtained immediately after adding 0.1 ml of vitamin K_1 hydroquinone in ethanol (10 mg ml^{-1}) to DMPO (100 mM final concentration) in 0.4 ml of 66% ethanol:water mixture. The ethanol had previously been gassed with CO_2 and O_2. Field 3385 G, frequency 9.50 GHz, power 30 mW, modulation 1 G, time constant 0.5 x, scan rate 0.4 Gs^{-1}, gain 5 x 10^4.

should be inhibited by SOD. Indeed carboxylation of both the pentapeptide substitute and prothrombin precursor by rat liver microsomes is inhibited by 64% and 35%, respectively, by SOD (10 mg ml^{-1}) (Ref. 10). However SOD at 1 mg ml^{-1} inhibited 60% of the carboxylation of the pentapeptide by calf liver microsomes (Table 2). Catalase was similarly more effective in the calf system compared to the rat; the reason for the increased sensitivity of the calf system is so far unexplained. It has been suggested that the relative insensitivity of the rat carboxylation reaction to SOD may be due to the inaccessability of the enzyme to the superoxide generating site in the carboxylase complex. A similar situation arises in the cytochrome P-450 dependent hydroxylation reactions, which are not inhibited by SOD if intact microsomes are used, but are inhibited when a reconstituted hydroxylation system is studied.[35]

Richter et al.[36] have shown that hydroxylation reactions in intact microsomes may be inhibited by a copper tyrosine complex known to have SOD activity. The effect of three copper (II) complexes with known dismutase activity has been tested on the carboxylation of the prothrombin precursor by vitamin K deficient rat liver microsomes[37] (Figure 5). To indicate that the inhibition is related to the dismutase activity of the complexes the K_i values of the copper complexes for the carboxylation reaction are tabulated with K_i for the inhibition of formazan production from nitroblue tetrazolium, a process also mediated by superoxide (Table 2). The effect of nickel (II) glycinate, which has no SOD activity, was also studied and, as can be seen from the results, the complex did not inhibit the

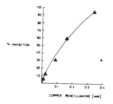

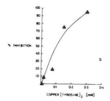

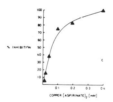

Fig. 5. The inhibition of ^{14}C incorporation into the trichloroacetic acid insoluble (protein) fraction of vitamin K-deficient rat liver microsomes by various copper complexes. Reactions were initiated by the addition of vitamin K_1. The incubation mixture contained 1 ml of microsomes suspended in 0.25 M sucrose, 80 mM KCl in 25 mM imidazole-HCl buffer, pH 7.5 containing 2% Triton X-100, dithiothreitol (60 µM), NADH (1 mM), NaH$^{14}CO_3$ (10 µCi), vitamin K_1 (100 µg) in a total volume of 1.25 ml. (a) plus copper penicillamine; (b) plus copper tyrosine; (c) plus copper aspirinate.

carboxylation reaction at a concentration of 830 µM. In the experiment with calf liver microsomes (Table 2), the carboxylation of the pentapeptide substrate was completely inhibited by 650 µM copper aspirinate. The radical trap (I), used to detect superoxide and hydroxyl radical production in the microsomes, has a K_i at least three orders of magnitude higher than the complexes (Table 3) and is a relatively poor inhibitor of both carboxylation and formazan production. Girardot et al.[38] obtained 50% inhibition at 5 mM (I), but their experimental conditions were considerably different from those used here.

The glutathione peroxidase/glutathione reductase system is a potent inhibitor of carboxylation according to Larson and Suttie.[14] However, we find that glutathione reductase alone inhibits carboxylation by about 30%, and that our preparations of glutathione peroxidase cause about 20% inhibition in the absence of other

TABLE 2

INHIBITION BY VARIOUS AGENTS OF PENTAPEPTIDE CARBOXYLATION IN CALF
LIVER MICROSOMES[a]

Conditions[b]	ml^{-1} DPM[c] microsomes	% Inhibition
Microsomes + vitamin K_1 + 1 mg BSA hydroquinone (0.012 mM)	3,234	0
+ 1 mg SOD (0.025 mM)	1,305	60
+ 1 mg CAT (0.003 mM)	1,788	45
+ 0.34 mg Cu aspirinate (0.65 mM)	120	96

[a]Incubations contained 1 ml microsomes, pentapeptide (2 mM), $NaH^{14}CO_3$ (10 µCi), NADH (1.4 mM), dithiothreitol (60 µM), vitamin K_1 (100 µg) in a volume of 1.25 ml.
[b]BSA = bovine serum albumin; CAT = catalase.
[c]DPM = disintegrations per minute.

TABLE 3

COMPARISON OF SUPEROXIDE DISMUTASE ACTIVITY AND INHIBITION OF ^{14}C INCORPORATION
INTO RAT PROTHROMBIN PRECURSORS BY METAL (II) COMPLEXES, DMPO AND SUPEROXIDE
DISMUTASE[a]

	Carboxylation K_i (µM)	Formazan production K_i (nM)
Copper penicillamine	135	140
Copper tyrosine	100	145
Copper aspirinate	65	125
Nickel glycinate	N.I. at 830[b]	N.I. at 8300[b]
DMPO	155×10^3	30×10^6
Superoxide dismutase	600[c]	1

[a]Incubations for the carboxylation assays contained 1 ml microsomes, NADH (1.4 mM), dithiothreitol (60 µM), $NaH^{14}CO_3$ (10 µCi), vitamin K_1 (100 µg) in a volume of 1.25 ml. Superoxide dismutase activity was assayed by inhibition of formazan production as in Ref. 23.
[b]N.I. = no inhibition.
[c]Estimated from Esnouf et al. (Ref. 10).

added components. Inhibition in these experiments may be due to small amounts of endogenous glutathione peroxidase and reductase present in the microsomal preparation.

DISCUSSION

The adrenochrome assay and the e.p.r. measurements both show that there is a marked vitamin K_1 stimulated release of superoxide from rat liver microsomes, confirming earlier observations.[10,15,16] The spin trapping measurements with the purified sample of vitamin K reductase demonstrate that the vitamin K_1 dependent production of superoxide in microsomes may originate from the NADH dependent vitamin K reductase couple presumed to be located in the carboxylase complex.

The release of superoxide seen in the absence of vitamin K_1 in the microsomes probably arises from the microsomal NADPH dependent cytochrome c reductase[39] and the microsomal cytochrome P-450. The increased rate of superoxide release seen when the microsomes are suspended in buffered KCl without sucrose may arise from the disruption of enzyme complexes which utilise superoxide. Strobel and Coon[35] have observed that microsomal hydroxylation reactions catalysed by cytochrome P-450 are influenced by the concentration of sodium chloride in the medium. Conversely SOD will only inhibit the carboxylation reaction if it is able to reach the site of superoxide production and/or utilisation in the microsomal membrane. Therefore it is not unexpected that a large concentration of SOD is required before effective inhibition of carboxylation occurs.

The small copper (II) complexes should have greater access to the carboxylase complex and it is most likely that their inhibitory action is due to their ability to dismutase superoxide. It is possible however that they may act as general radical scavengers; they have no reported catalase activity.

Glutathione peroxidase, which is very effective at inhibiting carboxylation at low enzyme concentrations, is normally associated with biomembranes[40] and under these experimental conditions might easily approach the carboxylase complex. Its site of action may be either the vitamin K_1 hydroperoxide as suggested by Larson and Suttie[14] or free hydrogen peroxide, or even peroxomonocarbonate (HCO_4^-) which is itself a hydroperoxide.

It has been proposed[10,15,16] that a synthetic step in the carboxylation process is the reaction of superoxide, or hydrogen peroxide derived from it, with carbon dioxide to yield one or more carbon containing intermediates (e.g. CO_2^-, CO_3^-, CO_4^-, HCO_4^-, $C_2O_6^{2-}$) which might act as a substrate for the carboxylase. This proposal should now be extended to include the possibility of CO_2 reacting directly with the hydroperoxide adduct of the vitamin (formed by the reaction of superoxide or hydrogen peroxide) to yield a vitamin K_1 peroxocarbonate, this species providing the carbon source for the reaction.

On the basis of the evidence presented here it is reasonable to suggest that

the superoxide radical anion is an intermediate in the vitamin K dependent carboxylation of glutamyl residues in precursor prothrombin.

ACKNOWLEDGEMENTS

S.J.W. acknowledges a grant from the Medical Research Council, A.I.G., M.J.O.-Z.
and M.R.G. are supported by the Science Research Council. Financial assistance
was provided by the Medical Research Council and the British Hearth Foundation.
M.P.E. and H.A.O.H. are members of the Oxford Enzyme Group.

REFERENCES

1. Stenflo, J. (1974) J. Biol. Chem. 249, 5527-5535.
2. Esmon, C.T., Sadowski, J.A. and Suttie, J.W. (1975) J. Biol. Chem. 250, 4744-
   4748.
3. Jones, J.P., Gardner, E.J., Cooper, T.G. and Olson, R.E. (1977) J. Biol.
   Chem. 252, 7738-7742.
4. Olson, R.E. and Suttie, J.W. (1977) Vitamin. Horm. 35, 59-108.
5. Stenflo, J. and Suttie, J.W. (1977) Annu. Rev. Biochem. 46, 157-172.
6. Martius, C., Ganser, R. and Viviani, A. (1975) FEBS Lett. 59, 13-14.
7. Sadowski, J.A., Esmon, C.T. and Suttie, J.W. (1976) J. Biol. Chem. 251, 2770-
   2776.
8. Patel, K.B. and Willson, R.L. (1973) J. Chem. Soc. Faraday Trans. I, 69,
   814-825.
9. Cadenas, E., Boveris, A., Ragan, C.I. and Stoppani, A.O.M. (1977) Arch.
   Biochem. Biophys. 180, 248-257.
10. Esnouf, M.P., Green, M.R., Hill, H.A.O., Irvine, G.R. and Walter, S.J.
    Biochem. J. (1978) 174, 345-348.
11. Matschiner, J.T., Bell, R.G., Amelotti, J.M. and Knauer, T.E. (1970) Biochim.
    Biophys. Acta, 201, 309-315.
12. Saito, I., Otsuki, T. and Matsurra, T. (1979) Tetrahedron Lett. 19, 1693-1696.
13. Feiser, L.F., Tishler, M. and Wendler, N.L. (1940) J. Amer. Chem. Soc. 62,
    2866-2871.
14. Larson, R.E. and Suttie, J.W. (1978) Proc. Nat. Acad. Sci. USA, 75, 5413-
    5416.
15. Esnouf, M.P., Green, M.R., Hill, H.A.O., Irvine, G.B. and Walter, S.J.
    (1979) in Oxygen free radicals and tissue damage, Ciba Found. Symp. 65
    (New Ser.), Exerpta Medica, Amsterdam, pp. 187-197.
16. Esnouf, M.P., Burgess, A.I., Walter, S.J., Green, M.R., Hill, H.A.O. and
    Okolow-Zubkowska, M.J. (1979) in Proc. 8th Steenbock Symp., Vitamin K meta-
    bolism and vitamin K-dependent proteins, Suttie, J.W., ed., in press.
17. Esmon, C.T. and Suttie, J.W. (1976) J. Biol. Chem. 251, 6238-6243.
18. Bonnet, R., Brown, R.F.C., Clark, V.M., Sutherland, I.O. and Todd, A. (1959)
    J. Chem. Soc., 2094-2102.
19. Buettner, G.R. and Oberley, L.W. (1978) Biochem. Biophys. Res. Commun. 83,
    69-74.
20. Sorenson, J.R.J. (1976) J. Med. Chem. 19, 135-148.
21. Brigelius, R., Hartmann, H.-J., Bors, W., Saran, M., Lengfelder, E. and
    Weser, U. (1975) Hoppe-Seylers Z. Physiol. Chem. 365, S.739-745.
22. McAuliffe, C.A. and Perry, W.D. (1969) J. Chem. Soc. A, 634-636.
23. Beauchamp, C. and Fridovich, I. (1971) Anal. Biochem. 44, 276-287.
24. Hosada, S., Nakamura, W. and Hayashi, K. (1974) J. Biol. Chem. 249, 6416-
    6423.
25. Millar, R.W. and Rapp, U. (1973) J. Biol. Chem. 249, 6084-6090.
26. Janzen, E.G. (1971) Account. Chem. Res. 4, 31-40.
27. Largercrantz, C.J. (1971) J. Phys. Chem. 75, 3466-3475.
28. Janzen, E.G. and Lui, J.I.P. (1973) J. Mag. Res. 9, 510-512.
29. Harbour, J.R., Chow, V. and Bolton, J.R. (1974) Can. J. Chem. 52, 3549-3553.

30. Janzen, E.G., Nutter, D.E., Davis, E.R., Blackburn, B.J., Poyer, J.L. and McCay, P.B. (1978) Can. J. Chem. 56, 2237-2242.
31. Lai, C.-S. and Piette, L.H. (1978) Arch. Biochem. Biophys. 190, 27-38.
32. Fridovich, I. (1970) J. Biol. Chem. 245, 4053-4057.
33. Tyler, D.D. (1975) Biochem. J. 147, 493-504.
34. Fridovich, I. (1979) in Oxygen free radicals and tissue damage, Ciba Found. Symp. 65 (New Ser.) Exerpta Medica, Amsterdam, pp. 77-93.
35. Strobel, H.W. and Coon, M.J. (1971) J. Biol. Chem. 246, 7826-7829.
36. Richter, C., Azzi, A., Weser, U. and Wendel, A. (1977) in Superoxide and Superoxide Dismutases, Michelson, A.M., McCord, J.M. and Fridovich, I., eds., Academic Press, New York, pp. 375-385.
37. Esnouf, M.P., Green, M.R., Hill, H.A.O. and Walter, S.J. (1979) FEBS Lett. 107, 146-150.
38. Giradot, J.M., Mack, D.O., Floyd, R.A. and Johnson, C.B. (1976) Biochem. Biophys. Res. Commun. 70, 655-663.
39. Bartoli, G.M., Galcotti, T., Palombini, G., Parisi, G. and Azzi, A. (1977) Arch. Biochem. Biophys. 184, 276-281.
40. Flohé, L. (1979) in Oxygen free radicals and tissue damage, Ciba Found. Symp. 65 (New Ser.), Exerpta Medica, Amsterdam, pp. 95-122.
41. Esnouf, M.P. (1977) Brit. Med. Bull. 33, 213-219.

OXYGEN SPECIES IN PROSTAGLANDIN BIOSYNTHESIS IN VITRO AND IN VIVO

C. DEBY AND G. DEBY-DUPONT
Laboratory of Applied Biochemistry, University of Liege, 4020-Liege,
Belgium

Prostaglandin endoperoxide synthetase, which acts as a dioxygenase, introducing two oxygen molecules in arachidonic acid (AA), to form PGG_2 and PGH_2 (Figure 1), exhibits also a peroxidase activity, since it can accept H_2O_2 as substrate, in the presence of a suitable hydrogen donor.[1,2] However, the relationship between prostaglandin biosynthesis and oxygen metabolism is not well defined. Assays of prostaglandin biosynthesis in vitro in the presence of H_2O_2, or O_2^- scavengers, give contradictory results. Some authors[3] claim that H_2O_2 enhances this process, other workers find an inhibitory effect.[1,2] According to Panganamala et al.,[4] superoxide dismutase exerts a stimulatory action on prostaglandin biosynthesis, whilst other workers find an inhibition[5] or no effect.[6] Also according to Panga-mala et al.,[7] hydroxyl radical is the directly involved oxygen species, in the oxidation of arachidonic acid, during synthetase activity. Nevertheless, numerous papers report a stimulatory effect of OH· scavengers on prostaglandin biosynthesis.[6,8,9]

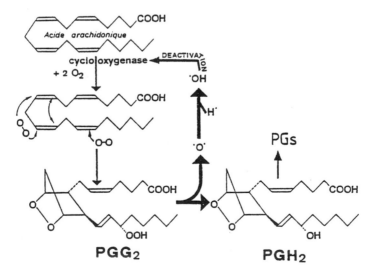

Fig. 1. The first steps of prostaglandin synthesis. During the conversion of PGG_2 to PGH_2, one atom of oxygen leaves the prostaglandin molecule and becomes OH·. If OH· is not scavenged, it inactivates cyclo-oxygenase (Ref. 9).

We have tried to explain these discrepancies, using three groups of techniques:
(a) In vitro assays: conversion of ^{14}C-arachidonic acid to prostaglandins by
bull seminal vesicles, according to a method previously described.[10]
(b) Study of stomach strip contractions induced by arachidonic acid converted
to prostaglandins by the gastric tissues. Prostaglandin synthetase is present in
the gastric tissues. Arachidonic acid provokes contractions only if it is converted
to prostaglandins. Pretreatment of a stomach strip with an inhibitor of prostaglandin
synthetase (aspirin, indomethacin) renders it completely insensitive to arachidonic
acid. Arachidonic acid induced stomach strip contractions are thus good indicators
of prostaglandin synthetase activity.
(c) Modifications of arachiodonic acid hypotension in the rabbit. Intravenously
injected arachidonic acid induces a fall of the arterial pressure, by conversion
into prostacyclin. Arachidonic acid sensitivity of the rabbit can be enhanced
twenty times by the combined effect of heparinization and perfusion of autologous
hemolyzed blood.[11] We define an AA_{50} as the intravenous arachidonic acid dose
necessary to obtain an hypotension representing 50% of the initial blood pressure.
To estimate the activating or inhibitory role of a factor, we use an arachidonic
acid activation coefficient, which is the number by which the AA_{50} must be multiplied
after the application of the modifying agent. Thus, this coefficient must be greater
than unity if an inhibitor is used, or less than unity if a stimulating factor is
injected in the rabbit before the arachidonic acid test.   More representative,
when an activator of the arachidonic acid effect is assayed, is the sensitization
coefficient or the reciprocal of the activation coeffecient. If the AA_{50} is lowered,
the sensitization coefficient will be greater than unity.

 The role of hydrogen peroxide.

(1) Effect of H_2O_2 on in vitro prostaglandin biosynthesis. Hydrogen peroxide
exerts a significant enhancing effect on arachidonic acid conversion by seminal
vesicle microsomes, at the optimal concentration of 5×10^{-6} M; above this concentration, H_2O_2 becomes a strong inhibitor (Figure 2). During the same time that
the inhibition appears, a production of hydroxyl radical occurs, revealed by the
addition to the enzymatic system of methional, which is converted in ethylene by
OH˙ (Figure 2). Prostaglandin synthesis inhibition and ethylene production vary
proportionally. The stimulating action of H_2O_2 is greatly potentiated by the
presence of a hydrogen donor, able to pass to a radical state, such as riboflavin.[12]
Catalase inhibits the H_2O_2 activation.[12] The biphasic aspect of the curve represented in Figure 2 explains why discrepancies are found in the literature about
H_2O_2 effects on prostaglandin biosynthesis.
    Rahimtula and O'Brien,[5] and van der Ouderaa et al.,[2] using respectively H_2O_2
concentrations of 10^{-5} M, and 2.2×10^{-4} M, find an inhibition when hydrogen
peroxide is added to the prostaglandin generating system. On the other hand,
Panganamala et al.[3] and Rahimtula and O'Brien[5] observe inhibition of prostaglandin

biosynthesis by catalase, even without the presence of H_2O_2. Those authors,[2,5]
who have found an inhibitory effect of H_2O_2, have used H_2O_2 concentrations in the
inhibitory part of the curve of Figure 2. There is thus no contradiction between
the two groups of results.

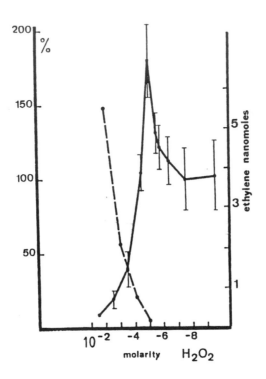

Fig. 2. Role of H_2O_2 in the conversion of ^{14}C-arachidonic acid to prostaglandins
by bull seminal vesicles, in presence of glutathione (1.5 x 10⁻⁴ M) and hydroquinone
(4 x 10⁻⁵M,) (Ref. 10). After extraction with diethylether and separation on
TLC plates, the areas corresponding to PGE_2 and $PGF_{2\alpha}$ were automatically eluted
and counted together with a beta spectrometer. Several experiments were performed
in presence of methional, giving ethylene when OH· is generated (Ref. 27). Estimation of ethylene was made by GLC, using Carbosieve as stationary phase. Left
ordinate: radioactivity yield, in percent of control assays (———); right
ordinate: ethylene generation (––––).

(2) <u>Effect of H_2O_2 on stomach strip contraction induced by arachidonic acid</u>.
Hydrogen peroxide provokes by itself strip contractions, even at concentrations
as weak as 10⁻⁶ M. These effects are abolished by indomethacin or by eicosatetraenoic
acid (TYA), two inhibitors of arachidonic acid enzymatic conversion to prostaglandins,
TYA being a highly competitive inhibitor. Arachidonic acid induced contractions

are also inhibited by indomethacin or TYA, whilst PGE_2 induced contractions are unchanged (Figure 3). These facts prove that stomach strip contractions are due to PGE_2 only. Administered together, H_2O_2 and arachidonic acid potentiate each other,[12] their effects being strongly diminished by previous treatment with indomethacin or TYA. The inhibition of H_2O_2 effects by a prostaglndin biosynthesis inhibitor as specific as TYA suggests a key-role for H_2O_2 in tissue prostaglandin biosynthesis.

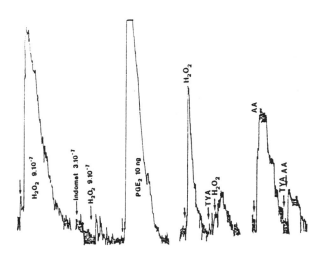

Fig. 3. Stimulation of stomach strip contractions by H_2O_2, illustrating inhibition by indomethacin (Indomet) and eicosatetraenoic acid (TYA). Contractions due to arachidonic acid are also impaired by TYA. Stomach strip assays were performed in Tyrode solution buffered at pH 7.6. The bath volume was 6 ml; Tyrode was automatically renewed every two minutes. Chemical reagents remained in contact with the strip for 30 seconds.

(3) Role of H_2O_2 on the arachidonic acid induced hypotension in the rabbit.
Hydrogen peroxide enhances the arachidonic acid hypotension only if it is administered 3 minutes before arachidonic acid injection. Otherwise, it is ineffective (Figure 4). This activation by H_2O_2 is abolished by the simultaneous administration of catalase (Figure 4). The delay of 3 minutes required by H_2O_2 to stimulate arachidonic acid hypotension may be due to the slowness of H_2O_2 penetration across the cell wall, or to a process, initiated by H_2O_2, giving an excited oxygen species. Aminotriazole, an inhibitor of catalase, given alone, increase the hypotensive effect of arachidonic acid (Figure 4). This phenomenon brings another proof for

the intervention of H_2O_2 in prostaglandin generation, in the animal.

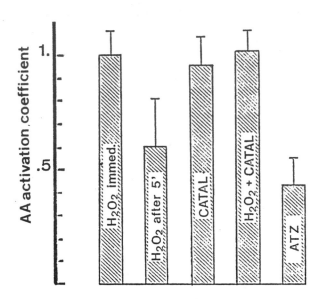

Fig. 4. Effect of arachidonic acid on rabbit blood pressure. If arachidonic acid is administered immediately after a perfusion of H_2O_2 (2×10^{-4} mole/kg) no modification of the effect of arachidonic acid on blood pressure is observed. But if the arachidonic acid assay is done 5 minutes after H_2O_2 injection, the hypotensive action of arachidonic acid is enhanced. Catalase inhibits this sensitization. Aminotriazal (ATZ) (2.5×10^{-4} mole/kg), an inhibitor of catalase, strongly sensitizes the rabbit to arachidonic acid. The rabbits received, one hour previously 1 ml of autologous hemolyzed blood, and 10 minutes before the assays, 10 mg/kg haparin (Refs. 11, 25). Ordinate: coefficient of arachidonic acid activation as defined in the text.

The role of superoxide anion, SOD and SOD analogs. Since H_2O_2 in living organisms arises in superoxide dismutation, it is logical to study the role of superoxide generating systems and of superoxide dismutating agents on prostaglandin biosynthesis.

(1) Effects on the in vitro biosynthesis. The classical system, hypoxanthine plus xanthine oxidase, was used, varying hypoxanthine concentration. Xanthine oxidase is inactive on prostaglandin biosynthesis, by itself. The prostaglandin biosynthesis rate has the same aspect under the effect of this O_2^- generating system, as under the effect of H_2O_2 (Figure 5). It is biphasic with an optimal point. In the presence of a standard concentration of 0.015 U/ml of xanthine oxidase, an inhibition began at hypoxanthine concentrations higher than 5×10^{-5} M.

Hypoxanthine alone was slightly stimulatory. This was explained by the finding of a weak endogenous xanthine oxidase activity, in the microsomal suspension. The positive and negative effects exerted by the superoxide generating system are strongly enhanced by the addition of SOD (100 μg/ml). This fact can be explained by the accelerated dismutation of O_2^- to H_2O_2, mediated by SOD, which provides the conditions of the experiment described on Figure 2.

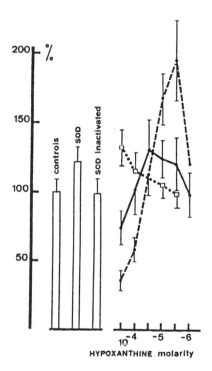

Fig. 5. In vitro biosynthesis of prostaglandins (conditions as for Fig. 2), in the presence of hypoxanthine (HX) plus xanthine-oxidase (XOS). Ordinate: radioactivity yield of TLC areas corresponding to PGE and PGF, counted together, in percent of controls. Bars: controls, no HX, no XOX, no SOD; SOD, controls plus SOD (100 μg/ml); SOD inactivated, controls plus heat-inactivated SOD (100 μg/ml). (●——●) Rate of prostaglandin biosynthesis as function of HX molarity in presence of XOX (0.015 U/ml); (●----● as in the presence of 100 μg/ml SOD; (●·······●) controls plus HX alone.

Panganamala et al.[4] reported that SOD (100 μg/ml) enhanced prostaglandin biosynthesis, whilst Rahimtula and O'Brien[5] described an inhibition with 50 μg SOD. The paradoxical effect of SOD on prostaglandin biosynthesis reconciles these opposite results (Figure 6). At low concentration (25 μg/ml), SOD inhibits prostaglandin production, whilst at concentrations higher than 100 μg/ml, it is stimulatory; this phenomenon is discussed further below and in the legend to Figure 6.

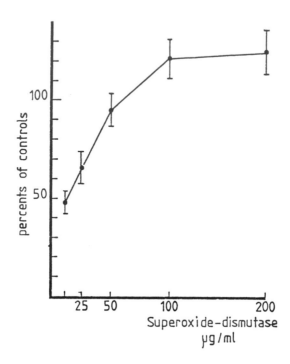

Fig. 6. Paradoxical effect of increasing concentrations of superoxide dismutase (SOD) on _in vitro_ prostaglandin biosynthesis, in the presence of hypoxanthine (10^{-6} M) plus xanthine oxidase (0.015 U/ml). Ordinate: radioactivity yields, in percent of control assays. Low SOD concentrations inhibit biosynthesis, whilst higher concentrations stimulate it significantly. In the first case, some O_2^- remain together with H_2O_2 giving $OH^{\cdot}$ in presence of Fe^{2+}: $OH^{\cdot}$ is an inactivator of cyclo-oxygenase (Fig. 1); in the second case, all O_2^- are dismutated to H_2O_2.

Entirely different is the action of the superoxide scavenger, copper-tyrosine, on _in vitro_ prostaglandin biosynthesis (Table 1). This agent is always an inhibitor, especially for PGE_2 synthesis. We have obtained similar results with copper acetate.

(2) Effect on rat stomach strip contractions exerted by arachidonic acid. The use of hypoxanthine is difficult, because it is poorly soluble at physiological pH, in Tyrode solution. The effects of the O_2^- generating reaction, acetaldehyde plus xanthine oxidase, were studied, showing a stimulatory effect[12] which is not modified by SOD. While SOD does not exert any modification on strip reactions to arachidonic acid, copper-tyrosine enhances greatly these responses. In fact, copper-tyrosine does not increase arachidonic acid conversion to prostaglandin, by stomach tissues, since prostaglandin contractions of stomach strips are also

enhanced by this agent. Copper-tyrosine seems to be an inhibitor of prostaglandin catabolism.

TABLE 1

EFFECTS OF Cu-Tyr ON PROSTAGLANDIN BIOSYNTHESIS IN VITRO[a]

SOD equiv.[b]	PGE_2	$PGG_{2\alpha}$
5.8 U/ml	96% ± 11	100% ± 9.6
58 U/ml	78% ± 8.4	95.4% ± 7.4
580 U/ml	30% ± 3.1	52.1% ± 5.3

[a]The conditions of prostaglandin biosynthesis are described in the legend to Fig. 2.
[b]The SOD-activity of Cu-Tyr was determined by Dr. P. Van Caneghem (Laboratory of Applied Biochemistry, University of Liege) by a polarographic method.

(3) Effect on rabbit arachidonic acid hypotension. We have observed[12,13] that intravenously injected hypoxanthine plus xanthine oxidase, administered before arachidonic acid, sensitizes rabbits by a coefficient of 3.2. Superoxide dismutase, simultaneously injected with the O_2^- generating system, is without effect, whilst the superoxide scavengers, copper-tyrosine and copper-penicillamine, enhance appreciably the hypotensive action of arachidonic acid (Figure 7). Copper acetate acts in the same way, but more moderately.

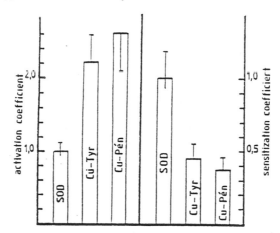

Fig. 7. Comparison of the effects of superoxide dismutase (SOD : 5900 U/kg), copper tyrosine (Cu-Tyr : 5770 U/kg) and copper-penicillamine (Cu-Pen : 5890 U/kg) on the hypotensive action of intravenously injected arachidonic acid in the rabbit. SOD is ineffective, whilst copper complexes are potent sensitizers.

The role of hydroxyl scavengers.

(1) On prostaglandin biosynthesis. Egan et al.[9] have reported a stimulation of prostaglandin biosynthesis by seminal vesicle microsomes, exerted by phenol and methional. They explain the self-deactivation of the prostaglandin synthetase by the production of OH$\cdot$ during its activity (Figure 1). This view is confirmed by our finding that OH$\cdot$ generation is parallel to prostaglandin synthesis inhibition, after addition of H_2O_2 to seminal vesicle microsomes (Figure 2). Collier et al.[8] have reported a positive effect of ethanol, OH$\cdot$ scavenger, on prostaglandin biosynthesis. Dimethylsulfoxide, another OH$\cdot$ scavenger, is also a stimulator of prostaglandin biosynthesis in vitro (unpublished observation).

(2) On stomach strip. The arachidonic acid induced contractions are 2.4 times higher if ethanol, at 4×10^{-3} M, is added.[12] At this concentration, ethanol is completely ineffective by itself.

(3) On rabbit hypotension. Hydroxyl scavengers are stimulating agents for hypotension (Figure 8), if intravenously administered, together with arachidonic acid.

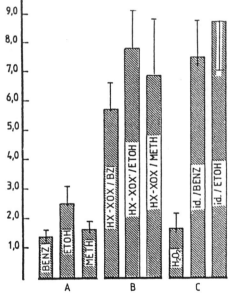

Fig. 8. Role of hydroxyl scavengers on the hypotensive response of the rabbit, to intravenously injected AA. The reagents were injected before the arachidonic test. Ordinate: sensitization coefficient as defined in the text. A: Effect of sodium benzoate (BENZ : 1.3×10^{-4} mole/kg), ethanol (ETOH : 1.3×10^{-4} mole/kg) and methional (METH : 1.3×10^{-4} mole/kg). B: Effect of successive intravenous injections of xanthine oxidase (XOX) (0.24 U/kg), hypoxanthine (HX) (4.5×10^{-5} mole/kg) and OH$\cdot$ scavengers (the same concentration as in A) on the response to arachidonic acid. The sensitization coefficient for the system xanthine oxidase plus hypoxanthine with OH$\cdot$ scavengers is 2.7. C: Effect of H_2O_2 combined with OH$\cdot$ scavengers. Arachidonic acid was injected 3 minutes later.

The role of carbon dioxide and bicarbonate anions. The recent work of Hodgson and Fridovich,[14] and Michelson et al.[15,16] have underlined the importance of carbonate and bicarbonate anions, and of carbon dioxide, in oxygen metabolism. These compounds appear to be able to react under certain conditions, either with O_2^- or with $OH^.$, according to the following equations:

$$CO_3^= + OH^. \longrightarrow CO_3^{.-} + OH^- \qquad (14)$$

$$HCO_3^- + O_2^. \longrightarrow OH^- + CO_2^{.-} + O_2 \qquad (15)$$

$$CO_2 + O_2^. \longrightarrow O_2 + CO_2^{.-} \qquad (15)$$

In these different ways, formate radicals are produced, which can be more specific reactants than $OH^.$. These radicals are potent oxidants for unsaturated lipids, but they undergo dimerization into oxalic acid, losing their radical state and their reactivity.[15] On the other hand, the generation of CO_2^- and $CO_3^{.-}$, from CO_2, HCO_3^-, and $CO_3^=$ perhaps a means to stop the radical reaction chain. Carbon dioxide and HCO_3^- may be efficient radical scavengers for $OH^.$ and O_2^-, in particular. It is for this reason, that we have proceeded to some experiments on prostaglandin biosynthesis, using a range of CO_2 and HCO_3^- concentrations, alone or in combination with an O_2^- generating system. The results are shown in Figure 9.

Carbon dioxide is, by itself, ineffective on prostaglandin biosynthesis, at concentrations lower than 10^{-3} M. On the contrary, if added to the superoxide generating system, hypoxanthine (5×10^{-5} M) plus xanthine oxidase, it enhances significantly prostaglandin biosynthesis, beginning at low concentrations. In the case of HCO_3^-, an optimal concentration is observed at 10^{-3} M.

Conclusions. Direct reactions between triplet oxygen and arachidonic acid (singlet) are forbidden.[17] Two possibilities can be considered:

(a) Oxygen is activated, losing its triplet configuration and/or an oxygen centered radical is generated;

(b) Transition metals, particularly iron, produce radical complexes with arachidonic acid, which are able to react with triplet oxygen;[18] cyclo-oxygenase contains hemin as co-factor, the iron of which can play such a role.

There is now experimental proof that the successive steps of oxygen metabolism can interact with prostaglandin biosynthesis in living tissues. All the biological pathways producing O_2^- and H_2O_2 are thus modifying factors of this mechanism.

Role of H_2O_2 and O_2^-. Catalase does not stop prostaglandin biosynthesis in all cases. For this reason, Vanderhoek and Lands[19] claim that H_2O_2 does not appear to be an obligatory intermediate in the synthesis of prostaglandins but our experiments afford some evidence that it can play an important role in many biological circumstances. The peroxidative property of cyclo-oxygenase is another argument for an active role of H_2O_2. Together, H_2O_2 and O_2^-, in the presence of iron, which is a universal constituent of tissues, generate $OH^.$, by a modified Haber-Weiss reaction.[20] The role of $OH^.$ is considered below.

94

The superoxide scavengers, copper-tyrosine and copper-penicillamine, seem to act as inhibitors of in vitro prostaglandin biosynthesis, by another mechanism than a SOD-like effect, since their action is always inhibitory. To the in vitro inhibitory effects of copper-tyrosine, are opposed the potent stimulation of stomach strip arachidonic acid contraction and of rabbit arachidonic acid hypotension, produced by this SOD-like agent. On the other hand, the peroxidative attack of H_2O_2 by prostaglandin synthetase in the presence of a flavin, as hydrogen donor, is perhaps the first requirement to produce the lipoperoxidation associated with prostaglandin biosynthesis,[21] the second being the radical state of the flavin, able to take a hydrogen from the unsaturated chains (Figure 10).

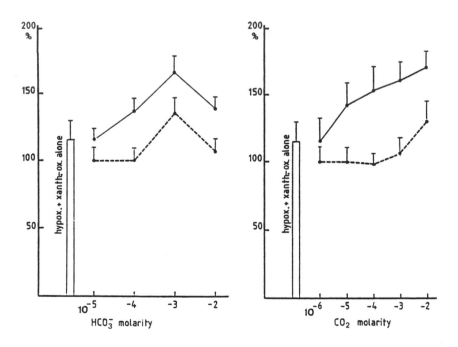

Fig. 9. Prostaglandin biosynthesis performed in the same manner as in Fig. 2, in presence of the O_2^- generating system (HX - XOX) : hypoxanthine (5×10^{-6} M) plus xanthine oxidase (0.015 U/ml). On the left, experiments performed with addition to the medium of various concentrations of HCO_3^-; and on the right, various concentrations of CO_2. CO_2 concentration was determined with a specific CO_2 gas electrode (Orion model 95-02). Ordinates show radioactivity yield in percent of control assays (without HX-XOX and without CO_2 or HCO_3^-). Continuous lines : HCO_3^- or CO_2 with HX-XOX; Dotted lines : HCO_3^- or CO_2 with HX-XOX. In each part of the figure, the bar shows the percentage of radioactivity in the presence of HX-XOX without CO_2 or HCO_3^-.

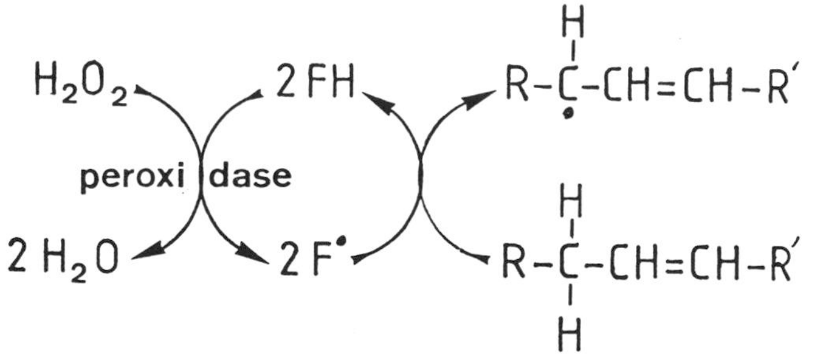

$$H_2O_2 \quad\longrightarrow\quad 2\,FH \quad\longrightarrow\quad \underset{\displaystyle \overset{\displaystyle H}{R-\overset{\bullet}{C}-CH=CH-R'}}{}$$

peroxi|dase

$$2\,H_2O \quad\longleftarrow\quad 2\,F^{\bullet} \quad\longleftarrow\quad \underset{H}{\overset{H}{R-C-CH=CH-R'}}$$

Fig. 10. The hypothetic role played by the peroxidative activity of cyclo-oxygenase, in the presence of flaving (FH), to generate free radical (F$^{\bullet}$) which can return to the FH state by removing a hydrogen atom from unsaturated fatty acid (R-CH$_2$-CH=CH-R'). This latter becomes a free radical oxidizable by triplet oxygen.

Role of singlet oxygen. Singlet oxygen is perhaps the oxygen species produced during the lipoperoxidation of parallel unsaturated chains, in phospholipids. After a Russell's reaction,[22] giving a tetroxide bridge, the decomposition of the latter occurs releasing singlet oxygen.[23] The hypothesis of an intervention of singlet oxygen during prostaglandin biosynthesis is only based on experiments demonstrating the inhibitory effect of some singlet oxygen scavengers in vitro[3,5,6] and in vivo.[13] Further work, using for instance spin labelled molecules, is required.

Role of OH$^{\bullet}$. Hydroxyl radical is in all cases an inactivator of prostaglandin synthetase[9] and contributes to its self-deactivation (Figure 1). It is certainly not the activated oxygen species involved in prostaglandin synthesis, as previously proposed.[7] The paradoxical effect of SOD on in vivo prostaglandin biosynthesis is probably due to the acceleration of O$_2^-$ dismutation, giving H$_2$O$_2$, but in an insufficient rate to eliminate all the O$_2^-$ radicals. The coexistence of H$_2$O$_2$ and O$_2^-$, in biological media containing Fe^{2+}, gives OH$^{\bullet}$, which impairs prostaglandin production. At higher SOD concentrations, O$_2^-$ is completely dismutated to H$_2$O$_2$, which reaches stimulating concentrations.

Hydroxyl radical can also arise from the conversion of PGG$_2$ to PGH$_2$ (Ref. 9). This might explain the inflammatory phenomena accompanying prostaglandin generation in tissues.

All OH$^{\bullet}$ scavengers stimulate prostaglandin biosynthesis in vitro.[6,9,8,24] Our own experiments with stomach strip assays and arachidonic acid hypotension show that OH˙ scavengers, particularly ethanol, enhance the in vivo arachidonic acid conversion into vasoactive prostanoids.

Role of HCO_3^- and CO_2. The stimulatory role played by CO_2 and by HCO_3^-, during prostaglandin biosynthesis, may be due to a scavenging effect of these compounds (see Refs. 14, 15). It may be underlined that CO_2 and HCO_3^- intervene to stimulate prostaglandin biosynthesis, mainly if an O_2^- generating system is present. These universal components of the biological media are perhaps regulators of prostaglandin biosynthesis in living tissues, in connection with the superoxide production. A misunderstood phenomenon, the prostaglandin biosynthesis stimulated by hypoxia, may be explained by the combined effect of CO_2 accumulation and enhancement of purine metabolism. We have in effect observed that a short hypoxia increases hypotensive responses of the rabbit to intravenous arachidonic acid.[25] Now, in hypoxia, the plasma level of hypoxanthine is greatly enhanced,[26] O_2^- generation increasing after the return to normal breathing. Thus, when hypercapnia accompanies hypoxia, the two factors, CO_2 accumulation and O_2^- production, are simultaneously present and prostaglandin biosynthesis is stimulated and arachidonic acid hypotension increases.

REFERENCES

1. O'Brien, P.J. and Rahimtula, A. (1976) Biochem. Biophys. Res. Commun. 70, 832–838.
2. Van der Ouderaa, E.J., Buytenhek, M., Nugteren, D.H. and Van Dorp, D.A. (1977) Biochim. Biophys. Acta, 487, 315–331.
3. Panganamala, R.V., Sharma, H.M., Sprecher, H., Geer, J.C. and Cornwell, D.C. (1974) Prostaglandins, 8, 3–11.
4. Panganamala, R.V., Brownlee, N.R., Sprecher, H. and Cornwell, D.G. (1974) Prostaglandins, 7, 21–27.
5. Rahimtula, A. and O'Brien, P.J. (1976) Biochem. Biophys. Res. Commun. 70, 893–899.
6. Marnett, L.J., Wlodawer, P. and Samuelsson, B. (1975) J. Biol. Chem. 250, 8510–8517.
7. Panganamala, R.V., Sharma, H.M., Heikkila, R.E., Geer, J.C. and Cornwell, D.G. (1976) Prostaglandins, 11, 599–607.
8. Collier, H.O.J., McDonald-Gibson, W.J. and Saeed, S.A. (1975) Lancet, 1, 702.
9. Egan, R.W., Paxton, I., Kuehl, F.A. (1976) J. Biol. Chem. 251, 7329–7335.
10. Deby, C., Descamps, M., Binon, F. and Bacq, Z.M. (1971) C.R. Soc. Biol. 165, 2465–2468.
11. Deby, C., Deby-Dupont, G. and Bacq, Z.M. (1979) Arch. Int. Physiol. 87, 149–155.
12. Deby, C. (1979) Tissue Reactions, 1, 611.
13. Deby, C. (1977) Drug. Exp. Clin. Res. 2, 105–113.
14. Hodgson, E.K. and Fridovich, I. (1976) Arch. Biochem. Biophys. 172, 202–205.
15. Michelson, A.M. and Durosay, P. (1977) Photochem. Photobiol. 25, 55–63.
16. Puget, K. and Michelson, A.M. (1976) Photochem. Photobiol. 24, 499–501.
17. Hamilton, G.A. (1974) in Molecular Mechanism of Oxygen Activation, Hayashi, O., ed., Academic Press, New York, pp. 405–451.
18. Peterson, D.A., Gerrard, J.M., Rao, G.H.R., Krick, T.P. and White, J.G. (1978) Prostaglandins Med. 1, 304–318.

19. Vanderhoek, J.Y. and Lands, W.E.M. (1978) Prostaglandins Med. 1, 251-264.
20. Koppenol, W.H., Butler, J. and van Leeuwen, J.W. (1978) Photochem. Photobiol. 28, 655-660.
21. Cook, H.W. and Lands, W.E.M. (1976) Nature, 260, 630-632.
22. Howard, J.A. and Ingold, K.U. (1968) J. Amer. Chem. Soc. 90, 1056-1058.
23. Nakano, M., Takayama, K., Shimuzu, Y., Tsuji, Y., Inaba, H. and Migita, T. (1976) J. Amer. Chem. Soc. 98, 1974-1975.
24. Glenn, M., Bowman, B.J. and Rohloff, N.A. (1977) Agent. Action. 7, 513-516.
25. Deby, C., Noël, F., Chapelle, J.P., Van Caneghem, P., Deby-Dupont, G. and Bacq, Z.M. (1978) Bull. Acad. Roy. Belg. Clin. Sci. 54, 644-648.
26. Saugstad, O.D., Østrem, G.B.T. and Aasen, A.O. (1977) Eur. Surg. Res. 9, 23-33.
27. Beauchamp, C. and Fridovich, I. (1970) J. Biol.Chem. 245, 4641-4646.

HYDROGEN PEROXIDE AND SINGLET OXYGEN IN LIPID PEROXIDATION

FULVIO URSINI, ALBERTO BINDOLI, LUCIA CAVALLINI AND MATULDE MAIORINO
Institute of Biological Chemistry, University of Padova and Centro Studio
Fisiologia Mitocondriale CNR, 35100 Padova, Italy

INTRODUCTION

Lipid peroxidation is a tissue injury, that can be promoted by dioxygen, its partially reduced forms and singlet oxygen.[1] The initiating step is generally followed by a free radical chain which causes the breakdown of polyunsaturated fatty acids (PUFA).[2] In particular superoxide anion produced by a number of biological systems can give rise to the other toxic forms. Among these are:

(1) H_2O_2 formed by spontaneous or catalyzed dismutation;

$$O_2^- + O_2^- + 2 H^+ \longrightarrow O_2 + H_2O_2$$

(2) $OH^{\cdot}$ radical and singlet oxygen generated through the Haber and Weiss reaction.

$$O_2^- + H_2O_2 \longrightarrow OH^- + OH^{\cdot} + {}^1O_2$$

In the present paper we are concerned with the lipid peroxidation depending on H_2O_2, which may generate ${}^1O_2$ through spontaneous disproportion,[3] and depending on ${}^1O_2$ photochemically generated.[4] We have demonstrated that H_2O_2 and ${}^1O_2$ form essentially lipid hydroperoxides (LHP). On the other hand, this barbituric acid reactive material (TBA-RM) is produced only when LHP breakdown is promoted by hematin.

MATERIALS AND MATHODS

Albino rats of the Wistar strain, fed al libitum on laboratory chow were used. Microsomes were prepared by differential centrifugation from livers homogenized in 0.25 M sucrose, 10 mM Tris HCl pH 7.5 and 5 mM EDTA. Microsomes were washed with the same buffer without EDTA. Boiled microsomes were obtained by incubating for 3 minutes in a boiling water bath the microsomes suspended in buffer in the test tubes; all the other assay reagents were added after cooling. Rat brain phospholipid liposomes were prepared according to Gutteridge.[5] Linolenic acid was from Merck, superoxide dismutase from Sigma. Crystalline beef liver catalase, from Boehringer Mannheim, was freed of thymol by passing through a Sephadex G-25 column equilibrated with 50 mM phosphate buffer, pH 7.5. Catalase activity was assayed polarographically, and the concentration determined spectrophotometrically. Hematin was prepared from beef red blood cells according to Shemin.[6] The formation of TBA-RM was determined by the thiobarbituric acid

(TBA) test according to Buege and Aust.[7] To the 1 ml assay samples 20 µl of butylated hydroxy toluene (10 mg/ml in ethanol) were added before the TBA reagent.

The formation of LHP was estimated by the KI oxidation (iodine) test according to Buege and Aust.[7] The photoactivation experiments were performed in vessels at $25^{\circ}C$; the reaction was started by illumination with a 200 W incandescent Lamp at a distance of 5 cm with a 1% K-bichromate solution filter. Oxygen uptake was detected with a Clark electrode. All chemicals were of analytical grade.

RESULTS AND DISUCSSION

TBA-RM is produced from rat liver microsomes incubated with H_2O_2 and hematin (Figure 1). There is no TBA-RM production when boiled microsomes, liposomes or linolenic acid dispersions are used, so indicating that an enzymatic component is required in the phenomenon. Since rat liver microsomes exhibit a catalase activity which can account for the complete breakdown of added H_2O_2 in 10-15 minutes under the employed conditions, we have tested the effect of the addition of catalase on boiled microsomes. As can be seen in Figure 2, 1 nm

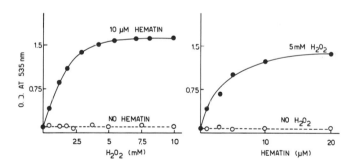

Fig. 1. Characteristics of H_2O_2 and hematin dependent microsomal lipid peroxidation. One ml samples of a mixture containing 0.1 M Tris-HCl pH 7.5 and 1 mg/ml microsomal proteins, were incubated in the presence of H_2O_2 and hematin at the indicated concentrations, at 20°C for 30 min. The reaction was stopped by addition of 20 µl of 1% ethanolic solution of butylated hydroxy-toluene and 2 ml of TBA reagent. Estimation of TBA-RM was performed as indicated under MATERIALS AND METHODS.

added catalase is able to restore TBA-RM production from boiled microsomes incubated with H_2O_2 and hematin. Addition of high concentrations of catalase to
boiled microsomes or of 5 nM catalase to fresh microsomes are inhibitory on
TBA-RM formation, probably because of too fast decomposition of H_2O_2. Also in
a linolenic acid dispersion, incubated in the presence of H_2O_2 and hematin,
catalase promotes TBA-RM production. One mM EDTA, 0.1 M mannitol, 0.1 M ter-
butanol, 10 µg/ml superoxide dismutase have no inhibitory effect; 10 mM histidine causes a 50% inhibition (Table 1). These properties indicate that $OH^{\cdot}$ and
O_2^{-} are not involved in the process, while $^1O_2$, generated non-enzymatically
from H_2O_2[3] or LHP breakdown,[8] can play some role. Boiled catalase or cytochrome
c and horse-radish peroxidase, in concentrations ranging from 1 nM to 10 µM
cannot replace native catalase.

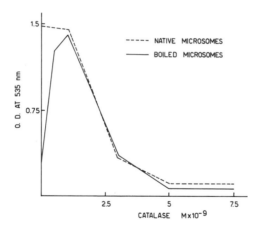

Fig. 2. Effect of catalase on TBA-RM production from native and boiled microsomes in the presence of H_2O_2 and hematin. One ml samples of a mixture containing 0.1 M Tris-HCl pH 7.5, 1 mg/ml microsomal proteins, 5 mM H_2O_2, 10 µM
hematin and catalase at the indicated concentrations were incubated at 20°C
for 30 min. For TBA test see legend to Figure 1.

Figure 3 shows the time course of the LHP formation from linolenic acid
incubated with : H_2O_2, hematin, H_2O_2 plus hematin, H_2O_2 plus hematin and catalase. TBA values at the end of the incubation are reported in the inset. LHP
are formed following the addition of H_2O_2 alone reaching a peak at about 20
minutes. Then their content starts to decrease slowly. Essentially no TAB-RM
can be found at the end of the 30-minute incubation period. The results with
H_2O_2 plus hematin are similar. When also catalase is added, TBA-RM is produced
at a great extent, but LHP rise appears to be modest. Hematin alone sparks a
small increase both of TBA-RM and LHP, probably acting on small amounts of
preexisting LHP.[9]

TABLE 1

TBA-RM FORMATION FROM LINOLENIC ACID[a]

				TBA Test
				O.D. at 535 nm
Linolenic Acid				0.080
"	+ H_2O_2			0.098
"	"	+ Hematin		0.130
"	+ Hematin			0.330
"	+ H_2O_2	+ Hematin + Catalase		0.970
"	"	"	"	+ EDTA	0.950
"	"	"	"	+ Mannitol	0.990
"	"	"	"	+ t-butanol	1.050
"	"	"	"	+ SOD	1.070
"	"	"	"	+ Histidine	0.480

[a]Incubation conditions: 0.1 Tris-HCl pH 7.5; 1 mg/ml linolenic acid; 5 mM H_2O_2; 10 µM hematin; 1 nm catalase; 1 mM EDTA; 0.1 M mannitol; 0.1 M ter-butanol; 10 γ/ml superoxide dismutase; 10 mM histidine; 1 ml samples were incubated at 20°C for 30 min. The TBA test was performed as indicated under MATERIALS AND METHODS

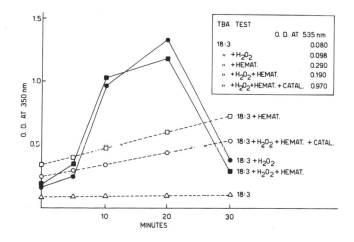

Fig. 3. Time course of LHP formation from linolenic acid in the presence of H_2O_2, hematin, H_2O_2 plus hematin and H_2O_2 plus hematin and catalase. Ten ml of a mixture containing 0.1 M Tris-HCl pH 7.5 and 1 mg/ml linolenic acid, were incubated in the presence of: (●—●) 5 mM H_2O_2, (□--□) 10 µM hematin, (■—■) 5 mM H_2O_2 plus 10 µM hematin, (O--O) 5 mM H_2O_2 plus 10 µM hematin and 1 nm catalase; (△--△) Control. Iodine test was performed on 1 ml aliquots withdrawn at different times. TBA test was performed on 1 ml aliquots at 30 min, as indicated under MATERIALS AND METHODS

From these results it appears that H_2O_2 alone gives rise to LHP, which can decompose spontaneously without TBA-RM formation. TBA-RM appears to be produced only when LHP decomposition is catalyzed by an electron transfer reaction (i.e., in the presence of hematin), but not when a spontaneous intramolecular rearrangement takes place.[10] Since the production of TBA-RM is evident only when H_2O_2, hematin and a definite catalase activity are present together, a competition between H_2O_2 and LHP for hematin can be postulated. The effect of catalase could be attributed to the slow progressive removal of H_2O_2, which can increase the availability of hematin for LHP breakdown. So the lipid peroxidation process studied in this work can be divided in two steps: LHP formation by the action of H_2O_2 and TBA-RM formation through LHP decomposition by the action of hematin, following the removal of H_2O_2. This is supported by the result of the experiment reported in Table 2. LHP are formed from linolenic acid in the presence of H_2O_2. Hematin forms TBA-RM provided that H_2O_2 is decomposed by a

TABLE 2

TWO STEPS IN THE FORMATION OF TBA-RM FROM BOILED MICROSOMES: EFFECT OF H_2O_2 HEMATIN AND CATALASE[a]

	30 min incubation	15 additional min incubation
	TBA test O.D at 535 nm	TBA test O.D at 535 nm
Boiled Microsomes	0.030	0.045
" " + H_2O_2	0.130	0.150
" " + Hematin	0.200	0.150
" " + H_2O_2 + Hematin	0.190	0.200
" " + H_2O_2	–	+ Catalase 0.185
" " + H_2O_2	–	+ Catalase + Hematin 0.675

[a]Incubation conditions: 0.1 M Tris-HCl pH 7.5; 1 mg/ml microsomal protein; 5 mM H_2O_2; 10 μM hematin; 20 nm catalase. TBA test was performed on 1 ml samples as indicated under MATERIALS AND METHODS

delayed addition of catalase by a two hour vigorous shaking in open air at room temperature ("aged" boiled microsomes) or by adding cumene hydroperoxide (CHP), hematin catalyzes TBA-RM production and addition of H_2O_2 is inhibitory (Table 3). Boiled microsomes were employed in these experiments to abolish catalase activity and to denaturate cytochrome P-450, which can promote peroxidation by interacting with CHP.

Since non-enzymatic disproportionation of H_2O_2 can give rise to $^1O_2$, which may account for the formation of LHP, we have looked for the direct identification of $^1O_2$ by using diphenylfuran, a specific $^1O_2$ reagent.[12] A clear-cut

TABLE 3

TBA-RM FORMATION FROM "AGED" BOILED MICROSOMES AND BOILED MICROSOMES PLUS CHP:
EFFECT OF HEMATIN AND H_2O_2 [a]

	TBA test O.D. at 535 nm
Boiled Microsomes + CHP	0.560
" " + Hematin	1.305
" " " + H_2O_2	0.540
Boiled Microsomes	0.030
"Aged" "	0.180
" " + Hematin	0.650
" " " + H_2O_2	0.230

[a] Incubation conditions: 0.1 M Tris HCl; 5 mM H_2O_2; 10 µM hematin; 3 mM CHP
(added as 200 mM ethanolic solution). "Aged" microsomes were prepared by
shaking for 2 hr the test tubes containing only boiled microsomes and buffer
before addition of the other reagents. 1 ml samples were incubated at 20°C
for 30 min. The TBA test was performed as described under MATERIALS AND METHODS.

demonstration of singlet oxygen formation was not obtained by this approach.
For this reason we have performed some experiments using rose bengal in the
presence of visible light, a well known system for generating $^1O_2$, to compare
the effect of $^1O_2$ with that of H_2O_2. Rose bengal, after light absorption, is
converted into an electronic excited state that can transfer its energy to
molecular oxygen. The reactive form of singlet oxygen, can react further with
different acceptor molecules (photoactivation system type II).[4] Consequently
an oxygen consumption can be recorded. As reported in Figure 4 the polarographic trace shows that the system composed of rose bengal and liposomes gives
rise to oxygen consumption when the light is switched on, while oxygen consumption stops in the absence of light. A similar behaviour is apparent when
dimethylfuran,[4] another $^1O_2$ reagent, is used in place of liposomes. Thus $^1O_2$
reacts with liposomes. LHP are formed from the interaction of $^1O_2$ and liposomes
(Table 4) and the addition of hematin when light is switched off can induce
TBA-RM production from LHP which disappear. Also in this case H_2O_2 abolished
the effect of hematin. It should be noted that when the photoactivation system
is employed, a small amount of TBA-RM is formed, also in the absence of hematin
addition. For this reason we cannot exclude the partial involvement of photoactivation system type I, which forms free radicals and can account for the
slight TBA-RM production. It is also possible that the direct insertion of
singlet oxygen on the PUFA, when it is produced in great extent by the photoactivation system, may form endoperoxidases which react with TBA.[13] In any
case hematin greatly increases the production of TBA-RM at the expense of LHP
formed by $^1O_2$.

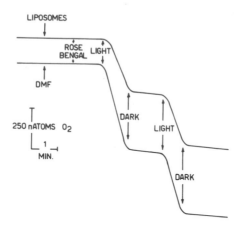

Fig. 4. Oxygen uptake by lipsomes or dimethylfuran induced by rose bangal and light. Liposomes (2 mg/ml) or 1 mM diemthylfuran were incubated in 0.1 M Tris-HCl pH 7.5 in the presence of 10 γ/ml rose bengal in a final volume of 3 ml. Experimental conditions are described under MATERIALS AND METHODS

TABLE 4

TBA-RM AND LHP FORMATION FROM LIPOSOMES IN THE PRESENCE OF ROSE BENGAL AND VISIBLE LIGHT: EFFECT OF HEMATIN AND H_2O_2 ON LHP DECOMPOSITION[a]

First Incubation Period			
15 min incubation		TBA test O.D. at 535 nm	Iodine test O.D. at 350 nm
Liposomes + RB	Dark	0.004	0.180
" "	Dark	–	–
" "	Light	0.210	0.810
" "	Light	–	–
" "	Light	–	–
Second Incubation Period			
15 min incubation		TBA test O.D. at 535 nm	Iodine test O.D at 350 nm
Liposomes + RB	Dark	0.010	0.130
" "	Dark + Hematin	0.095	0.090
" "	Dark	0.230	0.630
" "	Dark + Hematin	0.610	0.420
" "	Dark + Hematin + H_2O_2	0.250	0.930

[a]Incubation conditions: The liposomes (1 mg/ml) were incubated at 20°C for 15 min in the presence of 10 μl rose bengal (RB) in a 1 ml final volume. After the addition of hematin or hematin and H_2O_2 the incubation lasted for a further 15 min the the dark. TBA test and iodine test were performed as described under MATERIALS AND METHODS

The results reported here suggest the following conclusions: (i) H_2O_2 and $^1O_2$ promote essentially the formation of LHP; (ii) H_2O_2 acts probably through the formation of $^1O_2$; (iii) TBA-RM is produced at the greatest extent when free radicals are formed in the lipophilic phase from hematin catalyzed LHP breakdown; and (iv) there is a competition between H_2O_2 and LHP for hematin.

ACKNOWLEDGMENTS

Thanks are due to Professor Carlo Gregolin for helpful discussions through-out this work and to Mrs. Marina Valente for excellent technical assistance. We are also grateful to Mrs. Maurizia Cuccia for secretarial aid.

REFERENCES

1. Kellogg, E.W.,III and Fridovich, I. (1975) J. Biol. Chem. 250, 8812-8817.
2. Mead, G.F. (1976) in Free Radicals in Biology, Vol.1, Pryor, W.A., ed., Academic Press, New York, pp.51-68.
3. Smith, L.L. and Kulig, M.J. (1975) J. Amer. Chem. Soc. 98, 1027-1029.
4. Foote, C.S. (1976) in Free Radicals in Biology, Vol.2, Pryor, W.A., ed., Academic Press, New York, pp.85-133.
5. Gutteridge, J.M.C. (1977) Anal. Biochem. 82, 76-82.
6. Shermin, D. (1957) Methods Enzymol. 4, 643-651.
7. Buege, J.A. and Aust, S.D. (1978) Methods Enzymol. 52, 302-310.
8. Hawco, F.J., O'Brien, C.R. and O'Brien, P.J. (1977) Biochem. Biophys. Res. Commun. 76, 354-361.
9. Kaschnitz, R.M. and Hatefi, Y. (1975) Arch. Biochem. Biophys. 171, 292-304.
10. Pryor, W.A. (1976) In Free Radicals in Biology, Vol.1, Pryor, W.A., ed., Academic Press, New York, pp.1-49.
11. O'Brien, P.J. and Rahimtula, A.D. (1975) J. Agr. Food Chem. 23, 154-158.
12. King, M.M., Lai, E.K. and McCay, P.B. (1975) J. Biol. Chem. 250, 6496-6502.
13. Pryor, W.A. and Stanley, J.P. (1975) J. Org. Chem. 40, 3615-3617.

SUPEROXIDE RADICALS AND LIPID PEROXIDATION IN TUMOUR MICROSOMAL MEMBRANES

T. GALEOTTI[+], G.M. BARTOLI[+], S. BARTOLI[+] AND E. BERTOLI[++]
[+]Institute of General Pathology, Catholica University, Rome Italy; [++]Institute of Biological Chemistry, University of Bologna, Bologna, Italy

ABSTRACT

1. In microsomal membranes from tumours with a large difference in growth rate (Morris hepatomas 44 and 3924A, Novikoff hepatoma and Ehrlich Lettrè ascites cells) the content of the electron carriers (NADPH-cytochrome $\underline{c}$ reductase and cytochrome P-450) which generate reactive oxygen intermediates and the NADPH-induced production of superoxide (O_2^-) radicals are directly related to the degree of differentiation and hence inversely to the growth rate.

2. NADPH-dependent lipid peroxidation, measured as malonaldehyde production, shows the same behaviour as superoxide formation.

3. The tumour lipid behaviour is also decreased, with respect to liver microsomes, in the presence of an excess of superoxide anions produced by xanthine and xanthine oxidase or ascorbate.

4. The lipid to protein ratio and the degree of fatty acid unsaturation decrease gradually from normal to minimal- and maximal-deviation tumour microsomes. The different lipid composition is reflected also by differences in the physical environment of the bilayer, as suggested by data obtained with spin-labelled fatty acids.

5. The content of cytosolic superoxide dismutase (SOD) is diminished in all tumours examined. The lowest amounts were found in the tumours with the fastest growth rates.

The results indicate that, besides the low activity of the generators of oxygen radicals, changes in lipid composition and assembly make the tumour microsomal membranes more resistant to peroxidation. Such changes might be related to a sudden increase after tumour transformation in the lipoperoxidative attack of O_2^- radicals, owing to the decrease in the protective effect of cytosolic SOD. It is proposed that the low amount of peroxides subsequently produced by membranes altered in their lipid composition may be a factor responsible for the high mitotic activity of transformed cells.

INTRODUCTION

It has been reported previously that the non-enzymic lipid peroxidation induced by ascorbic acid is very low in tumour mitochondria.[1,2] Also a low rate

of the NADPH-dependent lipid peroxidation has been observed recently in mitochondria isolated from hepatoma D30.[3] Moreover, mitochondria and microsomal
membranes from poorly differentiated, fast-growing, hepatomas have a lowered
amount of phospholipids and a decreased ratio of polyenoic to monoenoic fatty
acids.[4-6] Work in our laboratory has demonstrated that fast-growing experimental
tumours of the rat and mouse lack the mitochondrial superoxide dismutase (SOD)[7]
and subsequent studies have established that the loss of the mitochondrial enzyme
is a universal feature of tumours whilst the cytosolic cupro-zinc enzyme is
usually, but not always, lowered in such tissues (see Ref.8 for review). The
whole of the above data has prompted us to focus attention on the possibility of
a correlation between growth rate and lipid peroxide production in tumour cells.
Indeed, if the hypothesis that lipid peroxides inhibit cell division[9-12] is valid,
the finding of a markedly decreased lipo-peroxidative activity in fast-growing
tumours with respect to progressively higher activities in slow-growing tumours
and normal tissues is indicative of the fact that in transformed cells lipid
peroxides become inadequate to control the mitotic activity in a proportion which
is directly related to the growth ability of the tumour.

The lowered lipid to protein and polyenoic to monoenoic ratios in tumour
membranes may be the expression of alterations consequent to the reduction of
the protective effect of SOD against the O_2^--induced lipid peroxidation. Membranes
already altered in their lipid composition and organization would then have much
less susceptibility to peroxidative agents.

In the present work we have studied several properties of microsomal membranes
(superoxide generation, lipid peroxidation, composition and fluidity) and determined the content of cytosolic SOD of tumours with a large difference in growth
rate. From the data obtained we are inclined to ascribe to the decreased SOD
content a role in determining the damage of tumour membranes. The consequent
reduced vulnerability of the membranes to peroxidative agents would result in
the production of an amount of lipid peroxides insufficient to control cell
division.

MATERIALS AND METHODS

Tumour and normal tissue preparations. Morris hepatomas 44 and 3924A (Ref.13)
were originally provided by Dr. H.P. Morris and maintained in our laboratory by
intramuscular and subcutaneous transfer into both hind legs of inbred rats of
the Buffalo and ACI/T strains, respectively. The tumours were excised 4-6 months
(hepatoma 44) and 3-4 weeks (hepatoma 3924A) after transplantation. Ehrlich
hyperdiploid Lettré and Novikoff hepatoma[14] ascites cells were grown i.p. in
albino Swiss mice and in Sprague-Dawley rates, respectively. The cells, harvested
6-8 days after the inoculation, were freed from contaminating erythrocytes by
differential centrifugations. Other details concerning the isolation of hepatomas

and preparation of ascites cells are reported elsewhere.[15,16] Normal male rats
of the Wistar and ACI/T strains, weighing 150-200 g, were used for rat liver
preparations. No significant differences in the parameters studied were found
between the two strains.

Isolation of microsomes. Liver and solid hepatomas were homogenized with 5
volumes of 0.25 M sucrose, 0.5 mM EGTA, 5 mM HEPES (pH 7.4) in a Potter-Elvehjem
glass-Teflon homogenizer. Ehrlich and Novikoff ascites cells were suspended
in the same medium and disrupted by homogenizing for 5 sec with an Ultra-Turrax
(Janke and Kunkel KG).[17] Microsomes were prepared by centrifugations at 8,000
and 18,000 x g for 10 min and 105,000 x g for 60 min.[18] The fractions were washed
once in 0.15 M KCl, 50 mM Tris-HCL (pH 7.5) by centrifugation at 105,000 x g
for 30 min and suspended in the same medium. Such preparations were virtually
free of any hemoglobin contamination as judged by difference absorption spectra of
microsomes saturated with CO versus aerobic microsomes. Proteins were estimated
by the biuret method.[19]

Assay methods. Lipid peroxidation was measured at $25^{\circ}C$ as malonaldehyde production in an oxygen-saturated medium containing 0.15 M KCl, 50 mM Tris-HCl
(pH 7.5), 4 mM ADP, 0.05 mM $FeCl_3$ and, when indicated, 0.33 mM xanthine. The
reaction was started either by 0.4 mM NADPH or by 50 μg/ml xanthine oxidase
(Sigma, Type I) or by 0.25 mM sodium ascorbate. Malonaldehyde was determined
at 535 nm by the thiobarbituric acid assay.[20]

The co-oxidation of epinephrine to adrenochrome induced by superoxide radicals[21]
was followed in a dual-wavelength/split-beam Aminco-Chance spectrophotometer at
480-575 nm. The reaction mixture contained 0.15 M KCl, 50 mM Tris-HCl, pH 7.5,
1 mM epinephrine. After saturation with O_2, 1 mg microsomal proteins per ml
were added and the reaction was started by addition of NADPH. The extinction
coefficient used for adrenochrome was 2.86 $mM^{-1}cm^{-1}$.

Analysis of cytochromes P-450 and b_5 was performed from difference absorption
spectra by the method of Omura and Sato,[22] using the dual-wavelength/split beam
spectrophotometer. The content of pigments was calculated from $\Delta\varepsilon(450-490)$ nm =
91 $mM^{-1}cm^{-1}$ for P-450 (Ref.22) and $\Delta\varepsilon(424-409$ nm) = 165 $mM^{-1}cm^{-1}$ for b_5 (Ref.23).
NADPH-cytochrome c reductase was determined by the method of Jones and Wakil.[24]

Lipids were extracted from the isolated membranes by the method of Folch et al.[25]
Phospholipid phosphorus was determined by the method of Bartlett[26] modified by
digestion with 70% $HClO_4$, according to Marinetti.[27] For analysis of fatty acids
by gas-liquid chromatography, methyl esters were prepared by interesterification
of lipid samples for 4h at $100^{\circ}C$ with 5% anhydrous HCl-methanol, then extracted
with petroleum ether and examined with a Perkin-Elmer F-II gas chromatograph.
The samples were injected onto a polyethyleneglycolsuccinate column operating
at $180^{\circ}C$. The column was calibrated with standard methyl esters of the different fatty acids.

Cytosolic SOD was extracted from 1.5–5 g of tissue by the procedure of McCord and Fridovich,[28] with the omission of the DE-32 column, as suggested by Sykes et al.[29] The enzyme content was measured in tissue extracts from the values giving 50% inhibition of autoxidation of epinephrine to adrenochrome, using a standard curve obtained with purified bovine blood enzyme (50% inhibition by 65 mg/ml at 30°C and pH 10.2).[30]

EPR spectroscopy. Microsomes were labelled with N-oxyl-4',4'dimethyloxazolidine derivatives of 5-ketostearic acid (5-NS) and 16-ketostearic acid (16-NS) purchased from Syna Assoc. (Palo Alto, Ca., U.S.A.). The spin labels were dissolved in ethanol and were evaporated to dryness with a stream of nitrogen. Microsomes (about 20 mg protein/ml) were then added to the label and vortexed at room temperature. The final concentration of the labels was 0.1 mM so that no liquid lines, due to the incorporated spin label, occurred.[31] Samples were held in a sealed Pasteur pipette and EPR spectra were recorded at room temperature using a Varian E-4 spectrometer. The freedom of motion of the spin-labelled fatty acids in microsomes was assessed from the order parameter (S), corrected by the polarity (a'n) according to the following equation (Ref.32):

$$ S = \frac{A_\parallel - A_\perp}{Azz - Axx} \cdot \frac{an}{a'n} $$

where Azz and Axx are the hyperfine principal values of the nitroxide radical and $A_\parallel$ and $A_\perp$ are measured as indicated in Figure 5, and from the motion parameter (τo), according to Ref.33, expressed in ns. Higher S and τo values are associated with a more restricted mobility of the hydrocarbon chains and therefore report the local fluidity of membrane lipids.[34]

RESULTS AND DISCUSSION

NADPH-dependent lipid peroxidation. The lipoperoxidative activity of liver and tumour microsomes, during 40-min incubation in the presence of NADPH, is shown in Figure 1. Malonaldehyde production is about 10% of control in membranes isolated from fast-growing tumours (hepatoma 3924A and Ehrlich ascites cells) and moderately higher in the slow-growing hepatoma 44 microsomes. Thus it appears that tumour transformation induces differential changes in the rate of the NADPH-dependent lipid peroxidation, depending on the degree of deviation from the normal tissue. Since it is known that lipid peroxidation induced by NADPH is initiated by reactive oxygen derivatives[35,36] generated by microsomal electron carriers, we expect to find also a decrease in the production of such chemical species in microsomes deriving from tumours. However, the possibility has to be considered also that changes in the membrane lipid composition and organization may be responsible for the lowered lipoperoxidative activity in tumours. In such a case low values of lipid peroxidation should be equally observed when the process is promoted by agents generated outside the membrane, such as O_2^- radicals or ascorbate.

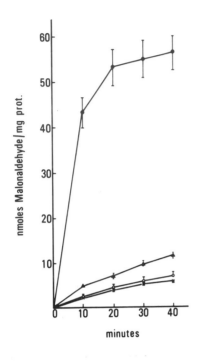

Fig. 1., NADPH-dependent lipid peroxidation of microsomal membranes from rat liver (●), Morris hepatomas 44 (▲) and 3924A (o) and Ehrlich ascites tumour cells (x). 0.2 mg/ml (rat liver and hepatoma 44) and 1 mg/ml (hepatoma 3924A and ascites cells) of microsomal proteins were suspended in the medium reported under MATERIALS AND METHODS. Values are given as means ± S.E. of 4-6 experiments.

Production of superoxide radicals and content of electron carriers. Microsomal membranes are also able to generate superoxide radicals in the respiratory system at the level of NADPH-cytochrome c reductase and cytochrome P-450 (Refs.37-40). Figure 2 shows that such production is substantially reduced in tumour membranes, with the lowest values present in the maximal-deviation tumours (Novikoff and 3924A hepatomas). Therefore, it appears that in tumours an inverse relationship exists between growth rate and another property of microsomal membranes, namely the generation of superoxide anions. The first column of Table 1 shows mean values of superoxide production in slowly and rapidly growing hepatomas. Such values range from a minimum of 15-20% in microsomes from the two poorly differentiated tumours to a maximum of about 40% with respect to liver in the well differentiated hepatoma 44. Difference spectra absorption spectra of tumour microsomes (Figure 3) show that cytochrome b_5 and P-450 are decreased in hepatoma 44 and practically undetectable in the 3924A and Novikoff hepatomas. The calculated values are reported in Table 1, together with the activities of NADPH-cytochrome c reductase. From such measurements it appears that in the membranes of the fast-growing tumours the low amount of adrenochrome formed is due to the generation of superoxide radicals, exclusively at the level of the flavoprotein.

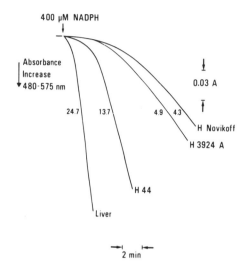

Fig. 2. Co-oxidation of epine-phrine to adrenochrome induced by superoxide radicals in micro-somes isolated from rat liver and various hepatomas (II). The num-bers along the traces indicate nmol adrenochrome formed per min/mg protein.

TABLE 1

RELATIONSHIP BETWEEN ADRENOCHROME FORMATION AND CONTENT OF ELECTRON CARRIERS IN MICROSOMAL MEMBRANES FROM RAT LIVER AND HEPATOMAS

	Adrenochrome	NADPH-cytochrome c reductase	Cytochrome b_5	Cytochrome P-450
	(nmol/min/mg protein)		(nmol/mg protein)	
Rat liver	$24.8\pm1.7^a(13)^{b,c}$	$83.1\pm3.4(4)$	$0.32\pm0.02(5)^c$	$0.56\pm0.06(6)^c$
Morris hepatoma 44	10.0 ± 1.1 (4)	$42.4\pm1.1(5)$	$0.15\pm0.01(3)$	$0.14\pm0.02(3)$
Morris hepatoma 3924A	3.4 ± 0.4 (4)	$17.8\pm1.2(5)$	$0.008\pm0.001(3)$	nil
Novikoff ascites hepatoma	4.7 (2)	25.1 (2)	nil	nil

[a] Mean ±S.E.

[b] Number in parentheses, number of observations.

[c] From Ref.40.

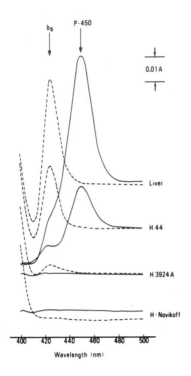

Fig. 3. Room-temperature difference absorption spectra of
cytochromes b_5 and P-450 in
microsomes from rat liver and
hepatomas (H). ----, NADPH-
treated minus untreated aerobic
microsomes; ———, $(Na_2S_2O_4+ CO)$-
treated minus $Na_2S_2O_4$-treated
microsomes. Suspensions were
1-3 mg protein/ml. NADPH, 330 µM.

Superoxide- and ascorbate lipid peroxidation. The observation that tumour mitochondria have a limited lipoperoxidative activity both in the presence of ascorbate
and NADPH[1-3] has suggested the possibility that a similar situation occurs in
tumour microsomal membranes. Microsomal membranes from tumours were exposed to
external lipoperoxidative agents and the rate of malonaldehyde formation was followed
during 40-min incubation. The results of such experiments are shown in Figure 4.
The activity, measured in the presence of xanthine and xanthine oxidase (A) or
ascorbate (B), decreases progressively from normal to hepatoma 44 and fast-growing
tumour microsomes. These results suggest that, besides the low rate of superoxide
generation within the membrane, the low peroxidative activity of tumour microsomes
should be ascribed to intrinsic features of the bilayer which seem to be altered
in a fashion related to the growth ability of the transformed tissue.

Lipid composition and fluidity. Table 2 reports the lipid content and fatty
acid composition of microsomal membranes isolated from rat liver and hepatomas
44 and 3924A. The lipid to protein ratio decreases considerably in the maximal-
deviation hepatoma whilst hepatoma 44 microsomes have a value closer to that of
rat liver. Furthermore, as indicated by the polyenoic to monoenoic fatty acid
ratio and by the double-bond index, the degree of unsaturation decreases in both

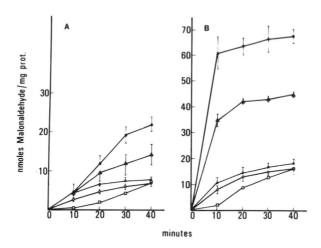

Fig. 4. Lipid peroxidation induced by superoxide radicals (A) and ascorbate (B) of microsomes isolated from rat liver (•), Morris hepatomas 44 (▲) and 3924A (o), Ehrlich (x) and Novikoff hepatoma (□) ascites cells. Suspensions were 0.2 mg protein/ml. The values are the mean of 3-5 experiments.

tumours with intermediate and minimal values in the 44 and 3924A hepatomas, respectively. The differences in lipid composition are reflected also by a difference in the physical environment of the bilayer, as indictaed by determina-tion of the lipid fluidity of the membranes by spin labelling method. The mobility of 5-nitroxide stearate (5-NS) in microsomes from normal and hepatoma cells is shown in the EPR spectra of Figure 5. Although there is a similarity in the spectra, the hyperfine splitting ($2T_{||}$) is different, indicating a less restricted mobility of the hydrocarbon chains or a less viscous environment around the spin probe in the control. Table 3 shows that $2T_{||}$ and the order (S) and motion (τ_0) parameters are higher in hepatoma 3924A microsomes, suggesting that the region in the proximity of the lipid polar head groups and in the lipid-core is more rigid than that of hepatoma 44 and rat liver. The above described modifications in the chemical and physical characteristics of tumour microsomal membranes might derive from a de-crease in the protective effect of SOD whose concentration has been reported to be lowered in tumour cells.[8] A support to this suggestion may be provided by the finding of a direct correlation between the degree of membrane perturbation and the entity of the decrease in SOD content. The experiments reported below seem to indicate the existence of such a correlation.

Content of cytosolic superoxide dismutase. Table 4 shows the values of cytosolic SOD in rat liver and Morris hepatomas 44 and 3924A. The enzyme content decreases

TABLE 2

LIPID CONTENT AND FATTY ACID COMPOSITION OF RAT LIVER AND HEPATOMA MICROSOMES

	Rat liver	Hepatoma 44	Hepatoma 3924A
Lipid/protein	0.66	0.56	0.32
		% hydrocarbon chains	
12.:0	0.2	0.5	0.7
12:1	0.4	0.9	0.7
13:0	0.2	0.2	0.3
13:1	0.2	1.3	0.5
14:0	1.3	2.3	2.8
14:1	0.2	1.0	1.7
15:0	1.0	1.8	1.0
15:1	0.2	0.8	1.9
16:0	25.9	23.4	24.8
16:1	2.6	4.5	6.2
17:0	1.2	0.7	0.7
17:1	0.3	0.9	0.3
18:0	18.9	14.9	13.3
18:1	14.3	12.0	18.4
19:0	traces	traces	traces
19:2	14.6	6.7	4.7
20:0	0.7	2.1	1.6
20:1	0.6	0.9	2.4
20:2	2.8	14.2	10.4
20:3	0.8	traces	traces
20:4	11.9	4.6	4.9
22:1	1.6	6.3	4.8
Polyenes/monenes	1.5	0.9	0.5
Double-bond index[a]	84	60	50

[a]The double-bond index was caluclated as the sum of the values obtained by multiplying the percentage of the unsaturated fatty acid by the number of doublebonds in that fatty acid.

progressively in the tumours examined with respect to the corresponding normal tissue, being still 25% in hepatoma and only 5% in hepatoma 3924A. Low values, similar to those of hepatoma 3924A, have been found for cupro-zinc SOD in other fast-growing tumours of the rat and mouse.[41,42] In relation to the proposed role of the diminished SOD content in the damage of tumour membranes it seems relevant to mention that two other protective enzymes, glutathione peroxidase and catalase, have low activities in this tissue.[41,43] Both mitochondrial and cytosolic glutathione perodixase have been found to be drastically decreased in the fast-growing hepatoma 27 (Ref.42). However, no data are yet available on the activity of this enzyme in minimal-deviation tumours. Interestingly, Mochizuki et al.[44] have observed that in Mottis hepatomas of different growth rate the content of peroxisomes decreases with increasing growth rate.

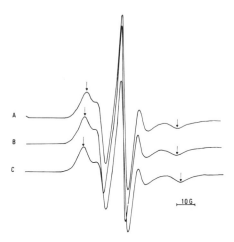

Fig. 5. EPR spectra of rat liver (A), Morris hepatoma 44 (B) and 3924A (C) microsomes labelled with 5-nitroxide stearate.

TABLE 3

THE FREEDOM OF MOTION OF LIPID SPIN LABELS IN MICROSOMES DERIVED FROM RAT LIVER AND MORRIS HEPATOMAS

	5-NS[a]		16-NS[b]
	$2T_{||}$ (G)	S	τ_0 (ns)
Rat liver	51.2	0.64	1.4
Hepatoma 44	51.8	0.67	1.7
Hepatoma 3924A	54.8	0.70	2.1

[a]5-NS, 5-nitroxide stearate

[b]16-NS, 16-nitroxide stearate

TABLE 4

CONTENT OF CYTOSOLIC SUPEROXIDE DISMUTASE IN RAT LIVER AND MORRIS HEPATOMAS

	Superoxide dismutase[a] (µg/g tissue)
Rat liver	207.5
Hepatoma 44	52.2
Hepatoma 3924A	10.7

[a]Each value represents the mean of 2-4 experiments performed in duplicate.

Conclusions. The experiments reported above demonstrate that the low lipo--peroxidative activity observed previously in tumour mitochondria[1-3] is also detectable in microsomal membranes. Since the phenomenon occurs both in the presence of internally generated (NADPH-induced) and of externally added (xanthine-xanthine oxidase or ascorbate) perodixative agents, it appears that the low NADPH-induced production of superoxide radicals within the membrane may not be the only factor responsible for such deficiency. Microsomal membranes have, in this respect, a lowered amount of lipids and a low degree of fatty acid unsaturation which easily justify the observed changes in lipid peroxidation. The loss of cytosolic SOD may represent a primary event which takes place after tumour transformation. The intracellular membranes exposed to a high concentration of oxygen radicals are damages in a way that they become much less susceptible to subsequent peroxidative injuries. The changes observed increase with increasing growth rate, suggesting that the low availability of lipid peroxides during tumour development may contribute to the maintenance of the uncontrolled mitotic activity of tumour cells.

ACKNOWLEDGEMENTS

We thank Dr.G. Rotilio for the kind gift of purified bovine blood superoxide dismutase. The work was supported by grants from Consiglio Nazionale delle Ricerche (78.02112.04) and Ministero Pubblica Istruzione 1977, Italy.

REFERENCES

1. Thiele, E.H. and Huff, J.W. (1960) Arch. Biochem. Biophys. 88, 208-211.
2. Utsumi, K., Yamamoto, G. and Inaba, K. (1965) Biochim. Biophys. Acta, 105, 368-371.
3. Player, T.J., Mills, D.J. and Horton, A.A. (1977) Biochem. Soc. Trans. 5, 1506-1508.
4. Feo, F., Canuto, R.A., Bertone, G., Garcea, R. and Pani, P. (1973) FEBS Lett. 33, 229-232.
5. Feo, F., Canuto, R.A., Garcea, R. and Gabriel, L. (1975) Biochim. Biophys. Acta, 413, 116-134.
6. Reitz, R.C., Thompson, J.A. and Morris, H.P. (1977) Cancer Res. 37, 561-567.
7. Dionisi, O., Galeotti, T., Terranova, T. and Azzi, A. (1975) Biochim. Biophys. Acta, 403, 292-300.
8. Oberley, L.W. and Buettner, G.R. (1979) Cancer Res. 39, 1141-1149.
9. Wolfson, N., Wilbur, K.M. and Bernheim, F. (1956) Exp. Cell. Res. 10, 556-558.
10. Cole, B.T. (1956) Proc. Soc. Exp. Biol. Med. 93, 290-294.
11. Glushchenko, N.N., Shestakova, S.V. and Danilov, V.S. (1975) Biol. Nauki. (Moscow), 18, 51-53.
12. Player, T.J., Mills, D.J. and Horton, A.A. (1977) Biochem. Biophys. Res. Commun. 78, 1397-1402.
13. Morris, H.P. and Wagner, B.P. (1968) Methods Cancer Res. 4, 125-152.
14. Novikoff, A.B. (1957) Cancer Res. 17, 1010-1027.
15. Russo, M.A., Galeotti, T. and van Rossum, G.D.V. (1976) Cancer Res. 36, 4160-4174.
16. Galeotti, T., Azzi, A. and Chance, B. (1970) Biochim. Biophys. Acta, 197, 11-24.

17. Bartoli, G.M., Dani, A., Galeotti, T., Russo, M. and Terranova, T. (1975) Z. Krebsforsch. 83, 223-231.
18. Werringloer, J. and Estabrook, R.W. (1975) Arch. Biochem. Biophys. 167, 270-286.
19. Layne, E. (1957) Methods Enzymol. 3, 450-451.
20. Hunter, F.E., Gebicki, J.M. Hoffsten, P.E., Weinstein, J. and Scott, A. (1963) J. Biol. Chem. 238, 828-835.
21. Misra, H.P. and Fridovich, I. (1972) J. Biol. Chem. 247, 3170-3175.
22. Omura, T. and Sato, R. (1964) J. Biol. Chem. 239, 2370-2378.
23. Garfinkel, D. (1958) Arch. Biochem. Biophys. 77, 493-509.
24. Jones, P.D. and Wakil, S.J. (1967) J. Biol. Chem. 242, 5267-5273.
25. Folch, G., Lees, M. and Sloane-Staney, G.H. (1957) J. Biol. Chem. 226, 497-505.
26. Bartlett, G.R. (1959) J. Biol. Chem. 234, 466-468.
27. Marinetti, G.V. (1962) J. Lipid Res. 3, 1-11.
28. McCord, J.M. and Fridovich, I. (1969) J. Biol. Chem. 244, 6049-6055.
29. Sykes, J.A., McCormack, F.X. and O'Brien, T.J. (1978) Cancer Res. 38, 2759-2762.
30. Misra, H.P. and Fridovich, I. (1972) J. Biol. Chem. 247, 3170-3175.
31. Butler, K.W., Tattrie, N.H. and Smith, I.C.P. (1974) Biochim. Biophys. Acta, 363, 351-360.
32. Seeling, J. (1976) in Spin Labeling, Theory and Applications, Berlines, L.J. ed. Academic Press, New York, pp.373-409.
33. Smit, J., Kamio, Y. and Nikaido, H. (1975) J. Bacteriol. 124, 942-958.
34. McConnell, H.M. and McFarland, B.G. (1970)Quart. Rev. Biophys. 3, 91-136.
35. Pederson, T.C. and Aust, S.D. (1975) Biochim. Biophys. Acta.385, 232-241.
36. Lai, C.S., Grover, T.A. and Piette, L.H. (1979) Arch. Biochem. Biophys. 193, 373-378.
37. Strobel, H.W. and Coon, M.J. (1971) J. Biol. Chem. 246, 7826-7829.
38. Aust, S.D., Roerig, D.L. and Pederson, T.C. (1972) Biochem. Biophys. Res. Commun. 47, 1133-1137.
39. Estabrook, R.W., Baron, J., Franklin, M., Mason, I., Waterman, M.and Peterson, J. (1972) in The Molecular Basis of Electron Transport, Schultz, J. and Cameron, B.F., eds., Academic Press, New York, pp.197-230.
40. Bartoli, G.M., Galeotti, T., Palombini, G., Parisi, G. and Azzi, A. (1977) Arch. Biochem. Biophys. 184, 276-281.
41. Bozzi, A., Mavelli, I., Finazzi-Agro, A., Strom, R., Wolf, A.M., Mondovi, B. and Rotilio, G. (1976) Mol. Cell. Biochem. 10, 11-16.
42. Peskin, A.V., Koen, Y.M., Zbarsky, I.B. and Konstantinov, A.A. (1977) FEBS Lett. 78, 41-45.
43. Greenstein, J.P. (1954) in Biochemistry of Cancer, Academic Press, New York.
44. Mochizuki, Y., Hruban, Z., Morris, H.P., Slesers, A. and Vigil, E.L. (1971) Cancer Res. 31, 763-773.

TUMOR CELLS AS MODELS FOR OXYGEN-DEPENDENT CYTOTOXICITY

G. ROTILIO, A. BOZZI, I. MAVELLI, B. MONDOVI AND R. STROM
Institutes of Biological Chemistry and Applied Biochemistry, University of
Rome, Rome, Italy

The State of the problem. The problem of cytotoxicity by reactive oxygen
derivatives has still many unsolved aspects. While the use of suitable sources
of active oxygen and of proper scavengers allow the investigation of the chemical
damage produced by oxygen on isolated biomolecules, such as enzymes, nucleic
acids or membrane lipids, the production and control of oxidative damage inside
intact cells cannot be used for straightforward solution to the cytotoxicity
problem. Firstly, such an approach requires knowledge of the cellular processes
which are potentially capable of producing and of metabolizing reactive oxygen
species and of the possible stimulatory mechanisms acting on them. Secondly,
addition of exogenous scavengers as the major interpretative tool is often un-
feasible with intact cells and only situations of natural deficiency of some
intracellular factors concerned with antioxidative defence can be used to test
their role provided that comparison is made with a "normal" type of response.

Concerning the first point, i.e. the knowledge of the metabolic processes
involved in production and/or scavenging of oxygen derivatives, it has repeatedly
been shown that a variety of cytotoxic agents act via autoxidation processes that
lead to the formation of superoxide, hydrogen peroxide and hydroxyl radicals.
A number of them are activated only in the presence of intracellular components
that are able to transfer one electron onto the exogenous molecule with formation of a radical. These agents include quinones of various types, such as
benzoanthraquinones (like the antitumor antibiotics adriamycin and daunomycine),[1]
N-heterocyclic quinones (mitomycin C and streptonigrin, which are also active as
antitumor antibiotics),[2] naphthoquinones (such as the trypanocidal drug lapa-
chone),[3] catechols (like the neurotoxic drug α-methyldopa),[4] and quaternary
bipyridyls (such as the herbicide paraquat).[5] In spite of their chemical diver-
sity, these drugs are all related to each other by the property of undergoing
redox cycling through a semiquinone form that is capable of reacting with oxygen,
with primary formation of superoxide and consequent production of hydrogen peroxide and hydroxyl radicals.[6] Reduction of the quinone drug to the semiquinone
radical can occur within the cells by a variety of mechanisms, and has indeed
been demonstrated to be catalysed by subcellular particles, either through the
mitochondrial[7,8] or through the microsomal[1,2,4-9] electron transport chain.

In any case the drug interacts with the system by apparently diverting electrons from the normal sequence of flow, and this results in an "uncoupling" of
the electron transport chain with increased oxygen uptake and superoxide formation. Superoxide, either spontaneously or enzymically, rapidly dismutes to
hydrogen peroxide, which is, in the absence of specific catalysts, much more
stable than its parent compound and can presumably diffuse out of the site of
production for a relatively long distance. It is likely that cytosolic enzymes
can interact with the peroxide produced, for instance, by the microsomal system[10]
and that non-catalytic reactions can occur with distant targets, such as the cell
membrane or the nuclear material, with consequent chemical and functional alterations.

As far as the second approach is concerned, i.e. the natural deficiency of
antioxidative defence pathways, it is reasonable to expect some correlation
between the extent of cytotoxicity of the quinone drugs mentioned above and the
cellular levels of enzyme activities related to the biological transformations
of oxygen derivatives. In particular the ratio between superoxide dismutase
and the enzymes decomposing hydrogen peroxide appears to be crucial.[11] In fact,
cytosolic and mitochondrial superoxide dismutases rapidly decrease the steady-
state concentration of superoxide that could have been liberated into the immediate environment by oxidases in special circumstances of incomplete coupling of
their molecular devices leading to water or hydrogen peroxide; at the same time
superoxide dismutases increase the cellular steady-state concentration of hydrogen peroxide. Both effects are potentially dangerous to the cell, as they increase the yield of hydroxyl radicals in the presence of redox metal ions.[12]
It should on the other hand be recalled that, while hydrogen peroxide is capable
of very efficient diffusion within and outside the cell[13] and is therefore not
restricted to interact with enzymes located at its intracellular generation
site, all cellular compartments can have defensive enzymes against it. In
liver cells, gluthathione peroxidase is present in the cytosol and in the mitochondrial matrix, and catalase is located in peroxisomes. Heme peroxidases
exist in specialized cells that utilize hydorgen peroxide for "useful" purposes:
typical examples are thyroid peroxidase, which plays a role in the synthesis of
thyroid hormones; myeloperoxidase, which is part of the oxidative machinery of
bacterial killing by neutorphils; lactoperoxidase, which is likely to have a
similar role in milk and saliva; and ovoperoxidase, which is involved in the
control of oocyte fertilization. These enzymes can cooperate in maintaining
low steady-state levels of hydrogen peroxide, in particular with respect to
the superoxide levels. We have thus shown[14] that tumor cells have lower specific superoxide dismutase activity than liver or red blood cells, but their content of catalase is practically zero and in some tumor cell types the decrease
of glutathione peroxidase relative to normal control cells is greater than that

of superoxide dismutase. In these cells, in other words, the metabolic flow
of oxygen derivatives is potentially driven toward accumulation of hydrogen
peroxide. Tumor cells may thus represent a simplified model to test the role
of hydrogen peroxide in some cases of oxygen-mediated cytotoxicity and also
because of the absence of catalase and the presence, in different cell strains,
of variable concentrations of glutathione peroxidase, to assess the importance
of the glutathione dependent pathway of hydrogen peroxide decomposition.

Experimental observations. In a preliminary set of experiments, different
strains of experimental ascites tumor cells were exposed to an extracellular
source of superoxide, namely illuminated FMN under aerobic conditions. The
results have been published[15] and only some conclusions relevant to the following discussion will be reported here:

(1) Out of four tumors tested, one – the Yoshida ascites hepatoma – was
fully resistant, as far as oxygen consumption, lactate production, cell permeability and cell viability were concerned, in conditions which affected to a
variable degree the other strains, and in particular caused an almost complete
inhibition – and subsequent cell death – of the metabolic activities of Ehrlich
ascites cells;

(2) Of the various scavengers competent for the different oxygen species,
only catalase was protective; accordingly addition of millimolar concentrations
of hydrogen peroxide reproduced the effects of aerobically illuminated FMN;

(3) The level of superoxide dismutase was lowest in the most sensitive strain,
but in the other types of cells, could not be correlated to the sensitvity to
illuminated FMN. On the other hand, glutathione peroxidase was highest in the
Yoshida cells and intracellular glutathione was 3-4 times more concentrated in
the resistant cells (Yoshida 0%, and Novikoff 10% sensitivity relative to that
of Ehrlich cells) than in the more sensitive Ehrlich and MC-1A cells.

The two extreme cases of response to extracellular hydrogen peroxide, i.e.
the resistant Yoshida and the sensitive Ehrlich cells, were then used in experiments where the oxygen uptake was correlated with varying concentrations of
representative quinone drugs, namely, daunomycin, streptonigrin and a water-
soluble analogue of menadione, 1,4-naphthoquinone-2-sulfonic-acid. The results
are reported in Figure 1, where also the response of isolated hepatocytes is
shown for comparison. Although there are significant differences in the effect
of the three compounds, it is generally observed that the Yoshida cells are
more resistant than the Ehrlich cells.

The specific resistance of Yoshida cells could be due to different rate of
drug penentration through the plasma membrane, which is known to possess differential structural and functional properties in different types of tumor cells.[16]
The experiments of Figure 1 were therefore repeated in the presence of lucensomycin, a polyenic antibiotic that has a rather general and non-specific effect

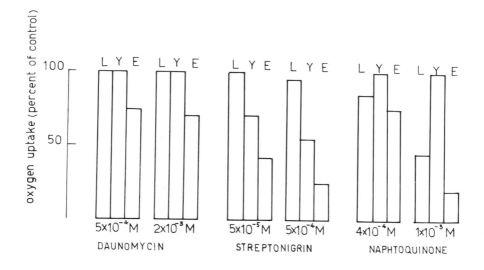

Fig. 1. Differential sensitivity of Ehrlich (E) and Yoshida (Y) ascites tumor cells and isolated heptocytes (L) to quinone drugs. Tumor cells were obtained and prepared as described in Ref.15. Hepatocytes were isolated according to Moldeus et al. (Ref.20). Oxygen consumption was measured in the presence of 15 mM succinate as exogenous substrate with a conventional Warburg apparatus. Cell concentration was 12×10^6/ml for E, 8×10^6/ml for Y and 4×10^6/ml for L in order to approximately normalize the value of oxygen consumption. The data reported refer to values recorded after 90 min incubation at 37°C.

of increasing the permeability of cell membranes. As shown in Table 1, the Yoshida cells are much more resistant than the Ehrlich cells also in these conditions, and again their response is comparable to that of liver cells. An interesting result is that lucensomycin protects both tumor cells from the toxic effect of streptonigrin; this effect may be ascribed to diffusion of the cyto-toxic drug semiquinone outside the cell in the presence of the permeability agent.

A puzzling point is the relatively greater sensitivity of liver cells to naphthoquinone sulfonate. This result may either be due to trivial experimental factors such as the most unfavourable drug-to-cell ratio in the case of hepa-tocytes, or point to additional mechanisms contributing to the overall effect of each drug.

In the absence of exogenous substrates, the effects of drugs on respiration are practically the same as in the presence of succinate. On the other hand glucose has a pronounced positive effect on the oxygen uptake of all types of cells in the presence of some quinone drugs, although in the case of Ehrlich

TABLE 1

EFFECT OF LUCENSOMYCIN ON OXYGEN UPTAKE OF TUMOR AND LIVER CELLS[a]

	Ehrlich	Yoshida	Liver
No drug	60	90	97
Daunomycin			
0.5 x 10^{-4} M	75	104	95
Streptonigrin			
5 x 10^{-5} M	128	143	98
5 x 10^{-4} M	180	189	99
Naphthoquinone sulfonate			
4 x 10^{-4} M	61	95	90
1 x 10^{-3} M	14	62	73

[a]Experimental conditions and expression of results as in Fig.1.

cells this increase is followed by inhibition at earlier times than with the Yoshida cells. Figure 2 reports the results obtained with Ehrlich cells and daunomycin. Apparently the drug stimulates a glucose-dependent pathway of oxygen reduction, such as the microsomal chain, and this stimulation overlaps in sensitive cells with inhibition at later times.

Paraquat, which had no effect on respiration and/or cell viability of either tumor type, was found to be unable to penetrate these cells. In the presence of lucensomycin some inhibition was observed only with Ehrlich cells.

A more direct evidence for the mechanism of activation of these quinone drugs and for the role of oxygen in the process came from EPR experiments on intact cells. When the various types of cells were incubated with the various drugs in a closed EPR sample container, a signal typical of a free radical appeared after some time and increased up to a maximum, the time scale of this phenomenon depending on the respiration rate of each particular cell type. Admission of air to the cell suspension led to disappearance of the signal; with Yoshida and liver cells, the redox cycle could be repeated by alternate exclusion and admission of oxygen, while in the case of Ehrlich cells, recovery of signal was poor already in the first cycle, indicating selective sensitivity of these cells to the reaction of the radical with oxygen. Some EPR spectra are reported in Figure 3, showing signals typical of immobilized radicals and significantly different for different drugs. The radicals are stable in the absence of oxygen. No signal is observed in cell-free supernatant. Paraquat, which does not penetrate the cells, gives no signal. On the other hand, addition of lucensomycin inhibits the appearance of the signal. All these facts demonstrate that the radical is formed from the drug inside the cell, in an environment that prevents its reaction with other radical molecules as well as its diffusion outside the

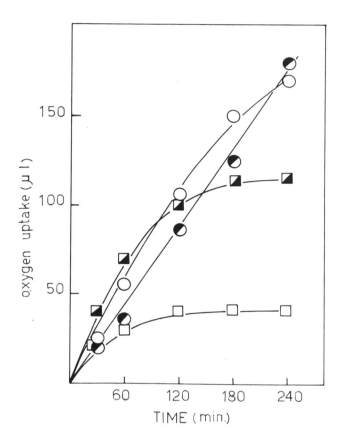

Fig. 2. Effect of glucose on oxygen uptake bt Ehrlich ascites tumor cells. (O) in the absence of exogenous substrates; (●) in the presence of 1.5 mM glucose as substrate (□) as (O), with 5 x 10⁻⁴ M daunomycin; (◪) as (●), with 5 x 10⁻⁴ M daunomycin. Other conditions as in Fig.1.

cell, where it would rapidly disappear via dismutation. It is also clear that no major reactions occur inside the cell beside that with oxygen. Therefore direct damage of nuclear material or membrane components by the radicals produced in the reductive activation of quinone drugs, as suggested by other authors,[9] seems unlikely. On the other hand, reaction with oxygen appears to be predominant and leads, in the case of Ehrlich cells, to selective cell damage.

The good correlation between the biochemical defence against oxidative stresses possessed by the cells and the sensitivity to the quinone drugs of the various cell types points to a sequence of events, upon penetration of a quinone drug

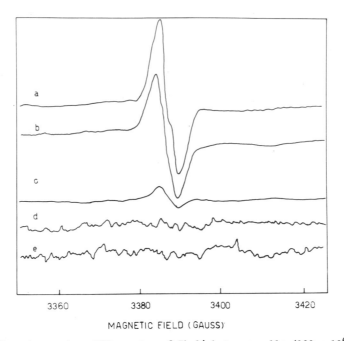

MAGNETIC FIELD (GAUSS)

— Fig. 3. Room temperature EPR spectra of Ehrlich tumor cells (130 x 10^6/ml).
a) In the presence of 1 x 10^{-3} M naphthoquinone sulfonate after 5 min incubation at room temperature. The signal amplitude was the same after 10 min.
b) In the presence of 1 x 10^{-3} M daunomycin after 30 min incubation at room
temperature. The signal amplitude was the same after 70 min. c) As b), in
the presence of lucensomycin (10 µg/10^6 cells). d) In the presence of 1 x
10^{-3} M paraquat. e) Supernatant of sample b) after gentle sedimentation of
cells. The spectra were recorded with an E-9 Varian instrument with the Varian
aqueous solution cell accessory. The amplification factor was x 4 for d) and
e).

inside an aerobic cell, that can be simplified as following: the exogenous

compound is reduced to its semiquinone, primarily by NADPH-dependent microsomal

enzymes. This process, which can be expected to be much less active in poorly

differentiated tumor cells of the maximally deviated type, such as those used

in this work, than in the hepatocyte[17] — as was confirmed by the different rate

of appearance of the radical in the EPR experiments — leads upon reaction of

the radical with oxygen, to the formation of superoxide. Superoxide dismutase

levels in Ehrlich and Yoshida cells, although lower than in the hepatocyte,[14]

are enough to dismute all the superoxide formed to hydrogen peroxide. The high

concentration of catalase and glutathione peroxidase of liver cell[14] easily

compensate for the greater amount of peroxide formed by the more developed

microsomal system. In tumor cells, instead, the fate of hydrogen peroxide,

even if produced in lower concentrations, is critically dependent on the levels

of the components of the glutathione cycle, because of the absence of catalase.

The intrinsic validity of this picture is supported by measurements of reduced and oxidized glutathione in Ehrlich and Yoshida cells after two hours incubation in the presence of succinate, with and without daunomycin. In the absence of quinone drug, the ratio GSH/GSSG is approximately 2.0 for Ehrlich and 5.0 for Yoshida cells, becuase of higher levels of glutathione reductase and its subsidiary enzyme glucose 6-phosphate dehydrogenase in Yoshida cells;[15] moreover the total amount of glutathione (oxidized plus reduced) is six times higher in Yoshida cells, reflecting the presence of a more efficient glutathione synthetase system. After treatment with daunomycin the ration changed to 0.45 in the Ehrlich and to 0.64 in the Yoshida cells, a fact which shows how the glutathione pool was subjected to oxidative challenge in both tumors under these conditions. However, the absolute amount of GSSG is at least four times higher in the Yoshida cells. Since the two types of tumors have comparable oxidation rates under the conditions used, the result indicates that the amount of hydrogen peroxide produced in the presence of daunomycin exceeds the potentialities of Ehrlich cells to counteract its noxious effects by the glutathione peroxidase system - a fact which accounts for the higher sensitivity of Ehrlich cells to the drug.

Conclusions. The data presented here and the related considerations are not intended to conclude that anticancer agents of the quinone type exert their therapeutic action primarily by their redox cycling in the microsomal or mitochondrial chain and the consequent formation of reactive oxygen derivatives. For several of them, it is indeed well known that they interact directly with DNA, their cytostatic effects being well correlated with this property.[18] On the other hand, the systems that we have considered above give unequivocal indication that these compounds elicit a stimulation of oxidative processes inside the cell. Under these conditions, certain tumor cells, which have a selectively modified pattern of enzymic defence against active oxygen species, respond in a way that appears to reflect the key role of hydrogen peroxide and of the hydrogen peroxide-removing enzymes, rather than the relatively small decrease, with respect to liver cell, of superoxide dismutase, which is present in the range of concentrations apt for establishing a "normal" steady-state level of superoxide inside the cell.[19]

ACKNOWLEDGEMENTS

This work was partly supported by a grant of Consiglio Nazionale delle Richerche, Progetto finalizzato "Controllo della Crescita Neoplastica", Contract No.78.02849.96.1159066.

REFERENCES

1. Goodman, J. and Hochstein, P. (1977) Biochem. Biophys. Res. Commun. 77, 797-803.
2. Handa, K. and Sato, S. (1975) Gann, 66, 43-37.
3. Docampo, R., Cruz, F.S., Boveris, A., Muniz, R.P.A. and Esquivel, D.M.S. (1978) Arch. Biochem. Biophys. 186, 292-297.
4. Dybing, E., Nelson, S.D., Mitchell, J.R., Sasame, H.A. and Gillette, J.R. (1976) Mol. Pharm. 12, 911-920.
5. Autor, A.P. (1977) Biochemical Mechanisms of Paraquat Toxicity, Academic Press, New York.
6. Misra, H.P. and Friodvich, I. (1972) J. Biol. Chem. 247, 188-192.
7. Iwamoto, Y., Hausen, I.L. Porter, T.M. and Folkers, K. (1974) Biochim. Biophys. Res. Commun. 58, 633-638.
8. Thayer, W.S. (1977) Chem. Biol. Inter. 19, 265-278.
9. Bachur, N.R., Gordon, S.L. and Gee, M.V. (1978) Cancer Res. 38, 1745-1750.
10. Jones, D.P., Thor, H., Andersson, B. and Orrenius, S. (1978) J. Biol. Chem. 253, 6031-6037.
11. Rotilio, G., Calabrese, L., Finazzi-Agro, A., Angento-Curu, M.P., Autuori, F. and Mondovi, B. (1973) Biochem. Biophys. Acta, 321, 98-102.
12. Rigo, A. and Rotilio, G., this publication, vol.1.
13. Chance, B., Sies, H. and Boveris, A. (1979) Physiol Rev. 59, 527-605.
14. Bozzi, A., Mavelli, I., Finazzi-Agro, A., Syrom, R., Wolf, A.M., Mondovi, B. and Rotilio, G. (1976) Mol. Cell. Biochem. 10, 11-16.
15. Bozzi, A., Mavelli, I., Mondovi, B., Strom, R. and Rotilio, G. (1979) Cancer Biochem. Biophys. 3, 135-141.
16. Schultz, J. and Block, R.E. (1974) Membrane Transformations in Neoplasia, Miami Winter Symp., Vol.8., Academic Press, New York.
17. Galeotti, T., Bartoli, G.M., Bartoli, S. and Bertoli, E., this volume.
18. Di Marco, A., Arcamone, F. and Zunino, F. (1975) in Mechanism of Action of Antimicrobial and Antitumor Agents, Cockran, J.W. and Hahn, F.M., eds., Springer Verlag, Berlin, pp. 101-128.
19. Tyler, D.D. (1975) Biochem. J. 147, 493-504.
20. Moldeus, P., Hogberg, J. and Orrenius, S. (1978) Methods Enzymol. 52, 60-71.

THE INFLUENCE OF FREE RADICALS ON AN IN VIVO MICROVASCULAR MODEL SYSTEM

R.F. DEL MAESTRO,[+] K.-E. ARFORS,[++] J. BJORK [++] AND M. PLANKER[+++]

[+]Laboratory for Experimental Brain Research, Victoria Hospital, Department of Neurosurgery, University of Western Ontario, London, Ontario, Canada, N6A 4G5; [++]Department of Experimental Medicine, Pharmacia AB, Uppsala, Sweden; [+++]Department of Medicine A, University of Dusseldorf, West Germany

INTRODUCTION

The activation of polymorphonuclear leukocytes, macrophages and monocytes on exposure to appropriate stimuli, such as bacteria, produces the "respiratory burst".[3] This respiratory burst is a coordinated series of metabolic events one of which involves the reduction of O_2 to the superoxide anion radical (O_2^-) (Ref. 5) by a surface bound enzyme called NADPH oxidase.[13] A significant portion of this generated O_2^- is released into the extracellular space surrounding the inflammatory cell[30] where spontaneous dismutation can occur with the formation of H_2O_2 and O_2 (Ref. 19). The release of O_2^- and subsequent generation of H_2O_2 in the phagosome, containing ingested foreign material, is essential for some bactericidal killing since patients with a genetic defect (chronic granulomatous disease) have polymorphonuclear leukocytes unable to reduce O_2 to O_2^- and suffer recurrent bacterial infestation from which they may succumb.[4] Babior et al.[6] and Rosen and Klebanoff,[31] using a substrate-xanthine oxidase radical generating system, have demonstrated bactericidal effects associated with O_2^- generation. Cellular membrane disruption,[23] lipid peroxidation[22] and hyaluronic acid degradation[9,10,23] have also been demonstrated using similar systems. It has therefore been suggested that some of the changes seen during inflammation may be related to O_2^- release from inflammatory cells.[11,29] The mechanisms by which O_2^- release results in both the bactericidal effects and tissue injury are poorly understood, but it is generally considered to be related to the further generation of more reactive activated oxygen species such as hydroxyl radical (OH·) and/or singlet O_2 ($^1\Delta g$) (Refs. 9, 29). Inflammatory processes are characterized by a series of alterations in the microcirculation which include decreased integrity of endothelial cell barriers, changes in vessel diameter and the margination and emigration of inflammatory cells.[36] It was considered important to elucidate the role of the extracellular generation of O_2 derived free radicals on an in vivo microcirculatory model system to assess their role, if any, in the changes observed during inflammation. In this communication the results of the in vivo experiments concerning microcirculatory permeability and polymorphonuclear leukocyte (granulocyte) adhesion are reported and a concept

of the roles of the individual radical species and their products is presented
in the hope that it may aid in the understanding and treatment of inflammatory
conditions.

MATERIALS AND METHODS

In vivo model. The hamster cheek pouch microvascular system has been used extensively to study alterations in both microvascular permability[33,34] and in vivo
granulocyte behaviour.[2] Except for its lack of lymphatics,[21] the cheek pouch
microvasculature may be considered representative of connective tissue microvascular
beds. Meticulous preparation and continuous superfusion allows observation of the
microcirculation by intravital microscopy (Figure 1) for over 3 hours without deterioration.[15,34] Macromolecular extravasation was assessed semi-quantitatively by
counting the number of leakage sites/cm^2 after injection intravenously of FITC-
dextran 150 (fluorescein-labelled dextran, mol. wt. 150,000; Pharmacia, Uppsala,
Sweden). Macromolecular leakage sites were observed as fluorescent areas found
in the interstitium originating from the vessel extravasating the intravascular
tracer (Figure 2). This method allows accurate delineation of the site of FITC-
dextran 150 extravasation. In studies in which granulocyte behaviour was assessed
the cheek pouch was scanned and a 30-45 μm venule was chosen. The venular segment
was monitored continuously using a television camera with the image displayed on
a television monitor with all information stored on a videotape recorder. Granulocyte rolling frequency was assessed as described by Atherton and Born[2] with the
rolling cells being monitored on the television monitor. Granulocyte velocity at
each time interval was calculated as the mean time taken by 10-20 rolling granulocytes to transverse a 100 μm venular distance.[2] Granulocyte adhesion was monitored
by counting the number of adherent granulocytes present in the 100 μm venular segment (Scale 0, no cells; 1+ representing 1-5; 2+, 6-10; 3+, 11-20 and 4+, >20 adhering granulocytes). Red cell velocity was measured in the venules studied as
described by Arfors et al.[1]

Substrate-xanthine oxidase free radical generation system. Free radical generation was induced by substrate-xanthine oxidase system which has been characterized
by Fridovich[18] and involves both a univalent pathway resulting in the reduction
of O_2 to O_2^- (reaction 1) and a divalent pathway resulting in the formation of H_2O_2
(reaction 2):

$$ENZ-H_2 + 2O_2 \rightarrow ENZ + 2O_2^- + 2H^+ \tag{1}$$

$$ENZ-H_2 + O_2 \rightarrow ENZ + H_2O_2 \tag{2}$$

In one group of experiments, the exogenous substrate hypoxanthine, was used while
in other experiments intrinsic cheek pouch substrates were sufficient to generate
a radical flux.[9] The substrate-xanthine oxidase system used during these studies
involves the generation and interactions of a group of activated O_2 species.

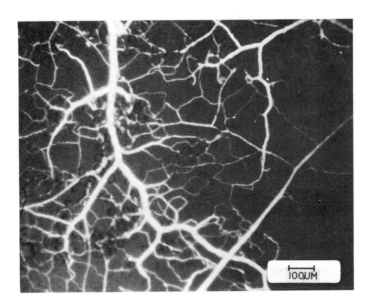

Fig. 1. Micrograph of cheek pouch microvasculature in fluorescent light at 35 x magnification prior to the application of 0.96 mM hypoxanthine and xanthine oxidase (0.05 units/ml).

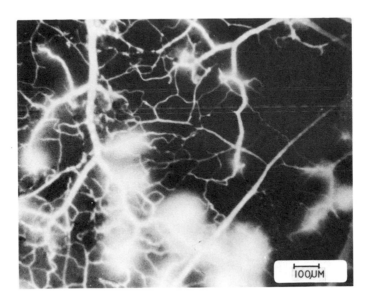

Fig. 2. The same region as in Fig. 1 2 min. following application of substrate and enzyme which demonstrates FITC-dextran 150 extravasation from postcapillary venules.

The role of individual compounds was assessed by using selective scavenging. A role for O_2^- was invoked if the incorporation of superoxide dismutase (SOD) in the reaction mixture resulted in an alteration from expected outcome. Similarly, the role of H_2O_2 was assessed by the ability of catalase (CAT) to influence expected findings. A role for OH· was suggested if specific OH· scavengers were effective. Dimethyl sulfoxide (DMSO) and L-methionine were employed as OH· scavengers.[14]

RESULTS

Microvascular permeability. The use of in vivo fluorescent microscopy at 35x magnification allowed continuous observation of the hamster cheek pouch microvasulature (Figure 1). FITC-dextran 150 appeared in all cheek pouch preparations 8-20 seconds following intravenous administration and was observed in all vessels within 35 seconds.

Table 1 shows the leakage sites/cm^2 seen before and after application of the test solution(s). During the 60 minute equilibration period the number of leakage sites seen varied among the individual preparations but no statistically significant differences were observed. Solutions containing only buffer, 0.96 mM hypoxanthine or 0.96 mM uric acid were associated with no observable alterations in the cheek pouch microvasculature and were not associated with significant increases in the number of leakage sites/cm^2. The applications of denatured enzyme caused no increase in macromolecular leakage. Figures 1 and 2 are typical micrographs taken prior and following application of 0.96 mM hypoxanthine and active enzyme to the cheek pouch reservoir. Figure 1 demonstrates no leakage sites immediately prior to application and Figure 2 demonstrates macromolecular extravasation from postcapillary venules (< 20 μm) 2 minutes after addition.

Macromolecular extravasation was associated with postcapillary venules and occasionally with larger venules but was not observed from arteries, arterioles or capillaries. Occasionally haemorrhages were also observed from postcapillary venules and these occurred in regions in which macromolecular leakage continued unabated during the observation period. During these investigations an alteration in granulocyte-endothelial interactions was observed, predominantly in postcapillary venules.

Since neither hypoxanthine, the substrate of aerobic oxidation by xanthine oxidase, nor uric acid, the product of the reaction, resulted in altered microvascular permeability, it was hypothesized that the changes seen could be related to an associated flux of radical species. A variety of compounds[8] can be acted on by xanthine oxidase to generate radical species and endogenous cheek pouch substances (purines, aldehydes) may be available to act as substrates for enzyme. The addition of active xanthine oxidase (0.05 units/ml) to cheek pouch homogenates resulted in an associated O_2^- flux (Table 2) which was measured as cyt c^{3+} reduction at 550 nm (Ref. 12) while the addition of denatured xanthine oxidase resulted in no measurable O_2^- formation.

TABLE 1

EFFECT OF TOPICAL APPLICATION OF TEST SOLUTION(S) ON MACROMOLECULAR LEAKAGE SITES[a]

Test Solution(s)	Pre-application	Time after topical application (min)				
		5	10	25	45	85
Bicarbonate buffer	1.5 ± 0.9	2.5 ± 1.6	1.7 ± 1.1	2.0 ± 1.1	1.7 ± 1.0	4.7 ± 3.8
Hyoxanthine (0.96 mM)	3.2 ± 2.7	4.8 ± 2.8	5.7 ± 3.3	10.3 ± 7.5	6.6 ± 4.3	6.6 ± 4.3
Uric acid (0.97 mM)	3.4 ± 1.6	4.9 ± 2.0	5.1 ± 2.8	3.7 ± 2.2	4.7 ± 2.8	8.8 ± 5.9
Xanthine oxidase (denatured) (0.05 unit/ml)	0.0 ± 0.0	0.7 ± 0.5	1.0 ± 0.6	0.0 ± 0.0	0.7 ± 0.5	2.5 ± 1.7
Xanthine oxidase (0.05 unit/ml)	1.5 ± 1.2	107.9 ± 38.2[b]	265.4 ± 36.7[b]	127.5 ± 52.2[b]	80.4 ± 39.4[b]	38.7 ± 15.2
Hypoxanthine (0.96 mM) and xanthine oxidase (0.05 unit/ml)	2.2 ± 1.1	207.9 ± 64.0[b]	295.6 ± 37.0[b]	164.0 ± 44.8[b]	93.7 ± 28.5[b]	89.7 ± 40.8[b]

[a] Values are mean number of leakage sites per cm^2 ± SEM for 6 hamsters in each group.

[b] Leakage sites values were compared at each time period against bicarbonate buffer as a control using the rank-sum test. Significant difference is indicated when α <0.025

These results suggest that endogenous cheek pouch substrates are available in the cheek pouch which can result in O_2^- generation with the addition of active xanthine oxidase to the reservoir surrounding the cheek pouch.

TABLE 2

RATE OF IN VITRO SUPEROXIDE FORMATION[a]

System[b]	O_2^- formation (nmoles/min/ml)
Hypoxanthine (0.96 mM) and xanthine oxidase (0.05 units/ml) (n=6)	12.0 ± 0.5
Denatured xanthine oxidase (0.05 units/ml) (n=6)	0
Cheek pouch homogenate and xanthine oxidase (0.05 units/ml) (n=8)	4.3 ± 0.7

[a]The assay mixtures contained 50 μM cyt c^{3+} in a bicarbonate buffer (pH 7.35) to which the test substances were added at 37°C. The increase in absorbance at 550 nm was monitored. Values represent mean ± SEM of O_2^- generation.
[b]n = number of experiments.

Role of individual radical species in increased permeability. The incorporation of SOD (10 μg/ml) in the cheek pouch reservoir prior to application of active xanthine oxidase was associated with only a minor decrease in mean leakage sites/cm^2 compared to the addition of xanthine oxidase alone (Figure 3). The addition of 50 μg/ml SOD was associated with a significant decrease in mean number of leakage sites/cm^2 (Figure 3). The addition of 50 μg/ml CAT to the reservoir fluid prior to addition of active enzyme resulted in a significant decrease in leakage sites/cm^2 (Figure 3). Both DMSO (10 mM) and L-methionine (10 mM) significantly decreased macromolecular extravasation induced by active enzyme.

Granulocyte-endothelial interactions. In experiments in which free radicals were enzymatically generated on the surface of the cheek pouch increased granulocyte adhesion and rolling were noted in postcapillary venules. The initial dynamic pattern of granulocyte behaviour and red cell velocity in the venules studied during the 33 minutes observation period were virtually identical. This consisted of an initial high red cell velocity, low granulocyte velocity and low granulocyte rolling frequency (Figures 4 and 5). After 12 minutes, red cell velocity decreased while

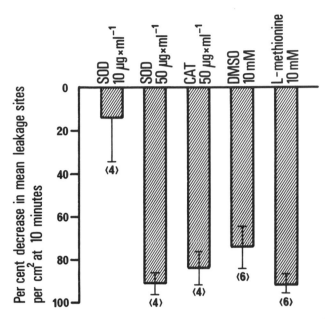

Fig. 3. Per cent decrease in the mean leakage sites/cm^2 ± SEM at 10 min. following application of 0.05 units/ml of xanthine oxidase in the presence of the substances tested as compared to the mean leakage sites/cm^2 in their absence. Number in parentheses = number of animals.

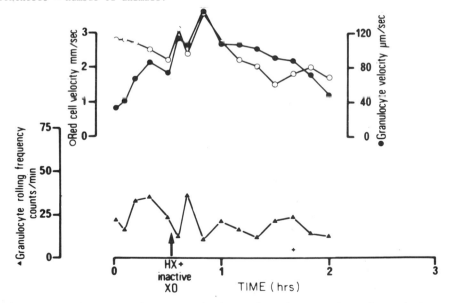

Fig. 4. Experiment in which 0.05 units/ml of inactive xanthine oxidase (XO) was added to the cheek pouch reservoir containing 0.96 mM hypoxanthine (HX) at 33 min for 1 min. Parameters were measured as described in MATERIALS AND METHODS.

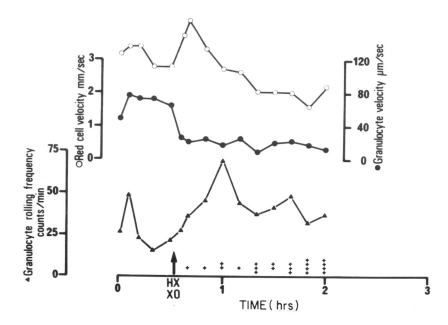

Fig. 5. Experiment in which 0.05 units/ml of xanthine oxidase (XO) was added to the cheek pouch reservoir containing 0.96 mM hypoxanthine (HX) at 33 min for 1 min. Parameters were measured as described in MATERIALS AND METHODS.

both granulocyte velocity and granulocyte rolling frequency increased. Thereafter granulocyte velocity and red cell velocity appeared to stabilize and granulocyte rolling frequency decreased to reach a rather constant low value between 20-30/min. Adherent granulocytes were quite unusual during the observation period. No significant differences in the parameters measured were noted among the various groups immediately prior to addition of active enzyme. An experiment in which the reservoir fluid was replaced with 0.96 mM hypoxanthine solution to which denatured enzyme (0.05 units/ml) was added can be seen in Figure 4. An initial increase in both red cell velocity and granulocyte velocity was observed with a direct association between these two parameters being maintained throughout the experimental period. No consistent increases in granulocyte rolling frequency were found and granulocyte adhesion was very uncommon. A distinctly different pattern was observed when active xanthine oxidase rather than the denatured enzyme was added to the hypoxanthine solution surrounding the cheek pouch. An initial increase (Figure 5) in red cell velocity was seen but this was associated with a significant decrease in granulocyte velocity which persisted during the observation period. Granulocyte rolling frequency was increased and many granulocytes became adherent.

<u>Role of individual radical species in altered granulocyte-endothelial interactions</u>. The incorporation of SOD (50 µg/ml) in the cheek pouch reservoir

(Figure 6) completely prevented the granulocyte-endothelial alterations seen in the absence of SOD (Figure 5). The dynamic association between red cell velocity and granulocyte velocity persisted and granulocyte adhesion was not seen. The addition of (CAT 50 µg/ml) or L-methionine (10 mM) to the reservoir fluid containing hypoxanthine prior to the application of active enzyme did not modify the results seen after addition of active enzyme to the reservoir fluid containing hypoxanthine alone.

CONCLUSIONS

The microvasculature of the hamster cheek pouch is markedly altered by the extra-cellular generation of free radical species by substrate-xanthine oxidase generating systems. A marked increase in leakage sites/cm^2 and an alteration of granulocyte behaviour was induced in postcapillary venules and in the larger venules studied. Arteries, arterioles and capillaries were not affected. Petechial haemorrhages were occasionally seen and these originated from postcapillary venules and were associated with both macromolecular extravasation and marked granulocyte adhesion in the region of the petechial haemorrhages. Since this permeability increase was decreased by SOD (50 µg/ml), CAT (50 µg/ml), DMSO (10 mM) and L-methionine (10 mM), a role for OH$^{\cdot}$ or more likely a diffusable factor generated by OH$^{\cdot}$ is suggested as a cause for the permeability increase. The mechanism by which OH$^{\cdot}$ is generated in this substrate-xanthine oxidase system has been extensively studied and the reaction sequence outlined in Figure 7 has been suggested.[9]

The results of addition of the substances tested can be explained in the light of this scheme. SOD by enzymatic scavenging of O_2^- prevents O_2^- mediated reduction of metal chelates[9,17] in the extracellular space and although the H_2O_2 concentration would increase, OH$^{\cdot}$ formation would not result via reaction 4.

$$O_2^- + Me^{n+} \rightarrow O_2 + Me^{(n-1)+} \tag{3}$$

$$H_2O_2 + Me^{(n-1)} \rightarrow OH^{\cdot} + O_2 + Me^{n+} \tag{4}$$

The presence of CAT would result in the removal of H_2O_2 and therefore the substrate (H_2O_2) of reaction 4 would be decreased resulting in less OH$^{\cdot}$ formation. DMSO and L-methionine by their ability to scavenge OH$^{\cdot}$ would directly decrease the reactive potential of any generated OH$^{\cdot}$. Since the normal extracellular concentrations of SOD and CAT are quite low,[29] OH$^{\cdot}$ may be generated in the extracellular space of the cheek pouch. Hydroxyl radicals once formed in the extracellular space may react, through radical intermediates such as the formate radical[27] with the polyunsaturated fatty acids of the plasmalemma. Lipid peroxide radical (ROO$^{\cdot}$) formation can occur resulting in further hydrogen abstraction from adjacent unsaturated fatty acids and hydroperoxide generation (ROOH) (Figure 7). Figure 8 is a schematic representation of this concept demonstrating how the O_2^- released from inflammatory cells may similarly interact to generated OH$^{\cdot}$ and result in tissue

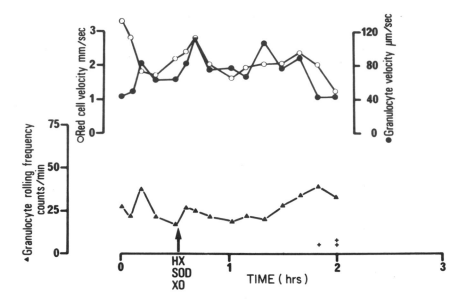

Fig. 6. Experiment in which 0.05 units/ml of xanthine oxidase (XO) was added to the cheek pouch reservoir containing 0.96 mM hypoxanthine (HX) and 50 µg/ml SOD at 33 min for 1 min. Parameters were measured as described in MATERIALS AND METHODS.

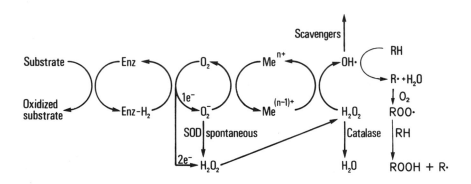

Fig. 7. Schematic illustration of the hypothesis in which the electron flux is from substrate to enzyme and then to O_2 forming O_2^-. Metal chelates (Me^{n+}) may be reduced by O_2^- and these ($Me^{(n-1)+}$) then react with H_2O_2 to generate $OH\cdot$. Hydroxyl radical may react with lipids generating radicals ($R\cdot$) that react with O_2 resulting in lipid peroxide radicals ($ROO\cdot$) and subsequently lipid hydroperoxides ($ROOH$).

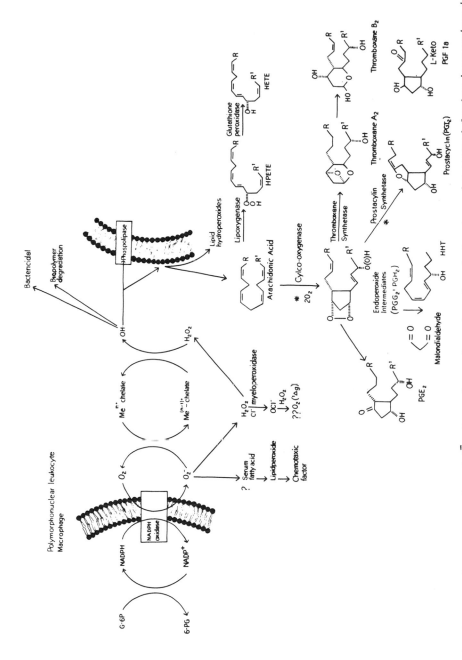

Fig. 8. Schematic representation of $\bar{O}_2$ release from polymorphonuclear leukocytes and suggested further interactions in the extracellular space.

injury.

Major consequences may result from the initiation of propagating radical chain reactions in plasmalemmal membranes. Extensive regions of plasmalemmal membrane may undergo peroxidative injury, resulting in disruption of the plasmalemmal and lyzosomal membranes[16] and eventually with the death of cells and extracellular release of destructive lyzosomal enzymes. A variety of toxic lipid hydroperoxides and possibly other fatty acid oxidative products may be released in the interstitial space during these interactions. Hydroxyl radicals or a generated product may activate phospolipase activity which would result in arachidonic acid release from lipid membranes and the generation of further active endoperoxides[32] (Figure 8). It would appear reasonable to suggest that if $OH^{\cdot}$ interactions with plasmalemmal membranes resulted in hydroperoxide formation, arachidonic acid release or the further generation of other lipid metabolites endothelial cell contraction may be induced resulting in the macromolecular leakage observed.

The alterations observed in granulocyte-endothelial interactions in the venules studied were increased granulocyte rolling frequency, a decreased granulocyte velocity and increased granulocyte adhesion. Our results suggest that the changes in granulocyte behaviour observed after addition of active enzyme to the hypoxanthine solution surrounding the cheek pouch are dependent on the generation of O_2^-. The incorporation of SOD (50 µg/ml) in the reservoir solution completely inhibited these changes in granulocyte behaviour while the incorporation of CAT (50 µg/ml) or L-methionine (10 mM) did not significantly alter the granulocyte response to the free radical flux induced by xanthine oxidase. Our results are consistent with the concept that an O_2^- dependent lipid hydroperoxide generated on the cheek pouch by the hypoxanthine-xanthine oxidase system markedly alters granulocyte behaviour in vivo (Figure 8). Similar alterations in granulocyte behaviour have been observed after the application of human serum exposed to an in vitro hypoxanthine-xanthine oxidase system to the hamster cheek pouch (unpublished observations). An O_2^- dependent factor has been isolated from plasma by McCord (personal communication) which is chemotaxic in vitro and when injected into rat's skin. Although O_2^- has not been implicated in lipid peroxidation[22,35] it can reduce metal chelates (reaction 3) which may then react with lipids resulting in the formation of chemotaxic hydroperoxides.[20] It has also been demonstrated that O_2^- reacts with hydroperoxides to release alkoxy radicals[35] which may have chemotaxic properties. A dissociation of the stimuli involved in the attraction and adhesion of polymorphonuclear leukocytes from those involved in bacterial killing,[6] increased permeability[9,11,12] and macromolecular degradation[10,24] appears evident since the former alterations are dependent on a O_2^- derived product (Figure 8).

The effectiveness of SOD in various experimental inflammatory models may be in part related to its ability to inhibit the formation of a O_2^- dependent chemotaxic factor. McCord and Wong[25] have demonstrated that SOD, chemically altered to increase its circulation time, decreased both the number of polymorphonuclear leukocytes seen and subsequent issue injury. Experimentally SOD appears to be effective in decreasing irradiation injury even when administered after irradiation in animal models[28] and clinically Menander-Huber and Huber[26] have reviewed the usefulness of intramuscular SOD given after irradiation of bladder and prostatic tumors in preventing irradiation injury sustained by the bladder and rectum. Since irradiation of tissue results in large fluxes of O_2^- (Ref. 7) part of the resultant initial inflammatory tissue injury may be associated with the production of an O_2^- dependent chemotaxic factor with subsequent leukocyte adhesion, emigration and activation.

Our results suggest that some of the permeability increase and granulocyte behaviour seen during the inflammation may be related to the release of O_2^- from inflammatory cells, possibly mediated through the interactions of $OH^{\cdot}$ with plasmalemmal fatty acids in the former case and directly related to an O_2^- derived product in the latter. A better understanding of the interrelationships between O_2 derived free radicals released from inflammatory cells and the other biochemical and pathological components of inflammation may be useful in developing new approaches to the understanding and treatment of disease.

ACKNOWLEDGEMENTS

We would like to thank Elizabeth Auld for excellent secretarial assistance. This study was supported in part by the Canadian Medical Research Council.

REFERENCES

1. Arfors, K.-E., Bergqvist, D., Intaglietta, M. and Westergren, B. (1975) Uppsala J. Med. Sci. 80, 27-33.
2. Atherton, A. and Born, G.V.R. (1972) J. Physiol. 222, 447-474.
3. Babior, B.M. (1978a) N. Engl. J. Med. 298, 659-668.
4. Babior, B.M. (1978b) N. Engl. J. Med. 298, 721-725.
5. Babior, B.M., Kipnes, R.S. and Curnutte, J.T. (1973) J. Clin. Invest. 52, 741-744.
6. Babior, B.M., Curnutte, J.T. and Kipnes, R.S. (1975) J. Lab. Clin. Med. 85, 235-244.
7. Benon, H., Bielski, J. and Gebicki, J.M. (1977) in Free Radicals in Biology. Vol. 3., Pryor, W., ed., Academic Press, New York, pp. 1-5.
8. Bray, R.C. (1963) in The Enzymes, Vol. 7, Boyer, P.H., Lardy, H. and Myrback, K., eds., Academic Press, New York, pp. 533-556.
9. Del Maestro, R.F. (1979) Diss. Acta Universitatis Upsaliensis, Uppsala, Abstr. 340.
10. Del Maestro, R.F., Arfors, K.-E. and Lindblom, R. (1979a) in 10th Europ. Conf. Microcirculation, Cagliari, 1978. Bibliotheca Anat. No. 18, Lewis, D., ed., Kaeger, Basel, pp. 132-135.
11. Del Maestro, R.F., Bjork, J. and Arfors, K.-E. (1979b) Microvasc. Res. 17, S104 (Abstract)
12. Del Maestro, R.F., Bjork, J. and Arfors, K.-E. (1980) in Active Oxygen and Medicine, Auror, A., ed., Raven Press (In press).
13. Dewald, B., Baggiolini, M., Curnutte, J.T. and Babior, B.M. (1979) J. Clin. Invest. 63, 21-29.

14. Dorfman, L.M. and Adams, G.E. (1973) Reactivity of Hydroxyl Radical in Aqueous Solutions, NSRDS-NBS No. 46, United States Department of Commerc, National Bureau of Standards, USA.
15. Duling, B.R. (1973) Microvasc. Res. 5, 423-429.
16. Fong, K.L., McCay, P.B., Poyer, J.L., Keele, B.B. and Misra, H. (1973) J. Biol. Chem. 248, 7792-7794.
17. Fong, K., McCay, P.B., Poyer, J.L., Misra, H.P. and Keele, B.B. (1976) Chem. Biol. Interact. 15, 77-89.
18. Fridovich, I. (1970) J. Biol. Chem. 215, 4053-4057.
19. Fridovich, I. (1978) Science, 201, 975-880.
20. Gutteridge, J.M.C. (1977) Biochem. Biophys. Res. Commun. 77, 379-386.
21. Handler, A.H. and Shepro, D. (1968)in The Golden Hamster, its Biology and Use in Medical Research, Hoffman, R.A., Robinson, P.F., Hagalhaes, H., eds., The Iowa State University Press, Ames, Iowa, pp. 195-201.
22. Kellogg, E.W. and Fridovich, I. (1975) J. Biol. Chem. 250, 8812-8817.
23. Kellogg, E.W. and Fridovich, I. (1977) J. Biol. Chem. 252, 6721-6728.
24. McCord, J.M. (1974) Science, 185, 529-531.
25. McCord, J.M. and Wong, K. (1979) in Oxygen Free Radicals and Tissue Damage (Ciba Found. Symp. 65). Elsevier/Excerpta Medica, North Holland, Amsterdam, pp. 343-360.
26. Menander-Huber, K.B. and Huber, W. (1977) in Superoxide and Superoxide Dismutases, Michelson, A.M., McCord, J.M. and Fridovich, I., eds., Academic Press, New York, pp. 537-549.
27. Michelson, A.M. (1977) in Superoxide and Superoxide Dismutases, Michelson A.M., McCord, J.M. and Fridovich, I., eds., Academic Press, New York, pp. 77-86.
28. Petkau, A. (1978) Photochem. Photobiol. 28, 765-774.
29. Salin, M.L. and McCord, J.M. (1977) in Superoxide and Superoxide Dismutases, Michelson, A.M., Mcord, J.M. and Fridovich, I., eds., Academic Press, New York, pp. 251-270.
30. Root, R.K. and Metcalf, J.A. (1977) J. Clin. Invest. 60, 1266-1279.
31. Rosen, H. and Klebanoff, S.J. (1979) J. Exp. Med. 149, 27-39.
32. Pryor, W.A. and Stanley, J.P. (1975) J. Org. Chem. 40, 3615-3617.
33. Svensjö, E. (1978) Doctoral thesis, University of Uppsala, Almqvist and Wiksell, Stockholm.
34. Svensjö, E. Arfors, K.-E., Arturson, G. and Rutili, G. (1978) Uppsala J. Med. Sci. 83, 71-79.
35. Thomas, M.J., Nehl, K.S. and Pryor, W.A. (1978) Biochem. Biophys. Res. Commun. 83, 927-932.
36. Thorgeirsson, G. and Robertson, A.L. (1978) Amer. J. Pathol. 93, 803-848.

THE POSSIBLE SIGNIFICANCE OF HYDROXYL RADICALS IN INFLAMMATION

P. PUIG-PARELLADA AND J.M. PLANAS
Department of Pharmacology, Faculty of Medicine, Casanovas 143, Barcelona, Spain

INTRODUCTION

Hydroxyl radicals (OH·) have been implicated in a number of biological phenomena. They are the primary free radicals that initiate the lysis of lysosomes
and peroxidation of cell membrane lipids.[1,2] The OH· radical reacts with nucleic
acids and may be responsible for the cellular damage induced by ionizing radiation.[3,4] Stimulated human neutrophils and monocytes generate OH·, which is
involved in its bactericidal activity.[5,6] Heikkila et al.[7] have demonstrated a
good correlation between capacity to scavenge OH· in vitro and prevention of
alloxan-induced diabetes. Cohen[8] has implictaed OH· in the sympathectomy induced
in peripheral sympathetic nerves by several toxic drugs. Panganamala et al.[9]
have demonstrated that OH· scavengers at high concentrations inhibit the biosynthesis of prostaglandins.

In the inflammatory process McCord[10] has shown that enzymatically generated
OH· depolymerizes bovine synovial fluid, being also capable of causing damage of
grave consequence to physiological systems.[11] Kuehl et al.[12] have demonstrated
that in the enzymatic oxidation of arachidonic acid an OH· is generated, which
is the major inflammatory agent. Ōyanagui[13,14] has implicated superoxide anions
at the prostaglandin phase of carrageenan foot-oedema; these results do not
exclude OH· as the main cause of inflammation because, as Halliwell[15] says,
"although superoxide is scarcely pleasant for the cell, its real danger seems
to lie in its ability to form OH· and singlet oxygen".

The results that we present here show the capacity of anti-inflammatory drugs
to inhibit the depolymerization of synovial fluid induced by the free radicals
generated in the enzymatic reaction between xanthine and xanthine oxidase. From
our own and other data it is suggested that the action of anti-inflammatory
drugs is due to their ability to scavenge OH·. This suggestion has been confirmed using the bleaching of p-nitrosodimethylaniline as a direct assay for OH·.

MATERIALS AND METHODS

Synovial fluid was obtained in the Barcelona slaughter house by intra-
articular extraction from the hind legs of cattle. The technique employed was
the one used by McCord[10]; after 30 min of enzymatic reaction a specimen was

taken and the relative viscosity tested. This was measured by recording the
time in seconds required for a given volume of the reaction mixture to drain by
gravity from the barrel of a plastic syringe through a needle of appropriate
size. The sample incubation and viscosity assay were performed at $37^{\circ}C$ (see
Ref.16 for more details).

For the detection of OH· we have employed the technique used by Bors[17] in which
the OH· generated in the oxidation of hypoxanthine by xanthine oxidase, in oxygenated solutions, bleaches the p-nitrosodimethylaniline (p-NDA). This reaction
is followed spectrophotometrically at 440 nm.

The reagents employed were: hypoxanthine, xanthine oxidase Type 1, superoxide
dismutase and cytochrome c 98% pure from Sigma; acetylsalicylic acid, acetaminophen
and EDTA from Scharlau, mefenamic acid from Parke-Davis; p-nitrosodimethylaniline
from Schuchardt; ethanol from Merck; indomethacin, niflumic acid and phenylbutazone USP grade.

In the p-NDA assay all the anti-inflammatory drugs were diluted in N,N'-
dimethylformamide 1.66% escept mefenamic acid and phenylbutazone that were diluted
at 3.33%. Concentrations are expressed as final concentrations. The comparison
between groups was done with the Student t test and the ID 50 (inhibitory dose
50) was calculated to Litchfield and Wilcoxon.[18]

RESULTS

Our results with synovial fluid (Table 1) show that the non-steroidal anti-
inflammatory drugs assayed at a concentration of 80 μM inhibited by more than
50% the decrease in viscosity induced by the free radicals generated in the
enzymatic reaction between hypoxanthine and xanthine oxidase. Acetysalicylic
acid is inactive. The decrease in relative viscosity is always significant
(P < 0.001) for all the anti-inflammatory drugs tested except for acetylsalicylic acid. The rank order of activity in this assay is equal to the one
obtained in the carrageenan foot-oedema by Vane[19] except for acetaminophen,
which is more active in the synovial fluid than in the foot-oedema assay.

In relation to the bleaching of p-NDA we have carried out experiments on its
specificity. Superoxide dismutase (0.033 to 0.66 μg/ml) inhibits the reduction
of cytochrome c[20] in proportion to its concentration, which permits us to
calculate an ID 50 of 0.066 μg/ml (0.01-3.3). The same enzyme at the same
concentration or even at concentrations ten times higher has no effect on the
bleaching of p-NDA. We have also assayed in both systems, p-NDA and ferricytochrome c, the activity of ethanol. At the concentrations tested (0.22 to
1.4 mM) this OH· scavenger is absolutely ineffective in the cytochrome c assay
and shows a dose response curve in the bleaching of p-NDA that gives an ID 50
of 0.7 mM (0.28-1.75).

TABLE 1

EFFECT OF SOME ANTI-INFLAMMATORY DRUGS ON THE DEGRADATION OF BOVINE SYNOVIAL
FLUID[a]

Non-steroidal anti-inflammatory drugs	Δ Viscosity (sec)[b]	Inhibition (%)
Control	8.28 ± 0.65[c]	100
Indomethacin	7.97 ± 1.36[c]	96.25
Niflumic acid	7.81 ± 1.49[c]	94.32
Acetaminophen	6.09 ± 1.55[c]	73.55
Mefenamic acid	5.25 ± 1.00[c]	63.40
Phenylbutazone	5.11 ± 1.38[c]	61.71
Acetylsalicylic acid	0.97 ± 0.72	11.71

[a]The synovial fluid degradation was induced by the following reaction mixture:
5 ml of bovine synovial fluid; 0.05 ml EDTA 0.1 M; 0.5 ml hypoxanthine 10 mM;
after 30 min preincubation at 37° enzymatic reaction was activated by the
addition of xantnine oxidase (0.10 U in 0.05 ml). The anti-inflammatory drugs
were added at time zero to the amount of 0.05 ml and final concentration of
80 μM. The final pH was 7.5.

[b]Values are: for control −mean of the sample differences ±95 per cent confidence
limits between samples of a complete reaction mixture without xanthine oxidase
and drug and samples of a complete reaction mixture without drug; for drugs −
mean of the sample differences ±95 per cent confidence limits between samples
of a complete reaction mixture without drug and samples of a complete reaction
mixture.

[c] $\underline{P} < 0.001$ of the sample differences.

The results obtained with the anti-inflammatory drugs assayed (Table 2) show
the capacity of these drugs to inhibit the bleaching of p-NDA. This inhibition
is proportional to the dose, giving a straight line, from which we calculate the
corresponding ID 50 and its confidence limits. The results obtained go from
0.0038 mM for niflumic acid, a very active drug, moving to 1.6 mM for indomethacin,
5.8 mM for mefenamic acid, 7.0 mM for acetaminophen, 7.2 mM for phenylbutazone
and 9.7 mM for acetylsalicylic acid, the least active of all. On considering
the decrease in optical density in absolute values we find significant differences between treated and control groups. The rank order of activity is equal
to that obtained in the prostaglandin synthetase assay[19] except for acetaminophen, which is more active in our system than in the prostaglandin system.

We have also checked that in a complete system without xanthine oxidase and
with an anti-inflammatory drug there is no alteration in the absorption at 440 nm.
Dimethylformamide, the diluting agent, was practically inactive at the concentrations employed.

DISCUSSION

The enzymatic reaction between hypoxanthine and xanthine oxidase generates
superoxide ion, a reduced and unstable form of oxygen.[20,21] Hydrogen peroxide
is produced in the dismutation reaction in which two molecules of superoxide

TABLE 2

EFFECT OF SOME ANTI-INFLAMMATORY DRUGS ON THE BLEACHING OF p-NITROSODIMETHYL-ANILINE[a]

Drug	Conc. (mM)	Bleaching (ΔA_{440}/min)	P	Inhibition (%)	ID 50 (conf.limits)
Control		0.0287 ± 0.0005		0	–
Indomethacin	8	0.0052 ± 0.0017	0.001	81.88	
	4	0.0100 ± 0.0000	0.001	65.15	1.6
	2	0.0127 ± 0.0008	0.001	55.74	(0.4–6.4)
	1	0.0153 ± 0.0001	0.001	46.68	
Niflumic acid	0.05	0.0012 ± 0.0012	0.001	95.81	0.0038
	0.01	0.0078 ± 0.0003	0.001	72.82	(0.001–0.011)
	0.005	0.0122 ± 0.0002	0.001	57.49	
Acetaminophen	8	0.0122 ± 0.0002	0.001	57.49	
	4	0.0196 ± 0.0003	0.001	31.70	7.0
	2	0.0228 ± 0.0006	0.001	20.55	(1.9–25.2)
Mefenamic acid	8	0.0118 ± 0.0007	0.001	58.88	
	4	0.0177 ± 0.0013	0.001	38.32	5.8
	2	0.0243 ± 0.006	0.001	15.33	(2.32–14.5)
Phenylbutazone	8	0.0137 ± 0.0006	0.001	52.26	
	4	0.0187 ± 0.006	0.001	34.94	7.2
	2	0.0237 ± 0.0022	0.05	17.42	(2.5–23.7)
Acetysalicylic acid	8	0.0161 ± 0.0003	0.001	43.90	
	4	0.0203 ± 0.0011	0.001	29.26	9.7
	2	0.0263 ± 0.0009	0.05	8.36	(1.4–63.0)

[a]The following reaction mixture was employed: p-nitrosodimethylaniline 53.5 mM; hypoxanthine 0.03 mM and phosphate buffer pH 7.8. Total volume 18.8 ml. The spectrophotometric cuvette contained: reaction mixture, buffer or anti-inflammatory drug and xanthine oxidase (0.0014 U/ml) to a total volume of 3 ml.

radical give rise to hydrogen peroxide either spontaneously or enzymatically.[22] Hydroxyl radical is generated by a reaction between the superoxide radical and hydrogen peroxide.[23] In the synovial fluid the fact that the degradative process was inhibited by either superoxide dismutase or catalase indicates that the actual depolymerizing species is the OH˙ radical.[10]

Our results show that the non-steroidal anti-inflammatory drugs inhibited the synovial fluid degradation because of their action on OH˙ produced in the xanthine oxidase reaction. It has been suggested[24] that the inhibition of erythrocyte hemolysis induced by anti-inflammatory drugs is due to their capacity to absorb or destroy free radicals or peroxides. More recently Kuehl et al.[12] have suggested that the OH˙ formed in the metabolism of PGG_2 is the inflammatory component liberated in the enzymatic oxidation of arachidonic acid.

Though from the present data we concluded that the OH· was responsible for the decrease in viscosity assayed, we became interested in confirming these conclusions in a more simple and specific system. We employed the bleaching of p-NDA as a more direct assay for OH·. This technique has some advantages such as the high efficiency of the reaction with OH· (Ref.25), its selectivity, as neither superoxide nor singlet oxygen reacts with the compound[25,26] and the ease of application, as one merely observes the bleaching of the sensitive absorption band at 440 nm. On the other hand, we have the lack of specificity for some recognized OH· scavengers, and the fact that no freely diffusible OH· is detected, but an OH·-analogous species is involved. It is suggested[17] that any OH· formed in biochemical systems is present only as an intermediary complex of H_2O_2, synonymous with biochemically generated OH·, this system being much more selective than OH· itself. Saeed et al.[27] have demonstrated that, in the oxidation of o-dianisidine catalyzed by horse-radish peroxidase-hydrogen peroxide, indomethacin acted by competing with o-dianisidine for peroxidase and hydrogen peroxide or peroxidase-hydrogen peroxide complexes.

Our results with p-NDA show the capacity of the anti-inflammatory drugs assayed to scavenge OH·. The ID 50 for the different drugs tested is of the same order of magnitude (in mM) as that obtained with ethanol, an efficient OH· scavenger. With regard to ethanol our data differ substantially from that obtained by Bors et al.[17] This difference is difficult to explain especially since the same experimental set up was used.

The rank order of activity obtained in our assay in relation to the prostaglandin assays permits us to speculate on a possible relationship between our results and those obtained by Kuehl et al.[12] Our results with acetominophen suggest that anti-inflammatory drugs acting on the central nervous system are more sensitive to OH·.

Another interpretation of our results would be possible if anti-inflammatory drugs had inhibited the enzymatic activity of xanthine oxidase, as colchicine does, or if they had been able to oxidize the p-NDA. The first possibility is difficult to confirm because most of these drugs at the concentrations employed absorb at the wavelength at which uric acid is detected. Any possible oxidation of p-NDA has been disregarded because in a complete system without xanthine oxidase there is no alteration in optical density.

The clinical value of our results must be based on the fact that deterioration of synovial fluid is a symptom which characterizes inflammatory types of arthritis. On the other hand, the phagocytosing polymorphonuclear leukocytes present in the synovial fluid produce superoxide radicals with attendant generation of hydrogen peroxide and hydroxyl radicals.[5,6] The above reaction has been suggested as the in vivo mechanism of synovial fluid degradation in inflamed joints. In addition, the same reagents that prevented the synovial fluid degradation were shown[11] to

protect phagocytosing leukocytes from premature death and release of hydrolytic
enzymes and chemotactic factors which play a role in perpetuating the inflammatory cycle.

SUMMARY

Bovine synovial fluid has been employed and its relative viscosity tested.
We demonstrated the capacity of anti-inflammatory drugs to inhibit the synovial
fluid depolymerization induced by the free radicals generated in the enzymatic
reaction between hypoxanthine and xanthine oxidase.

The possible implication of hydroxyl radicals in the synovial fluid degradation has been assayed in a more simple and specific system using the bleaching
of p-nitrosodimethylaniline as a direct assay for hydroxyl radicals. Our results
show the ability of anti-inflammatory drugs to scavenge hydroxyl radicals, this
mechanism being a possible interpretation of their action.

REFERENCES

1. Lai, C.-S. and Piette, L.H. (1977) Biochem. Biophys. Res. Commun. 78, 51-59.
2. Fong, K.-L., McCay, P.B. and Poyer, J.L. (1978) J. Biol. Chem. 248, 7792-
   7797.
3. Pryor, W.A. (1973) Fed. Proc. 32, 1862-1869.
4. Myers, L.S., Jr. (1973) Fed. Proc. 32, 1882-1894.
5. Tauber, A.I. and Babior, B.M. (1977) J. Clin. Invest. 60, 374-379.
6. Weiss, S.J., King, G.W. and LoBuglio, A.F. (1977) J. Clin. Invest. 60, 370-
   373.
7. Heikkila, R.E., Winston, B., Cohen, G. and Banden, H. (1976) Pharmacol. 25,
   1085-1092.
8. Cohen, G. (1978) Photochem. Photobiol. 28, 669-676.
9. Panganamala, R.V., Sharma, H.M., Heikkila, R.E., Geer, J.C. and Cornwell,
   D.G. (1976) Prostaglandins, 11, 599-607.
10. McCord, J.M. (1974) Science, 185, 529-531.
11. Salin, M.L. and McCord, J.M. (1975) J. Clin. Invest. 56, 1319-1323.
12. Kuehl, F.A., Jr., Humes, J.L., Egan, R.W., Ham, E.A., Beveridge, G.C., and
    Van Arman, C.G. (1977) Nature, 265, 170-174.
13. Oyanagui, Y. (1976) Biochem. Pharmacol. 25, 1473-1481.
14. Oyanagui, Y. (1976) Biochem. Pharmacol. 25, 1465-1472.
15. Halliwell, B. (1978) Cell Biol. Int. Rep. 2, 113-128.
16. Puig-Parellada, P. and Planas, J.M. (1978) Biochem. Pharmacol. 27, 535-537.
17. Bors, W., Michel, C. and Saran, M. (1979) Eur. J. Biochem. 95, 621-627.
18. Litchfield, J.T., Jr. and Wilcoxon, F.J. (1949) Pharmacol. Exp. Ther. 96,99.
19. Vane, J.R. in Advances in Biosciencies, Vol.9. Bergström, S., ed., Pergamon
    Press, Oxford, p.395.
20. Fridovich, I. (1970) J. Biol. Chem. 245, 4053-4057.
21. McCord, J.M. and Fridovich, I. (1968) J. Biol. Chem. 243, 5753-5760.
22. McCord, J.M. and Fridovich, I. (1969) J. Biol. Chem. 244, 6049-6055.
23. Beauchamp, C. and Fridovich, I. (1970) J. Biol. Chem. 245, 4641-4646.
24. Fujihira, E. and Otomo. (1970) Yakugaku Zasshi, 90, 1355.
25. Sharpatyi, V.A. and Kraljic, I. (1978) Photochem. Photobiol. 28, 587-590.
26. Kraljic, I. and Mohsni, S.El. (1978) Photochem. Photobiol. 28, 577-581.
27. Saeed, S.A. and Warren, B.T. (1973) Biochem. Pharmacol. 22, 1965-1969.

SUPEROXIDE AND INFLAMMATION

YOSHIHIKO ŌYANAGUI
Research Laboratories, Fujisawa Pharmaceutical Co. Ltd., 1-6,2-Chome, Kashima,
Yodogawa-Ku, Osaka 532, Japan.

 Participation of superoxide in carrageenan paw edema. Superoxide radical (O_2^-)
or hydroxyl radical (OH·) was demonstrated in vitro by McCord[1] to be involved in
synovial fluid degradation, and I have shown that superoxide participates in
rat carrageenan paw edema.[2] Repeated intravenous injections of 1 or 2 mg/kg
superoxide dismutase (SOD, from bovine blood, Sigma product) suppressed completely
the paw swelling of prostaglandin phase (2-4 hr after carrageenan injection into
hind paw), but did not influence histamine and serotonin phase swelling (0-2 hr).
Heat-inactivated SOD, catalase, D-mannitol, etc. showed no suppression (Figure 1).
I also tested the effect of SOD with agranulocyte rat which was prepared with
methotrexate treatment according to DiRosa et al.[3] The suppression of paw swelling
was more evident in agranulocyte rats than in normal rats, suggesting that macrophage is the principal participating cell in this experimental model. McCord[4]
verified this result with a single injection of a stable SOD derivative. He
demonstrated also the participation of active oxygen radicals in the migration
test of leukocytes and in the reverse passive Arthus reaction.

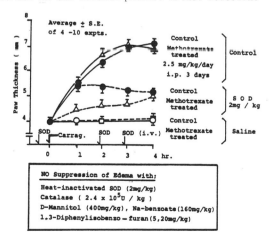

Fig. 1. Carrageenan paw edema in control and agranulocytic (methotrexate-
treated) rats.

Superoxide production by macrophages in vitro. A simple in vitro screening
method was established using oil-induced guinea pig peritoneal macrophages or
the xanthine oxidase reaction, and adapting LDH-bound NADH method og Chan and
Bielski[5] for superoxide detection. Other assay methods were not my choice
because of unpreferable pH, interferance of drugs, etc. As shown in Figure 2,
typical non-steroidal anti-inflammatory drugs inhibited superoxide production
by macrophages, but they failed to inhibit the xanthine oxidase reaction.[6,7]
Catalase, D-mannitol, sodium azide were negative in both systems as well as many
other therapeutic drugs. SOD, pyrogallol, ascorbate and chloropromazine are
superoxide scavengers and showed inhibition in both systems.

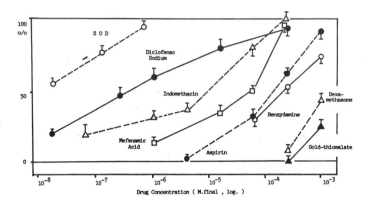

Fig. 2. Inhibition of superoxide radical production by macrophages with typical
anti-inflammatory drugs.

Many anions and cations were also examined. Phosphate(Pi), pyrophosphate
(PPi), ATP, ADP or sulfate was shown to be essential for superoxide production
by macrophages (Figure 3). These anions are reported to increase in synovial
fluid of rehematic patients, etc.[8,9,10] and can induce the paw swelling when
they are injected in the hind paw.[11]

Many metal cations are inhibitory or ineffective, but platinic ion (Na_2PtCl_6)
was very stimulatory (Figure 4)[12]. This effect is derived from the chain reaction of superoxide production as observed with paraquat or streptonigrin.
Platinic compounds are known as anti-tumour agents.[13] Bleomycin and adriamycin
are supposed to be effective by their capacities to form active oxygen radical(s).[14,15]

Physiological and pathological superoxide scavengers. Endogenous SOD protects
the host from the hazard of superoxide. Moreover, I observed comparatively
effective scavenging action of blood plasma or tumor ascites fluid.[16] Ceruloplasmin was recently reported as a superoxide scavenger.[17] Figure 5 shows the

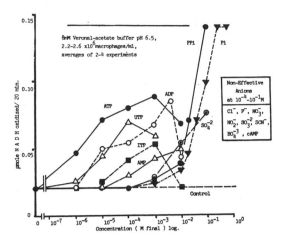

Fig. 3. Stimulation of macrophage superoxide radical production by anions.

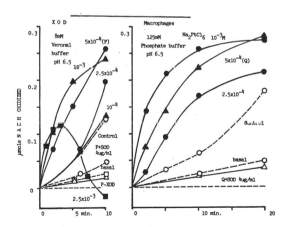

Fig. 4. Stimulation of superoxide radical production by platinic ion.

superoxide scavenging effect of inflammatory exudate fluid induced by peritoneal
injection of formalin solution. This scavenging capacity increased with time
and may serve to inhibit the development of inflammation.

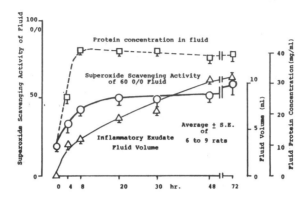

Fig. 5. Superoxide scavenging activity of inflammatory peritoneal fluid induced by formalin (2% in saline, 10 ml/kg, i.p.).

Participation of active oxygen radicals in immunity. Kreuger et al.[18] demonstrated that nitroblue tetrazolium (NBT) reducing macrophages appeared after culturing them with sensitized lymphocytes and specific antigen. The degree of NBT reduction (formazan particle forming macrophages/total macrophages) correlates well with the hypersensitivity skin test. Casein-induced peritoneal macrophages containing BCG sensitized lymphocytes of guinea pig were cultured with 5μg PPD/ml for 2 days in an 8-well Lab-Tek chamber slide. About half of macrophages manifested strong NBT reducing activity. SOD or D-mannitol prevented the appearance of NBT reducing macrophages (Figure 6). Catalase itself showed little effect, but when mixed with a less effective dose of SOD, the inhibition of NBT reducing activity became evident. These results indicate that the hydroxyl radical is responsible to render the macrophages metabolically active (increase of hexose monophsophate shunt pathway and of superoxide production). SOD added to an NBT reducing reaction of 2 hr duration, after 2 days culture, was not effective. Peritoneal exudate cells from normal animals with PPD, or from sensitized animals without PPD, did not reduce NBT.

Conclusion. The role of free active oxygen radicals in inflammation and killing of bacteria or tumor cells may be supposed to occur as shown in Figure 7 from the results of my experiments and of others.

Initial stimuli to macrophages or leukocytes may alter the cell surface structure which activates membrane-bound NAD(P)H oxidase. This enzyme produces superoxide and other active oxygen radicals in the presence of phosphate, pyro-

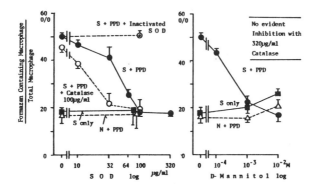

Fig. 6.   Inhibition of NBT reducing macrophage increase with SOD and D-mannitol
(S,N = Peritoneal exudate cells of BCG sensitized and normal guinea pigs,
respectively; PPD 5 μg/ml, 48 hr culture, direct method).

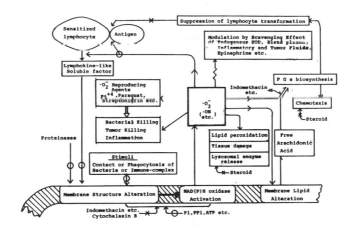

Fig. 7.   Supposed role of oxygen free radicals in inflammation and tumor or
bacterial killing.

phosphate, ATP, etc. Active oxygen radicals kill bacteria or tumor cells directly or after receiving amplification with Pt^{+4}, paraquat or streptonigrin. Active oxygen radicals provoke lipid peroxidation, tissue damage, lysosomal enzyme release, and prostaglandin (PG) biosynthesis[19,20] which lead to inflammation development. Some PGs and thromboxanes are reported as chemotaxants.[21,22] PGE_2 is suppressive to lymphocyte transformation and must be an end product to become a signal for ceasing inflammation. Amounts of active oxygens might be controlled by endogenous SOD, blood plasma or epinephrine, etc. in the physiological state. Inflammatory fluid can scavenge superoxide to stop the inflammation. Tumor ascites fluid may prevent tumor cells from the attack of active oxygens. Sodium periodate ($NaIO_4$) or galactose oxidase treatment of macrophages and/or lymphocytes resulted in the increase of lymphocyte-macrophage interaction and of lymphokine production.[23,24] Modification of cell surface carbohydrate by activated oxygen radicals is also supposed by these reports. Some proteinases are believed to be required for superoxide production by human leucocytes.[25]

SUMMARY

Repeated injections of superoxide dismutase (SOD) suppressed the prostaglandin phase swelling (2-4 hr) of rat carrageenan paw edema. Typical nonsteroidal anti-inflammatory drugs blocked the production of superoxide by guinea pig macrophages, but they did not scavenge superoxide produced by xanthine oxidase. Phosphate, pyrophosphate, ATP, ADP or sulfate was stimulatory for superoxide production by macrophages. Metal cations are inhibitory or non-effective except Pt^{4+} which is stimulatory. Pt^{4+} may amplify the amount of superoxide and serve to kill bacteria or tumor cells. Blood plasma proteins, tumor ascites and inflammatory fluid were proved as scavengers and their physiological meaning is considered. Nitroblue tetrazolium reducing macrophages appeared after incubation with BCG sensitized guinea pig lymphocyte and antigen PPD. This reduction was inhibited with SOD or D-mannitol, suggesting that hydroxyl radical (OH·) is involved in rendering macrophages metabolically activated. Prostaglandin(s), chemotaxis, lysosomal enzymes and proteases are discussed in connection with active oxygen radicals.

REFERENCES

1. McCord, J.M. (1974) Science, 185, 529-531.
2. Ōyanagui, Y. (1976) Biochem. Pharmacol. 25, 1465-1472.
3. DiRosa, M., Papadiumitroiou, T.M. and Willoughby, D.A. (1971) J. Pathol. 104, 239-245.
4. McCord, J.M., Proc. Symp. Active Oxygen and Medicine, Honolulu, Jan.30 - Feb.1, 1979. in press.
5. Chan, P.C. and Bielski, B.H.J. (1974) J. Biol. Chem. 249, 1317-1319.
6. Ōyangui, Y. (1976) Biochem. Pharmacol. 25, 1473-1481.
7. Ōyangui, Y. (1978) Biochem. Pharmacol. 27, 777-782.
8. Bennett, R.M., Lehr, J.R. and McCarty, D.J. (1975) J. Clin. Invest. 56,1571-1579.

9. Willoughby, D.A., Dunn, C.J., Yamamoto, S., Capasso, F., Deporter, D.A. and Giroud, J.P. (1975), Agent. Action. 5, 35-42.
10. Ghosh, P., Stephens, R.W. and Taylor, T.K.F. (1975) Med. J. Australia 1, 317.
11. Oyanagui, Y. (1977) Agent. Action. 7, 125-132.
12. Oyanagui, Y. (1977) Biochem. Pharmacol. 26, 473-476.
13. Rosenberg, B. and Van Camp, L. (1970) Cancer Res. 30, 1799-1802.
14. Ishida, R. and Takahashi, T. (1975) Biochem. Biophys. Res. Commun. 66, 1432-1436.
15. Gregory, E.M. and Fridovich, I. (1973) J. Bacteriol. 114, 1193-1197.
16. Oyanagui, Y. Proc. Symp. Active Oxygen and Medicine, Honolulu, Jan.30-Feb.1, 1979, in press.
17. Goldstein, I.M., Kaplan. H.B., Edelson, H.S. and Weissmann, G. (1979) J. Biol. Chem. 254, 4040-4045.
18. Krueger, G.G., Ogden, B.E. and Weston, W.L. (1976) Clin. Exp. Immunol. 23, 517-524.
19. Rahimtula, A. and O'Brien, P.J. (1976) Biochem. Biophys. Res. Commun. 70, 893-899.
20. White, J.G., Rao, G.H.R. and Gerrad, J.M. (1977) Amer. J. Pathol. 88, 387-402.
21. McClatchey, W. and Synderman, R. (1976) Prostaglandins, 12, 415-426.
22. Booth, J.R., Dawson, W., and Kichen, E.A. (1976) J. Physiol. 257, 47P.
23. Presant, C.A. and Parker, S. (1976) J. Biol. Chem. 251, 1864-1870.
24. Dixon, J.F.P., Parker, J.W. and O'Brien, R.C. (1976) J. Immunol. 116, 575-584.
25. Kitagawa, S., Takaku, F. and Sakamoto, S. (1979) FEBS Lett. 99, 275-280.

A ROLE FOR SUPEROXIDE IN GRANULOCYTE MEDIATED INFLAMMATION

JOE M. McCORD, DENIS K. ENGLISH[*] AND WILLIAM F. PETRONE
Department of Biochemistry, University of South Alabama, Mobile, Alabama 36688, USA

Phagocytically activated polymorphonuclear leukocytes (granulocytes) generate substantial quantities of superoxide free radical by an NADPH oxidase located on the plasma membrane.[1,2] During the ingestion of foreign particles the membrane invaginates to become the lining of the phagocytic vacuole. Superoxide is thus produced in the phagosome where it plays a major role in the intracellular killing of microorganisms.[3] A sizable fraction of this toxic oxgyen radical is detectable in the extracellular medium. Granulocytes from individuals with chronic granulomatous disease cannot produce superoxide, presumably because they lack the NADPH oxidase.[4] The microbicidal effectiveness of these granulocytes is thus greatly diminished, causing recurrent and ultimately fatal infections. The anti-inflammatory activity of bovine superoxide dismutase was discovered long before its enzymatic activity was known (for review see Refs.5,6). The findings summarized above permitted the development of an obvious rationale for the anti-inflammatory activity based on the catalyzed dismutation of the cytotoxic superoxide radical. Evidence presented here supports a new role for superoxide in the inflammatory process which goes beyond direct cytotoxicity. It reacts with a precursor molecule present in plasma to create a potent chemotactic factor for granulocytes, thereby providing a mechanism for cell-to-cell communication.

We have observed the anti-inflammatory properties of parenterally administered superoxide dismutase in three animal models of induced inflammation: the reverse passive Arthus reaction, carrageenan-induced foot edema, and immune-complex induced glomerular nephritis.[7,8] In all cases the enzyme dramatically reduces edema formation. An unexpected finding was noted during histological examination of the sites of the potential lesions. The enzyme-treated animals showed very little infiltration of granulocytes into the tissue, in contrast to the non-treated animals. This suggested that superoxide was somehow involved with the migration of cells into the inflammed areas. If superoxide dismutase could prevent the arrival of inflammatory cells, it could prevent all granulocyte-mediated tissue damage (i.e. oxidative damage by myeloperoxidase, hydrolytic damage by granular enzymes, etc.), not just that caused by the direct cytotoxicity of superoxide.

* Present address: Division of Pediatric Oncology-Hematology, Vanderbilt University Medical Center, Nashville, TN 37232, USA.

Due to the instability of the superoxide radical and the unlikelihood that it could serve directly as a chemoattractant, we proposed the existence in extra-cellular fluids of a "precursor molecule" capable of reacting with superoxide to produce a stable, active chemotactic factor. The transient production of superoxide by the first granulocyte to arrive at the locus of a developing inflammatory lesion could, by such a mechanism, be translated into a stable signal calling additional cells to the site. Upon arrival, the new cells would likewise be stimulated to produce the radical, assuming the stimulus had not been eliminated, and an amplified signal would go out to summon still more cells. This cycle would repeat until the stimulus were removed. Furthermore, we proposed that incubation of granulocytes with the putative chemotactic factor should not induce superoxide production in the cells. That is, the factor should serve only as a chemoattractant and not as a metabolic stimulus to the granulocyte. This is an important theoretical point. If the factor activated cells it would catalyze its own production; the inflammatory process, once initiated, would escalate to an explosive end.

To test the hypothesis we exposed normal human plasma to a superoxide gen-erating system (xanthine oxidase-xanthine), and assessed the chemotactic response of isolated human peripheral granulocytes toward this "activated" plasma. Chemo-taxis was examined using the "chemotaxis-under-agarose" method[9] as well as the Boyden technique.[10] With the former method there was an increase in the rate of migration of the cells toward the well containing plasma exposed to superoxide (Figure 1, plate A). The addition of superoxide dismutase (but not catalase) to plasma before exposure to superoxide prevented the formation of the chemotactic activity (Figure 1, plate B); the addition of superoxide dismutase after exposure to superoxide had occurred, however, had no effect on the chemotactic response of the granulocytes (Figure 1, plate A). Cells responding to a gradient of superoxide-exposed plasma migrated 67% father than those responding to plasma protected by superoxide dismutase (see Figure 1). Qualitatively identical results were obtained using the Boyden chamber technique[10] modified as previously described.[11] A millipore filter (3 micron pore size) separated a suspension of granulocytes from solutions to be tested for chemotactic activity. After incubation for 95 minutes at $37^{\circ}C$, the filters were removed and stained with hematoxylin. The average number of cells (per microscopic field) which had migrated through the filter toward the chemotactic substance was determined. These data are reported in Table 1. Again, plasma exposed to superoxide became potently chemotactic. If superoxide dismutase were present during the exposure no chemotactic activity was developed. Catalase, on the other hand, caused no significant inhibition of chemoattractant formation. Thus, the precursor molecule reacts directly with superoxide, and is not dependent upon the secondary generation of hydroxyl radical via a Haber-Weiss mechanism.

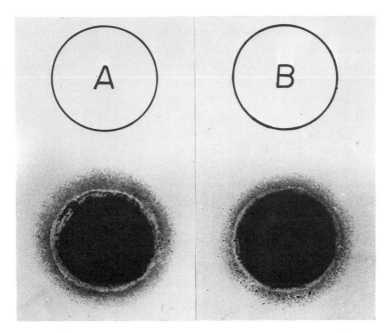

Fig. 1. Chemotaxis of human granulocytes under agarose. Agarose plates were prepared by the method of Nelson et al. (Ref.9) with wells 2.8 mm in diameter and 1.6 mm apart. Leukocytes were prepared by mixing 8 ml of heparinized human blood with 2 ml of 5% dextran in 0.15 M saline and allowing the red cells to sediment by gravity. The upper layer, rich in leukocytes, was centrifuged 10 min. at 250 x g. The cells were washed once with Hank's balanced salts solution, then resuspended in this medium to a concentration of 4×10^7 cells/ml. Ten μl of the cell preparation was added to each of the lower wells. Activated plasma was prepared by the addition of 0.006 U of xanthine oxidase and 0.1 ml of 1 mM xanthine to 1 ml of 10% plasma in Hank's medium. This mixture was incubated at $37^\circ C$ for 60 min., then 75 U of bovine superoxide dismutase was added. Ten μl of this mixture was placed in well (A). The plasma in well (B) was prepared identically, except that superoxide dismutase was present from time zero. The plate was incubated at $37^\circ C$ for 3 hr. in a humidified CO_2 incubator, then fixed and stained according to Nelson et al. (Ref.9).

To determine whether the chemotactic factor also served as a metabolic stimulant to granulocytes, the cells were incubated in 10% plasma which had been previously activated by exposure to superoxide. Superoxide production by the cells themselves was monitored by the cytochrome $\underline{c}$ reduction method.[1] The factor did not induce superoxide production by granulocytes, supporting our hypothesis concerning its role in the inflammatory process.

We next examined the ability of the superoxide-dependent chemotactic factor to provoke granulocyte infiltration _in_ _vivo_. Since it does not induce the "respiratory burst" associated with superoxide production by the cells, we predicted that there would be cellular influx without the usual signs of inflammation. This hypothesis was tested by the intradermal injection of rats

TABLE 1

DEMONSTRATION OF THE SUPEROXIDE-DEPENDENT CHEMOTACTIC ACTIVITY BY THE BOYDEN
TECHNIQUE.[a]

Experiment	Description	Cells/Field ± S.E.M.	% Inhibition
1	Normal plasma	4.3 ± 3.0	–
	Superoxide-treated plasma	16.2 ± 4.2	–
	omit xanthine	6.4 ± 2.5	87%
	omit xanthine oxidase	4.3 ± 1.6	100%
2	Normal plasma	1.3 ± 0.9	–
	Superoxide-treated plasma	20.7 ± 3.9	–
	plus SOD at t=0[b]	1.5 ± 1.4	99%
	plus SOD at t=60	19.0 ± 3.5	9%
3	Normal plasma	2.3 ± 1.4	–
	Superoxide-treated plasma	16.3 ± 3.4	–
	plus catalase at t=0[c]	14.2 ± 3.8	15%
	plus catalase at t=60	12.3 ± 5.5	28%

[a]Plasma was used at a concentration of 10% in Hank's balanced salts solution.
Superoxide-treated plasma was incubated for one hour at 37°C with 0.3 mM xanthine
and 0.01 unit/ml xanthine oxidase.
[b]Superoxide dismutase (SOD) was added at a concentration of 50 µg/ml at 0 min.
or after 60 min.
[c]Catalase (83 µg/ml) was added, as above.

with plasma which had been incubated with a xanthine-xanthine oxidase superoxide
generating system. During a five hour period following injection there was no
ostensive signs of either edema or erythema at the injection points. Histological examination of hematoxylin and eosin stained sections of this tissue,
however, revealed that there was indeed substantial migration of granulocytes
to the site (Figure 2, plate A). When the xanthine-xanthine oxidase system was
allowed to react to completion before the addition of plasma (thereby precluding
the formation of the chemotactic factor by temporally separating superoxide
generation from the addition of the precursor substance), almost no accumulation
of granulocytes resulted (Figure 2, plate B).

When superoxide-activated plasma was subjected to gel filtration and ion
exchange chromatography, the chemotactic factor co-eluted with albumin. Commercially purified albumin, however, (Cohn fraction V) is not chemotactic and
cannot be rendered so by exposure to superoxide. The possibility that the
chemotactic factor might be a lipid-albumin complex was supported by the observation that a chloroform extract of normal plasma could be redissolved in a solution of commercial albumin, yielding a complex which was itself of weak chemotactic activity, but which became potently chemotactic upon exposure to super-

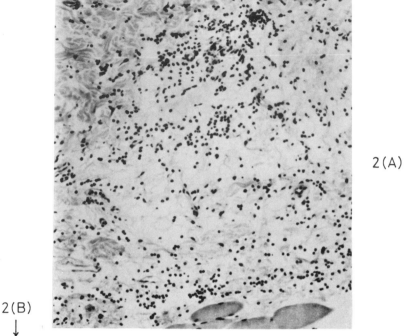

2(A)

2(B)
↓

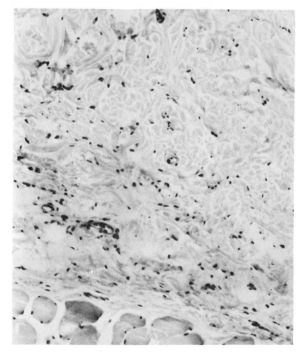

Fig. 2. The superoxide-
dependent chemotactic factor
induces granulocyte infiltration in vivo. Activated rat
plasma (A) was prepared by
incubating 0.1 ml of 1 mM
xanthine and 0.002 U of
xanthine oxidase with 0.5 ml
of heparinized rat plasma at
37°C for 2 hr. Control plasma
(B) was prepared by allowing
the same amounts of xanthine
and xanthine oxidase to react
to completion before the
addition of the plasma. Rats
were injected intradermally
with 50 µl aliquots of these
preparations at shaved, depi-
lated, laterally paired, dorsal
sites. After 5 hr., the animals
were sacrificed and the injection sites were excised, fixed,
sectioned, and stained with
hematoxylin and eosin. Slides
were photographed at 100X
magnification. The tissue receiving activated plasma (A)
shows heavy,diffuse infiltration
by inflammatory cells. Control
plasma (B) elicited little or
no information.

oxide. That is, the reconstituted complex of albumin plus chloroform extract behaves much like normal plasma.

Thomas et al.[12] have shown that superoxide is unreactive toward polyunsaturated fatty acids, but that it is reactive with fatty acyl hydroperoxides. Perez and Goldstein[13] have shown that the action of strong oxidizing radicals on arachidonic acid gives rise to chemotactic products. Thus, while the exact identity of the naturally-occurring precursor of the superoxide-dependent chemotactic factor is unknown at present, evidence suggests a complex of albumin with a lipid hydroperoxide. The prevention by superoxide dismutase of its activation in vivo may be the principal mode of anti-inflammatory action by the enzyme.

ACKNOWLEDGEMENT

This work was supported in part by a grant from the NIH (USPHS grant AM20527).

REFERENCES

1. Babior, B.M., Kipnes, R.S. and Curnutte, J.T. (1973) J. Clin. Invest. 52 741-744.
2. Dewald, B., Baggiolini, M., Curnutte, J.T. and Babior, B.M. (1979) J. Clin. Invest. 63, 21-29.
3. Johnston, R.B.,Jr., Keele, B.B.,Jr., Misra, H.P., Lehmeyer, J.E., Webb, L.S., Baehner, R.L. and Rajagopalan, K.V. (1975) J. Clin. Invest. 55, 1357-1372.
4. Curnutte, J.T., Whitten, D.M. and Babior, B.M. (1974) N. Engl. J. Med. 290, 593-597.
5. Huber, W. and Saifer, M.G.P. (1977) in Superoxide and Superoxide Dismutases, Michelson, A.M., McCord, J.M. and Fridovich, I., eds., Academic Press, New York, pp.537-550.
6. Menander-Huber, K.B. and Huber, W. (1977) in Superoxide and Superoxide Dimutases, Michelson, A.M., McCord, J.M. and Fridovich, I., eds., Academic Press, New York, pp.551-556.
7. McCord, J.M. and Wong, K. (1979) in Oxygen Free Radicals and Tissue Damage, Ciba Foundation Symp. 65 (New Series), Excerpta Medica, Amsterdam, pp.343-351.
8. McCord, J.M., Stokes, S.H. and Wong, K. in Proceedings of the International Congress of Inflammation, Paoletti, R., Samuelsson, B. and Weissmann, G., eds., Raven Press, New York, in press.
9. Nelson, R.D., McCormack, R.T. and Fiegel, V.D. (1978) in Leukocyte Chemotaxis, Gallin, J.I. and Quie, P.G., eds., Raven Press, New York, pp.25-42.
10. Boyden, S. (1962) J. Exp. Med. 115, 453-466.
11. Smith, C.W., Hollers, J.C., Patrick, R.A. and Hasset, C. (1979) J. Clin. Invest. 63, 221-229.
12. Thomas, M.J., Mehl, K.S. and Pryor, W.A. (1978) Biochem. Biophys. Res. Commun. 83, 927-932.
13. Perez, H.D. and Goldstein, I.M. (1979) Fed. Proc. 38, 1170.

EFFECT OF OXYGEN-DERIVED FREE RADICALS ON CONNECTIVE TISSUE MACROMOLECULES

ROBERT A. GREENWALD

Division of Rheumatology, Long Island Jewish-Hillside Medical Center,
New Hyde Park, New York 11042, and State University of New York at
Stony Brook, New York, USA

One of the areas where the biological effects of oxygen-derived free radicals
may have clinical implications is the field of arthritis. Our laboratory has been
conducting some studies of the effects of these radicals on connective tissue macromolecules (see Ref. 1), and this report will summarize some of our recent work.

One of the basic problems in the control of arthritis is the arrest of degradation of connective tissues. Although there are some exceptions, most forms of treatment for arthritis have no effect on the progression of the disease, and we cannot
prevent joint tissue destruction despite maximal anti-inflammatory therapy. Thus,
much research has been directed at the mechanisms by which connective tissues undergo
degradation. Most of that research has been directed at study of enzymatic processes
by which connective tissue molecules can be destroyed. The three major molecules
which have been the subject of study are collagen, proteoglycans, and hyaluronic
acid.

Collagen is found in cartilage and in the supporting structures of the joints,
and collagen is generally susceptible to degradation only by specific collagenases.
Collagenase could arise from leukocytes, synovium, or cartilage itself, but there
are many endogenous inhibitor molecules present in serum and synovial fluid which
could impair its action. Although it is probable that collagenase plays a role
in the destruction of cartilage, there are many questions which have not been re-
solved.

Considering hyaluronic acid as a substrate, the situation is even more unclear.
One of the characteristic features of inflammatory arthritis is the loss of viscosity of synovial fluid. Hyaluronic acid accounts almost entirely for the viscosity of this fluid. Despite the frequent finding of decreased viscosity in inflammatory arthritis, there is no convincing evidence of the existence of hyaluronidase
in pathologic joint fluids, in synovium, in cartilage, or in polymorphonuclear
leukocytes. Furthermore, there is no direct data demonstrating depolymerization
of hyaluronic acid in such fluids. Based on current knowledge, there is no way
to invoke an enzymatic mechanism to account for the loss of hyaluronic acid viscosity which accompanies inflammatory arthritis.

Accumulation of leukocytes is a characteristic feature of many forms of inflammatory arthritis. Since leukocytes can generate free radicals under a variety of stimuli, many of which probably take place in the inflamed joint, we have hypothesized that superoxide ion and related species could be involved in the degradation of connective tissue. We have therefore studied some possible effects of oxygen-derived free radicals on these macromolecules in vitro. This report will deal with some recent work involving two macromolecules, namely, collagen and hyaluronic acid.

In the collagen experiments, we tested the effect of xanthine oxidase on the gelation of soluble collagen. When acid soluble calf-skin collagen in dilute solution at pH 7.4 is warmed at $37^{\circ}C$, the collagen will gel. The progress of the gelation reaction can be monitored in a spectrophotometer by following the absorbance at 400 nm. To do these experiments, we took aliquots of a collagen solution and treated them with xanthine oxidase, with or without hypoxanthine and various inhibitors. After pre-treatment of the collagen in the superoxide generating system, the collagen solutions were moved to the spectrophotometer and brought to $37^{\circ}C$. Our finding, confirmed repeatedly, was that superoxide radical could alter the soluble collagen and prevent its normal gelation.

Figure 1 shows the results of an experiment performed with crude xanthine oxidase obtained from Sigma Chemical in early 1978. Panel (a) shows a normal gelation curve produced when untreated soluble collagen was brought to $37^{\circ}C$ in the spectrophotometer. If crude xanthine oxidase was added to this solution in the absence of hypoxanthine, there was some mild inhibition of gelation but the material did eventually gel and in fact, given enough time, it would reach the same final optical density as the control in collagen gelation (Figure 1). However, when hypoxanthine was present during the xanthine oxidase incubation, gelation was abolished. Panel (b) shows that the presence of either superoxide dismutase or catalase in the collagen solution prior to the treatment with xanthine oxidase and hypoxanthine prevented the inhibition of gelation. If superoxide dismutase and catalase were both used simultaneously, a completely normal gelation curve, superimposable on the control, was found. In panel (c), collagen solutions were treated with xanthine oxidase and hypo xanthine for 60 minutes at varying concentrations of xanthine oxidase. The more xanthine oxidase present, the greater the degree of inhibition of gelation. Panel (d) shows the same type of experiment performed with a 4 hour pre-incubation.

We then attempted to try to remove from the xanthine oxidase the contaminating protease which is known to co-purify with the active enzyme. We put some crude xanthine oxidase on a column of Sephadex G-100 and eluted the column with phosphate-salicylate-EDTA buffer at pH 7.8. Three peaks of protein were found (Figure 2). The first peak contained the protease activity, and the third peak had neither activity.

When collagen was incubated with the purified xanthine oxidase material from

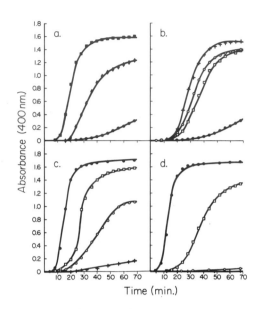

Fig. 1. Effect of unfractionated, crude xanthine oxidase on the thermally initiated gelation of collagen. a, Gelation of collagen preincubated (60 minutes) in 1 mM Hx without any enzymes (■); gelation of collagen preincubated with 0.0156 mg/ml crude XO but without Hx (◆); gelation of collagen preincubated with both XO and Hx at the same concentrations (●); b, Gelation of control sample containing Hx but no XO (+); inhibited gelation with the combination of XO and Hx (●); reversal of this inhibition with either SOD (60 μg/ml) (O), or catalase (Worthington, 15 μg/ml) (□) (When both catalase and SOD were added together, a normal gelation curve superimposable on (+) was oabtained); c, Gelation curves for collagen preincubated for 60 minutes with varying concentrations of XO in 1 mM Hx. No XO (●); collagen/XO ratio 40 (□); collagen/XO ratio 20 (O); collagen/ XO ratio 13 (+); d, Gelation curves for collagen samples preincubated at the same collagen/XO ratios as in c for 4 hours.

peak 1 in the absence of hypoxanthine (Figure 3), a completely normal gelation curve, reminiscent of the controls shown on the previous slides, was noted. When the protease material from the second peak of Figure 2 was added to the collagen, there was a slight retardation of gelation but the material did eventually gel and reached the same final optical density. However, if the peak material from the first peak, i.e. the purified xanthine oxidase, was allowed to act on the collagen in the presence of hypoxanthine, marked inhibition of gelation was obtained. Finally, if some second peak material was added, total abolition of gelation was obtained.

This suggested that collagen acted upon by superoxide might become susceptible to proteolytic degradation which would not be possible if the collagen were in its native state.

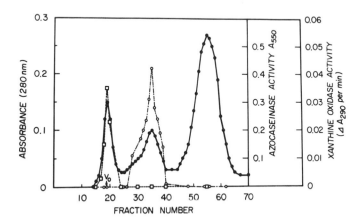

Fig. 2. Chromatographic profile of crude commercial xanthine oxidase on Sephadex G-100 eluted with phosphate-salicylate-EDTA buffer pH 7.8. Aliquots from each fraction containing protein (●) were assayed for their activity in conversion of hypoxanthine to uric acid (□) and for proteolytic activity (○).

Figure 4 shows the results from an experiment in which a solution of collagen and xanthine oxidase was placed in the cold room for several hours and "shots" of hypoxanthine were given at various intervals. At each interval, an aliquot of collagen was removed and tested for its ability to gel, while the remainder was allowed to remain under the action of the enzyme. The longer the collagen remained under the action of xanthine oxidase, the greater the degree of inhibition of gelation of the material over a six hour experiment. When superoxide dismutase was present in the sample pre-incubated for five hours with both xanthine oxidase and hypoxanthine, the collagen was totally protected from inhibition of gelation.

Figure 5 shows the results of a comparative inhibitor study using superoxide dismutase and copper-penicillamine complex. In these experiments, samples of a collagen solution containing 778 μg of collagen in 750 μg of buffer were incubated for four hours at 2°C. Each sample contained 91 μg of xanthine oxidase, 1.24 mM hypoxanthine, and either superoxide dismutase or copper-penicillamine at various concentrations. After pre-incubation in the cold room, the gels were moved to the spectrophotometer and gelation was induced. The superoxide dismutase concentrations on the left panel ranged from zero to 23 micrograms per ml, and the greater the concentration, the greater the degree of inhibition of the reaction. Copper-penicillamine on the right was present at concentrations ranging from 0.09 mM to 0.4 mM. Again, the greater the concentration of scavenger, the greater the degree of inhibition of the reaction.

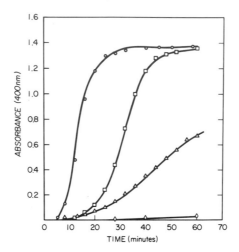

Fig. 3. Effect of various fractions from G-100 chromatography of commercial crude xanthine oxidase on collagen gelation. When enzyme from peak I (fractions that convert hypoxanthine to uric acid but have no effect on Azocoll) was added to collagen with Hx omitted, a normal gelation curve was observed (O). If Hx was added during the preincubation, substantial inhibition of gelation ensured (△). The protease material (peak II) delayed gelation (□) but did not inhibit the final plateau of absorbance. If enzymes from peak I and peak II were both present (◇) gelation never occurred, which suggests that collagen molecules acted upon by superoxide might have become susceptible to subsequent proteolytic degradation.

Gelation of neutral pH at 37°C is a physiologic property of normal soluble collagen. Degraded collagen will not gel properly, nor will collagen with major modifications of amino acid side chains which participate in noncovalent aggregation during the nucleation phase of the reaction. We therefore used thermal gelation of collagen as a test system to detect an effect of superoxide on collagen molecules. These studies indicate that the free radicals are capable of altering the normal behavior of the protein used in these experiments. Our results are in agreement with those found by investigators in India who reported that singlet oxygen would inhibit the gelation of tropocollagen, and that this was associated with lowering of the viscosity of the solution and loss of certain amino acid residues. We have not yet had an opportunity to perform amino acid analysis on our collagen preparations.

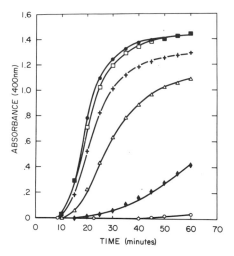

Fig. 4. Effect of sequential addition of "shots" of Hx to collagen samples preincubated with xanthine oxidase (56 milliunits/sample). The enzyme was added to a control collagen sample without Hx, and aliquots were removed at various times for determination of their ability to gel. Regardless of the time of pre-incubation, gelation was always normal even in the presence of the enriched enzyme if Hx was omitted (●). To another collagen sample incubated in parallel, "shots" of Hx were added as aliquots were removed such that the Hx concentration varied from 0.43 mM at the start to 0.88 mM by the end of the experiment. Aliquots removed from the Hx samples at various times showed progressive inhibition of gelation: 1 hour (+); 2 hours (△); 4 hours (◆); 5 hours (O). When SOD (50 μg/ml) was present in a sample preincubated for 5 hours with both XO and Hx, the collagen was protected from inhibition of gelation (□).

However, preliminary studies have indicated that polyacrylamide gel pattern of the collagen treated with superoxide is basically normal, although the material does have a lowered intrinsic viscosity. We believe that superoxide and related radicals can adversely affect collagen and that this effect might play a role in joint tissue destruction or repair.

The remainder of our studies pertain to hyaluronic acid. As indicated above, loss of synovial fluid viscosity is one of the hallmarks of inflammatory arthritis. Since there is no hyaluronidase in leukocytes or in joint tissues, the mechanism of this effect is unknown. We believe that polymorphonuclear leukocytes probably mediate the loss of viscosity through the action of superoxide on hyaluronic acid. I would now like to present some data to support this hypothesis.

Figure 6 shows the results of a simple experiment in which hyaluronic acid was incubated in a viscometer with various concentrations of crude xanthine oxidase. The contaminating protease is irrelevant in these experiments since hyaluronic acid is not susceptible to degradation by proteases. In the experiments shown here,

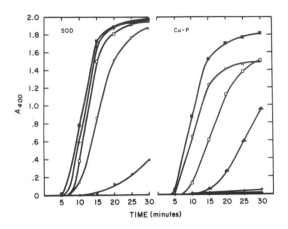

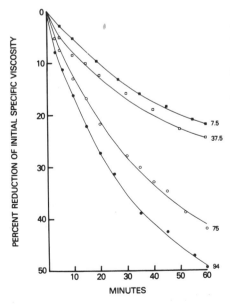

Fig. 5.   Comparative effectiveness of superoxide dismutase and copper-penicil-lamine complex in protecting soluble collagen from the inhibition of gelation induced by superoxide radical.   The test system contained 778 μg of collagen, 91 μg xanthine oxidase, and 1.24 mM hypoxanthine in 750 μl.   After preincubation for 4h at 2°C, the gels were moved to the temperature-controlled compartment of a spectrophotometer and gelation monitored at 400 nm as described in the text.   The two scavengers were present during the preincubation as follows: Superoxide dis-mutase (left panel): (●), 0; (+), 23.3 μg; (□), 5.9 μg; (O), 0.5 μg; (■), 140 μg but hypoxanthine omitted.   Copper penicillamine complex (right panel): (●), 0; +, 0.085 mM; (▲), 0.188 mM; (□), 0.255 mM; (O), 0.426 mM; (■), no hypoxanthine.

Fig. 6.   Effect of superoxide genera-tion on the viscosity of hyaluronic acid.   Hyaluronate (700 μg/ml) was incubated in a viscometer and the re-action was initiated by addition of xanthine oxidase to the final concen-trations (μg/ml) shown on the figure. At increasing enzyme concentrations, in the presence of excess hypoxanthine, a progressively greater degree of vis-cosity fall was noted.

hyaluronate at a concentration of 700 µg per ml was incubated with xanthine oxidase at concentrations ranging from 7.5 to 94 µg per ml. The greater the enzyme concentration, the greater the degradation of the hyaluronate as measured by reduction of initial specific viscosity.

Figure 7 shows the results of inhibitor studies. In panel (A) in the upper left superoxide dismutase was present at two different concentrations. Partial inhibition of the effect of xanthine oxidase on hyaluronate was noted. In panel (B), catalase was present at various concentrations, and again partial inhibition was noted. In panel (C), copper-penicillamine also proved to be an effective inhibitor of hyaluronate degradation. Finally, it is seen in panel (D), that the presence of the combination of trivial concentrations of both copper-penicillamine and catalase, at concentrations below that which would be effective for either component individually, were totally protective of the effect of superoxide on hyaluronic acid.

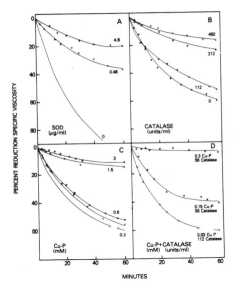

Fig. 7. Effect of free radical scavengers on the superoxide induced fall in viscosity of hyaluronate. Experiments were performed as in Fig. 1 with prior addition of the substances indicated above.

Figure 8 shows similar data involving histidine and mannitol. On the left are various concentrations of superoxide dismutase and mannitol. The combination of 1 µg of superoxide dismutase and 220 mM mannitol was totally protective of hyaluronate. On the right, 1 µg per ml of superoxide dismutase and 40 mM histidine was also totally protective. We have concluded from these studies that hyaluronic acid can be degraded by oxygen-derived free radicals and that the reactions can be inhibited by the known scavengers of these substances.

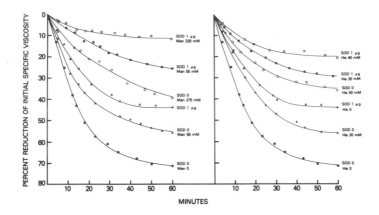

Fig. 8. Effect of histidine and mannitol on the superoxide induced fall in viscosity of hyaluronate. These two scavengers, at the indicated concentrations, were added to the test system with or without superoxide dismutase.

Finally, we have turned our attention to the action of polymorphonuclear leuko-cytes (PMNs) on hyaluronic acid. Human polymorphonuclear leukocytes were prepared from freshly drawn heparinized blood from normal volunteers. The PMNs were washed twice with normal saline and once with isotonic 5 mM HEPES buffer at pH 7.4. The PMNs were suspended in a hyaluronic acid solution containing 5 µg ferrous iron. Superoxide production was induced by addition of phorbol myristate acetate (PMA) at various concentrations on the order of several hundred nanograms per ml. The reactions were stopped by plunging the tubes into an ice water bath and centri-fuging down the cells. The hyaluronic solutions were then studied by viscometry, dialysis, and chromatography.

Figure 9 shows the results of a simple experiment in which PMNs were suspended in a hyaluronic acid solution. In the absence of a stimulus to superoxide produc-tion, there was no effect on the viscosity of the solution. In the presence of 5 million cells, 200 nanograms of PMA induced a mild reduction in the viscosity of the hyaluronic acid solution. When 13 million PMNs were present, a substantial effect on the viscosity was noted. Calibration of the system with cytochrome c revealed that 50 nanograms per ml of PMA would lead to the generation of 5.8 nm of superoxide per million cells per 15 minutes. Addition of mannitol and superoxide dismutase to the PMN system completely abolished the effect of PMY on hyaluronic viscosity.

We were unable to measure any increase in reducing sugars in the hyaluronic solutions treated either with xanthine oxidase or with polymorphonuclear leukocytes.

Most of these samples showed a mild to moderate loss of uronic acid upon dialysis after treatment with the superoxide generating systems.

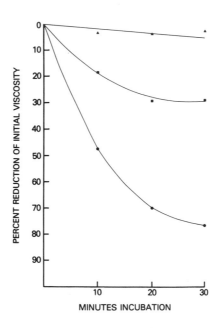

Fig. 9. Effect of superoxide generation by polymorphonuclear leukocytes on the viscosity of hyaluronate. Hyaluronic acid (1.74 mg/ml) was incubated with human PMNs with or without addition of phorbol myristate acetate (PMA) and the specific viscosity of the solution measured at various times. (▲), 13.6 million PMNs/tube, no PMA; (■), 4.9 million PMNs/tube, 200 ng PMA; (●), 13.6 million PMNs/tube, 600 ng PMA.

Figure 10 shows the results of chromatographic analysis on Sepharose CL-2B of the products of hyaluronate degradation produced by either the xanthine oxidase system, shown in panels (A)-(C), or the PMN system shown in panels (D)-(F). Panels (A) and (B) show the profiles of untreated hyaluronic acid, which always eluted in the void volume of the column. After treatment with either xanthine oxidase and hypoxanthine as in panel (B), or polymorphonuclear leukocytes as in panel (E), the material eluted within the included volume of the column, indicating a lowering of the molecular weight. Treatment of the hyaluronic acid with the superoxide generating systems in the presence of superoxide dismutase and mannitol resulted in material which still eluted in the void volume as shown in panels (C) and (F). We concluded from these studies that oxygen-derived free radicals could indeed lower the molecular weight of hyaluronic acid.

The degradation of hyaluronic acid by free radicals has been well known for many years. Other laboratories have studied this reaction involving superoxide.

170

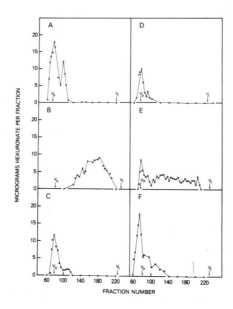

Fig. 10. Chromatographic profiles of hyaluronate on Sepharose CL-2B after
exposure to superoxide generation with the xanthine oxidase system (panels A-C)
or with PMNs stimulated by PMA (panels D-F). Elution profiles of control material
before superoxide exposure (panels A and D) show a single peak in the void volume
of the column. After superoxide exposure (panels B and E), carbazole-reactive
material appears in the included volume. Superoxide exposure in the presence of
free-radical scavengers (superoxide dismutase and mannitol, panels C and F) had
no effect; the material still eluted in the void volume served as the control.

We have extended these observations to demonstrate directly the effect of superoxide
generated by leukocytes on hyaluronate and to show analytic evidence of a lowering
of the molecular weight of the material. Time did not permit the presentation of
data concerning intrinsic viscosity, but suffice it to say that the intrinsic viscosity of hyaluronate treated with superoxide drops dramatically. We believe that
these studies clearly implicate superoxide as a possible mediator of the loss of
viscosity of synovial fluid which accompanies inflammatory arthritis.

It is a long way from the test tube to treatment of the patient. There is evidence suggesting that drugs which interfere with the superoxide system may be thera-
peutically effective in arthritis (see Refs. 2, 3). There have been uncontrolled
studies involving the injection of superoxide dismutase preparations into arthritic
animal models which have suggested that the material ameliorates the arthritis.
Penicillamine (as copper complex) and histidine, both of which are scavengers of
oxygen-derived free radicals, are known to have therapeutic efficacy in some inflammatory arthritic conditions. Steroids have been reported to inhibit the production of superoxide by leukocytes, and of course steroids are also highly anti-

inflammatory. We believe that oxygen-derived free radicals may indeed play a role in degradation of joint tissue macromolecules, and we believe that this is an area for future investigation.

REFERENCES

1. Greenwald, R.A. and Moy, W.W. (1979) Arthritis Rheum. 22, 251-259.
2. Menander-Huber, K.B., this volume.
3. Flohé, L., Biehl, O., Hofer, H., Kadruka, F., Kölbel, R. and Puhl, W., this volume.

ACTIVATION OF THE SUPEROXIDE GENERATING SYSTEM IN HUMAN GRANULOCYTES AND
MYELOID LEUKEMIA CELL LINES

PETER E. NEWBURGER, MARGARET E. CHOVANIEC AND HARVEY J. COHEN
Division of Hematology-Oncology, Children's Hospital Medical Center, and
Department of Pediatrics, Harvard Medical School, Boston, MA 02115, USA

INTRODUCTION

Phagocytic cells generate superoxide as a prime event of their function in
antimicrobial host defense.[1] In the resting state, these cells produce little
or no O_2^-, but upon exposure to soluble[2] or particulate[3] stimuli they initiate
a respiratory burst consisting of O_2^- production and associated O_2 consumption,
hexosemonophosphate shunt stimulation, and generation of other reactive oxygen
species such as peroxide, hydroxyl radical and perhaps singlet oxygen.[1] Continuous monitoring of O_2^- generation,[4,5] O_2 uptake,[6] and H_2O_2 production[7]
demonstrates a delay between contact with a stimulant and the onset of the
respiratory burst. This lag period represents the time taken by the activation
of the membrane-bound reduced pyridine nucleotide oxidase responsible for O_2^-
generation (see H.J. Cohen et al., this volume).

We have developed a continuous assay of granulocyte O_2^--dependent cytochrome
c reduction[4,5] and applied it to the study of the activation (measured as a
lag time) and activity (measured as the rate of O_2^- production) of the human
granulocyte O_2^- generating system. We have also examined these two processes
in two human myeloid leukemia-derived cell lines. One HL-60, derived from a
patient with acute promyelocytic leukemia, differentiates in response to
dimethyl sulfoxide,[8] thus providing an in vitro model for the study of O_2^-
generation during development from promyelocytes to functional mature granulocytes.[9] The other, line 120, derived from a patient with acute myeloblastic
leukemia, appears myeloid by morphologic and functional criteria,[10] but is
infected by Epstein-Barr virus,[11] a B-lymphoid characteristic.

MATERIALS AND METHODS

Human peripheral blood granulocytes were purified by Dextran and Ficoll-
Paque sedimentation as previously described.[5] Cell lines HL-60 and 120
(generously provided by Drs. R.C. Gallo and A. Karpas, respectively) were
maintained in culture as previously described.[10] O_2^- production was measured
quantitatively in pooled cells by a previously described continuous spectrophotometric assay of superoxide dismutase-inhibitable cytochrome c reduction.[4]

Cells were stimulated by phorbol myristate acetate (PMA) 1 µg/ml or by opsonized zymosan 4 mg/ml. Assays were performed at $37^\circ C$ in a double-beam spectrophotometer. Both sample and reference cells contained ferricytochrome c (50 nmol), cells (2.5×10^6), and either PMA (1 µg) or opsonized zymosan (4 mg) in 1 ml total volume of Krebs'-Ringer's phosphate buffer (pH 7.4). The reference cell contained, in addition, superoxide dismutase (0.01 mg). The rate of O_2^- production was calculated by dividing the linear change in A_{550} by the molar extinction coefficient for the reduction of ferricytochrome c ($\Delta\varepsilon_M$ = 21,000) (Ref.12). O_2^- generation was measured qualitatively in individual cells by the NBT slide test.[9] Cells (2.5×10^6/ml), NBT (0.1%), and stimulant were incubated in phosphate-buffered saline at $37^\circ C$ for 10 min., washed three times in cold normal saline containing 1 mM N-ethyl maleimide, resuspended in 5% human serum albumin, smeared on glass slides, and counterstained with safranin.

We adapted the method of Stossel et al.[13] for the measurement of hexosemonophosphate shunt activity. Cell suspensions (2.5×10^6 in 1 ml of Krebs'-Ringer's phosphate buffer) with or without various stimulants were incubated with 5 mM glucose containing 0.1 µCi/µmol of $(1-^{14}C)$-glucose and released $^{14}CO_2$ was trapped in hyamine hydroxide for scintillation counting.

Results for all complement-dependent assays are expressed as the difference between results obtained with particles opsonized in fresh versus heat-inactivated ($56^\circ C$ for 1 hour) human serum.

RESULTS

The typical pattern of response by phagocytic cells in the continuous spectrophotometric assay of cytochrome c reduction appears in Figure 1, curve A. After addition of the stimulus, there was an initial period with no change in A_{550}, then a gradual rise to a linear rate of change representing the activity of the superoxide generating system. As illustrated, the lag time (representing the activation process) was calculated from the intercept of the back-extrapolated linear portion of the curve with the pre-activation baseline of zero absorption change. Curve B (superoxide dismutase added) demonstrates the O_2^- dependence of the observed cytochrome c reduction, thus permitting the use of single-beam (i.e. no reference cuvette) assays for rates > 1 nmol O_2^-/min/10^6 cells. Curve C (stimulus omitted) shows the absence of measurable resting O_2^- generation. Similar curves were obtained with opsonized zymosan as the stimulant.

Normal human granulocytes. We examined the effects of physical and chemical perturbations on human granulocytes, and report here those changes that differentially affected the lag time and the rate of O_2^- production. The effect of temperature on PMA-stimulated O_2^- generation is illustrated in Figure 2. The rate reached a maximum at $37^\circ C$ and fell off sharply at higher and lower temperatures. In contrast, the lag time progressively shortened (i.e. activation proceeded with increasing rapidity) over the entire range.

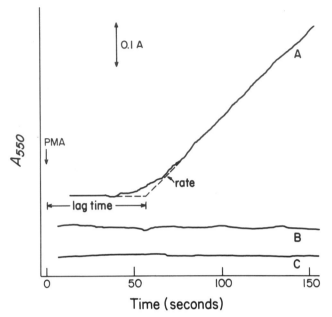

Fig. 1. Continuous assay for granulocyte O_2^- production. C: Granulocytes at
2.5×10^6/ml were incubated in 1 mol of KRP (pH 7.4) containing glucose (5 mM)
and ferricytochrome c (50 mM). B, A: In the presence (B) or absence (A) of
10 μg of superoxide dismutase, 1 μg of PMA was added at zero time (arrow).
Dashed lines illustrate the methods for deriving the rate of activity (slope of
the linear portion of the curve) and lag time for activation (back extrapolation
to zero absorbance change) of the O_2^- generating system.

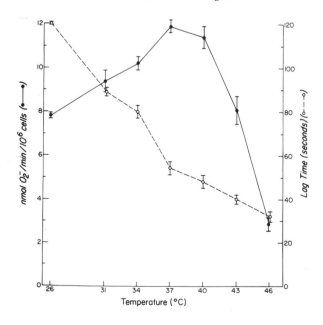

Fig. 2. The effect of
temperature on the rate
(●) and lag time (o) of
PMA-stimulated O_2^- production by human granu-
locytes. The temperature
for the standard assay
was varied as indicated
on the abscissa. Each
point and bar represents
the mean and S.D. of
triplicate determinations.

The addition of glucose to the assay mixture (Figure 3) enhanced the rate of O_2^- production in a dose-dependent manner but did not significantly alter the lag time. Other perturbations that affected the rate but not the lag time were variation in pH and inhibition by N-ethyl maleimide and 2-deoxyglucose.[5]

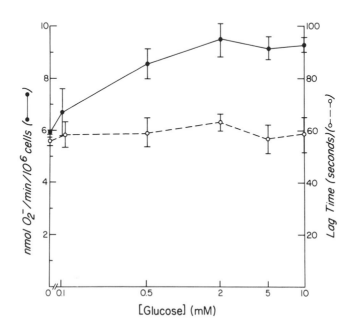

Fig. 3. The effect of glucose on the rate (●) and lag time (o) of PMA-stimulated O_2^- production by human granulocytes. Glucose at the concentrations indicated on the abcissa was added to the cytochrome c reduction assay mixture. Each point and bar represents the mean and S.D. of triplicate determinations.

Leukemia cell lines. As HL-60 cells differentiated during nine days of culture in 1.3% dimethylsulfoxide, they showed marked increases in their capacities for O_2^- generation, hexosemonophosphate shunt stimulation, ingestion, degranulation, and bacterial killing.[9] As shown in Figure 4, the rates of O_2^- production in response to PMA or opsonized zymosan increased 18-fold, to levels 1/2 to 2/3 those of normal granulocytes. Concurrently, the lag times shortened approximately 50%, achieving activation periods comparable to normal granulocytes.

Line 120 cells, which morphologically resemble myeloblasts, also generated O_2^- and increased hexosemonophosphate shunt activity in response to PMA (Table 1). As with immature HL-60 cells, the rate of O_2^- production was lower than that by PMN and the lag time much longer. However, the 120 cells failed to activate the respiratory burst in response to complement-opsonized particles. They were stimulated by non-opsonized particles such as latex beads. Line 120 cells have proven capable of ingestion of complement-opsonized oil particles and bacteria and of degranulation in response to complement-opsonized zymosan.[18]

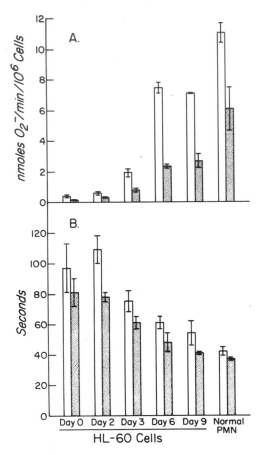

Fig. 4. Superoxide generation by HL-60 cells and normal PMN. HL-60 cells were cultured for the indicated times in 1.3% dimethyl-sulfoxide. (A) shows linear rates of O_2^- production. (B) shows the lag time for the activation of the O_2^- generation system. Cells were stimulated by PMA (clear bars) or opsonized zymosan (stippled bars). Bar heights represent means of triplicate determinations; error lines, SEM.

Nitroblue tetrazolium (NBT) slide tests, which permit examination of individual cells for ingestion and NBT reduction,[9,14] confirmed the defect in 120 cells. The ingested complement-opsonized zymosan, immunoglobulin-opsonized bacteria, and non-opsonized latex beads; but only the latter two showed associated NBT reduction. Normal PMN showed both ingestion and NBT reduction in response to all three.

TABLE 1

RESPIRATORY BURST ACTIVITY OF LINE 120 CELLS AND PMN IN RESPONSE TO SOLUBLE AND
PARTICULATE STIMULI

		O_2^- Generation		Hexosemonophosphate
		Rate[a]	Lag time[b]	shunt activity[c]
Line 120 cells:	PMA	0.786 ± 0.035	92 ± 1	620 ± 50
	OpZ[d]	<0.005	>500	2 ± 1
	Latex	ND[e]	ND	63 ± 10
PMN:	PMA	9.33 ± 0.49	54 ± 2	2270 ± 260
	OpZ	4.74 ± 0.43	33 ± 2	2140 ± 80
	Latex	ND	ND	734 ± 170

[a] nmol O_2^-/min/10^6 cells

[b] seconds

[c] cpm/30 min/10^6 cells

[d] OpZ = opsonized zymosan

[e] ND = not determined

DISCUSSION

The continuous assay for O_2^- generation permits the dissection of normal
granulocyte O_2^- production into separable processes of activation and activity,
measured respectively as the lag time and rate of cytochrome c reduction. The
two processes respond differently to manipulations involving temperature, energy
metabolism (glucose enhancement and 2-deoxyglucose inhibition), pH, and N-ethyl
maleimide inhibition. It appears that separate enzymatic processes activate
the O_2^- generating system and produce its activity (the reduction of O_2 to O_2^-).
These processes differ in temperature and pH maxima, dependence on ongoing
energy metabolism, and susceptibility to the sulfhydryl reagent N-ethyl maleimide.
These findings in human peripheral blood granulocytes differ markedly from those
previously reported in guinea pig peritoneal exudate granulocytes, in which
activation was far more sensitive to perturbation.[15]

The studies with the HL-60 cell line suggest that the process of myeloid
differentiation involves not only a quantitative increase in the amount of the
O_2^--generating oxidase per cell, but also a maturation of the activation process.
We do not know whether or not in vivo differentiation follows a similar course.
NBT slide testing has indicated semi-quantitatively that activity matures at
the metamyelocyte level[16] in normal human bone marrow but there are no data on
the development of the activation process.

The dissociation between complement-opsonized particle phagocytosis and
respiratory burst function in line 120 provides some insight into the triggering
of thr activation process. The respiratory burst of the cells in response to

PMN, latex, and immunoglobulin-opsonized particles demonstrates the presence of the O_2^- generating system. Their ability to ingest and degranulate in response to complement opsonized particles[18] (but not zymosan or oil particles incubated in heat-inactivated serum) shows that the previously demonstrated[11] line 120 cell complement receptor is present and connected to the cytoskeleton. However, the line 120 cells fail to activate O_2^- generation in response to complement-opsonized particles, even while ingesting them. Figure 5 illustrates a model for this disconnection. The complement receptor triggers O_2^- generation by a mechanism (blocked or absent in line 120 cells) separate from that membrane signal initiating ingestion and degranulation. Other stimuli activate O_2^- generation, ingestion, and degranulation <u>via</u> receptors independent of that for complement or by direct membrane stimulation. A selective defect in activation similar to that in line 120 cells has been reported in a patient with a clinical picture of chronic granulomatous disease.[17]

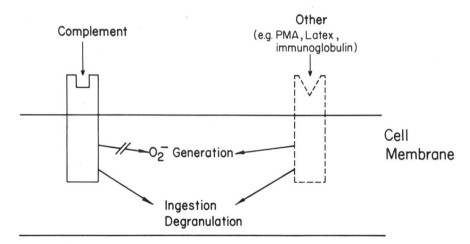

Fig. 5. A model for the triggering of O_2^- generation, ingestion, and degranulation. The interaction of complement-opsonized particles with a complement receptor results in activation of the O_2^- generating system (cross lines indicate the defect in line 120 cells) as well as ingestion and degranulation (intact in line 120 cells). Other stimuli interact with the cell membrane independently from the complement recpetor. The dashed lines indicate that they may do so <u>via</u> receptors or by direct action on the membrane.

Thus, normal granulocytes define the separable characteristics of the activation and activity of the O_2^- generating system, differentiating HL-60 cells show the development of the processes during myeloid maturation, and line 120 cells help to describe the relationship between membrane triggering and the activation process. Defects at any step could produce the clinical syndrome of chronic granulomatous disease.

ACKNOWLEDGEMENTS

Supported by a Basil O'Connor Starter Grant from the National Foundation-
March of Dimes and NIH grants CA26033 and CA26506. Dr. Newburger is an American
Cancer Society Junior Faculty Clinical Fellow; Dr. Cohen is supported by Research
Career Development Award 1K04-00311.

REFERENCES

1. Babior, B.M. (1978) N. Engl. J. Med. 298, 659-668.
2. Graham, R.C.,Jr,. Karnovsky, M.J., Shafer, A.W., Glass, E.A. and Karnovsky,
    M.L. (1967) J. Cell Biol. 32, 629-647.
3. Babior, B.M., Kipnis, R.S., Curnutte, J.T. (1973) J. Clin. Invest. 52,
    741-744.
4. Cohen, H.J. and Chovaniec, M.E. (1978) J. Clin. Invest. 61, 1081-1087.
5. Newburger, P.E., Chovaniec, M.E. and Cohen, H.J. (1978) Blood, in press.
6. Segal, A.W. and Coade, S.B. (1978) Biochem. Biophys. Res. Commun. 84,
    611-616.
7. Root, R.K., Metcalf, J., Oshino, N. and Chance, B. (1975) J. Clin. Invest.
    55, 945-955.
8. Collins, S.J., Ruscetti, F.W., Gallagher, R.E. and Gallo, R.C. (1978)
    Proc. Nat. Acad. Sci. USA, 75, 2458-2462.
9. Newburger, P.E., Chovaniec, M.E., Greenberger, J.W. and Cohen, H.J. (1979)
    J. Cell Biol. 82, 315-322.
10. Greenberger, J.S., Newburger, P.E., Karpas, A. and Moloney, W.C. (1978)
    Cancer Res. 38, 3340-3348.
11. Karpas, A., Hayhoe, F.G.J., Greenberger, J.S., Barker, C.R., Cawley, J.C.
    Lowenthal, R.M. and Moloney, W.C. (1977) Leuk. Res. 1, 35.
12. Massey, V. (1959) Biochim. Biophys. Acta, 34, 2555-256.
13. Stossel, T.P., Murad, F., Mason, R.J. and Vaughan, M. (1970) J. Biol. Chem.
    245, 6228-6234.
14. Newburger, P.E., Cohen, H.J. Rothchild, S.B., Hobbins, J.C., Malawista, S.E.
    and Mahoney, M.J. (1979) N. Engl. J. Med. 300, 178-181.
15. Cohen, H.J. and Chovaniec, M.E. (1978) J. Clin. Invest. 61, 1088-1096,
16. Zakhireh, B. and Root, R.K. (1979) Blood, 54, 429-439.
17. Harvath, L. and Anderson, B.R. (1979) N. Engl. J. Med. 300, 1130-1135.
18. Newburger, P.E., Pagano, J.S., Greenberger, J.S., Karpas, A. and Cohen, H.J.,
    in preparation.

LOCALIZATION AND ACTIVATION OF THE GUINEA PIG GRANULOCYTE NAD(P)H-DEPENDENT
SUPEROXIDE GENERATING ENZYME: KINETICS OF ACTIVATION AND EFFECT OF DETERGENTS

HARVEY J. COHEN, MARGARET E. CHOVANIEC AND WAYNE E. DAVIES
Department of Medicine, Division of Hematology and Oncology, Children's Hospital
Medical Center, Boston, Ma 02115, USA

INTRODUCTION

As a consequence of the interaction of phagocytic cells with appropriately
opsonized particles and certain soluble stimuli a series of metabolic changes
occur. These include an increase in oxygen consumption,[1] the stimulation of the
hexosemonophosphate shunt[2] and the generation of reactive oxygen species such
as superoxide (O_2^-), (Ref.3) hydrogen peroxide,[4] hydroxyl radical[5,6] and perhaps
singlet oxygen.[7,8] These metabolic events are related to the activation and
activity of an O_2^- generating system. We have previously reported that when the
soluble stimulant, digitonin, is added to guinea pig granulocytes (PMN), O_2^- is
produced.[10] Employing a continuous assay, we have demonstrated that the production of O_2^- commences after an initial lag of about 30 seconds followed by an
increasing rate until a linear rate occurs after 1-2 minutes. This rate continues for 5-10 minutes after which the rate decreases and finally O_2^- production
ceases. We investigated the effects of pH, temperature, N-ethyl maleimide,
divalent cations and inhibitors of glycolysis and oxidative phosphorylation on
both the rate of activation and the activity of PMN O_2^- production.[10,11]

It has been shown that human PMN stimulated by opsonized zymosan contain a
reduced pyridine nucleotide oxidase[12] that generates O_2^- (Ref.13). This enzyme
is found in a sedimentable fraction of PMN homogenates[12,13] and has recently been
localized to the plasma membrane.[14] The localization and nature of the O_2^-
generating system in guinea pig PMN is not known. It has been suggested that
a soluble NADH oxidase is responsible for O_2^- production by these cells.[15,16]
We report evidence that demonstrates that O_2^- production by guinea pig PMN is due
to an enzyme which can utilize either NADH or NADPH as a substrate. We utilized
a technique previously developed for obtaining macrophage plasma membrane-bound
vesicles - podosomes[17] to determine if O_2^- production occurs in association with
the plasma membrane. The O_2^- generating enzyme is enriched in these podosomes
that are equally enriched for plasma membrane markers. The onset of PMN O_2^-
production correlates with the activation of this enzyme and manipulations that
affect whole cell O_2^- production alter the specific activity of this enzyme.

We also report that the same O_2^- generating system is present in phagocytic vesicles (PV) obtained from these PMN. The apparent activity of the podosome but not PV enzyme is enhanced by low concentrations of detergents which presumably increase the permeability of the membrane to reduced pyridine nucleotides. We, therefore, feel that in the guinea pig PMN, O_2^- production is a result of the activation and activity of a NAD(P)H-dependent enzyme located on the inner surface of the plasma membrane which translocates to the PV with ingestion.

MATERIALS AND METHODS

Superoxide production by digitonin-stimulated PMN was quantitated by a previously described continuous spectrophotometric assay of superoxide dismutase inhibitable cytochrome c reduction.[10]

Podosomes were prepared by slight modification of a previously described procedure.[17] 1.0×10^8 PMN were incubated at 37°C in glass conical centrifuge tubes with or without 50 μg of digitonin in 5 ml of Krebs-Ringer's phosphate (KRP)(or KRP with 5 mM glucose). Except where indicated, the time of incubation was 3-4 minutes. After incubation the suspension was placed in a room temperature sonication bath for 6-8 seconds, then placed in an ice bath. The samples were centrifuged at 70 x g for 5 minutes in a Sorvall centrifuge Model RC-3 refrigerated at 4°C. The resulting cloudy supernatant was placed in 15 ml Corex no.844 centrifuge tubes and centrifuged at 12,000 x g for 10 minutes in a Sorvall centrifuge Model RC-2B refrigerated at 4°C. The resulting pellet was suspended in 0.6-1.0 ml of 0.34 M sucrose at 1-2 mg protein/ml. PV were made by the technique of Stossel et al.[18]

Superoxide production by podosomes and PV was assayed for by examining the rate of O_2^--dependent cytochrome c reduction in the presence of reduced pyridine nucleotides. In standard assays, cytochrome c (50 nmol), NAD(P)H (1.0 μmol) and NaCN (0.5 μmol) were added to both the sample and reference compartments of a double-beam spectrophotometer in a total volume of 0.9 ml. In addition, superoxide dismutase (20 μg) was added to the reference compartment. The assays were started with the addition of 0.1 ml of the podosomes or PV to both compartments and the absorbance change at 550 nm followed. Cyanide was necessary in this assay since there was contamination of the podosomes with some mitochondria and rexodiation of the cytochrome c occurred in the absence of cyanate. The use of cyanide required that the superoxide dismutase concentration to be 20 μg/ml to ensure complete scavenging of the O_2^- formed. The sensitivity of this assay is such that it will detect rates of O_2^- production as low as 0.05 nmol/min. NADPH oxidation was measured under similar conditions in the absence of cytochrome c and superoxide dismutase. The absorbance change at 340 nm was followed.

β-glucuronidase was assayed by measuring the formation of p-nitrophenol (A_{410}) from p-nitrophenyl β-D-glucuronide catalysed by PMN, homogenates, podosomes or supernatant fluid in the presence of 0.05% Triton X-100 (Ref.19). Lysozyme assays measured the change in light scattering (OD_{450}) resulting from the lysis of <u>Micrococcus</u> <u>lysodeikticus</u>[20] in the presence of 0.05% Triton X-100. The peroxidase assay was a previously described assay[21] modified by continuously monitoring the initial linear increase in A_{450} upon the addition of sample to a cuvette containing hydrogen peroxide (80 μmol), 3,3'-dimethyoxybenzidine (0.32 μmol) in 1.0 ml of 0.1 M citrate, pH 5.5, containing 0.05% Triton X-100. 5'-nucleotidase was assayed for by measuring phosphate release from AMP in the presence of tartrate, an inhibitor of nonspecific phosphatase as previously described.[22] Inorganic phosphate was assayed for by the method of Lowry and Lopez.[23]

Adenylate cyclase activity was assayed for by incubating 20 μl samples of podosomes or homogenates with 50 μl of substrate to a mixture of the following composition: ATP, 1 mM, containing 100,000 cpm of ^{14}C-labelled ATP (45 Ci/mol); $MgCl_2$, 5 mM; cAMP; 2 mM; theophylline, 10 mM; bovine serum albumin, 0.1%; NaF, 10 mM; Tris-HCl, 50 mM; pH 7.4. Incubation were carried out for 1 hour at $37^{\circ}C$ and the reaction terminated by the addition of 20 μl of EDTA, 0.1 M containing 3000 cpm of ^{3}H-cAMP (37.7 Ci/mol) which served as a means of calculating recovery of cAMP in subsequent steps. Assay tubes were placed into a boiling water bath for 2 minutes and coagulated protein pelleted at 1000 x g, 10 min. Separation of cAMP from its precursor and other labelled nucleotides was effected according to the method of White and Zenser,[24] using columns of neutral alumina and 50 ml Tris-Cl buffer, pH 7.4.

RESULTS

Superoxide generation in podosomes. Podosomes from resting and digitonin-stimulated PMN were examined for their ability to generate O_2^- in the presence of reduced pyridine nucleotides. Table 1 shows that using either NADPH or NADH very little O_2^- is formed from podosomes obtained from resting cells. Podosomes made from PMN stimulated for three to four minutes with 10 μg/ml digitonin produce O_2^- in the presence of either reduced pyridine nucleotide at rates that are about twenty times that of podosomes from resting cells. Podosomes in both instances contain the same amount of protein. In addition, the sodium/dodecyl/sulfate (SDS) gel-electrophoresis protein profile of podosomes in both 5% and 10% acrylamide was identical whether the podosomes were made from resting or activated cells. PV made from opsonized lipopolysaccharide particles also contain both an NADH- and NADPH-dependent O_2^- generating system. In both preparations the addition of both nucleotides results in no further increase in O_2^- production above that obtained for either alone.

Guinea pig PMN contain 5'-nucleotidase in their plasma membrane.[25] Adenylate cyclase is also present in these cells and has been used as a plasma membrane

TABLE 1

NAD(P)H-DEPENDENT SUPEROXIDE PRODUCTION BY PODOSOMES AND PHAGOCYTIC VESCILES

	nmol O_2^-/min/mg[a]		
	Podosomes		Phagocytic vesicles
	Resting	Stimulated	
NADPH (1 mM)	0.74 ± 0.29 (12)	13.6 ± 2.6 (12)	34.5 ± 5.8 (12)
NADH (1 mM)	0.54 ± 0.23 (12)	13.6 ± 2.9 (12)	27.8 ± 5.9 (10)
NADPH + NADH	–	14.2 ± 3.1 (4)	38.2 ± 5.7 (4)
0	–	0 (6)	0 (4)

[a]Number of experiments in parentheses

marker for other cells.[26] We determined whether the podosomes obtained were enriched for these enzymes and the O_2^- generating system. In addition we determined the relative enrichment of granule enzymes in these particles. There is a 3.1-fold enrichment in specific activity of NADPH-dependent O_2^- production in the podosomes. The same enrichment is also seen with NADH (data not showm). Enrichment for adenylate cyclase (3.2-fold) and 5'-nucleotidase (2.8-fold) are of the same magnitude. The granule enzymes, myeloperodixase, lysozyme and β-glucuronidase are enriched only 1.5-fold in these podosomes.

In other experiments, the mitochondrial enzyme succinate-cytochrome c reductase and the cytoplasmic enzyme glucose-6-phosphate dehydrogenase were found not to be enriched for in podosomes. In addition, as can be seen in Table 2, except for the O_2^- generating enzyme, the specific activtities of the other enzymes do not differ in podosomes made from either activated or resting cells. Thus, the only difference we have found in podosomes obtained from resting cells and activated cells is their ability to catalyze NAD(P)H-dependent O_2^- production. These podosomes are devoid of whole cells microscopically and are composed of small vesicles as previously shown for macrophages[17] Podosomes made from resting cells cannot be activated with digitonin even in the presence of KRP, ATP, and reduced pyridine nucleotides. In addition, in podosomes made from digitonin-treated cells, no O_2^- is generated in the absence of reduced pyridine nucleotides. Using either NADH or NADPH as the electron door, homogenates of activated granulocytes can generate O_2^- at about 40% the rate generated by whole cells. Podosome activity accounts for about 50% of the homogenate activity.

Substrate specificity. The affinity of the podosome and PV enzymes for reduced pyridine nucleotides is shown in Table 3. The apparent Michaelis constant (K_m^{APP}) for NADPH is one-tenth that for NADH but the maximum rates of O_2^- production are similar. The rate of O_2^- production is unaffected by the addition of both NADPH and NADH over that obtained for each separately at 1 mM concentrations of each. Thus, it appears that a single enzyme system is responsible for both NADPH- and NADH-dependent O_2^- production in both podosomes and PV. It has

TABLE 2

COMPARISON OF ENZYMATIC ACTIVITIES IN HOMOGENATES AND PODOSOMES[a]

	Homogenates Resting	Homogenates Activated	Podosomes Resting	Podosomes Activated	Activated Podosomes / Activated Homogenates	Activated Podosomes / Resting Podosomes
NADPH-dependent O_2^- generating enzyme (nmol O_2^- min/mg)	0.26	4.21	0.83	12.9	3.1	15.5
Adenylate cyclase (pmol/ min/mg)	–	363	–	1170	3.2	–
5'-nucleotidase (mg P/30 min/mg)	16.3	18.0	39.1	50.1	2.8	1.3
Myeloperoxidase (ΔA_{450}/ min/mg)	6.61	6.31	10.43	9.53	1.5	0.9
Lysozyme (ΔOD_{450}/min/ mg)	0.74	0.79	1.01	1.23	1.6	1.2
β-glucuronidase (units/ mg)	5.20	5.64	8.26	8.57	1.5	1.0

[a]Homogenates and podosomes were made as described in MATERIALS AND METHODS from PMN incubated at 37°C for 4 minutes in the presence (activated) and absence (resting) of 10 μg/ml digitonin. Assays for enzyme activity and protein were performed on samples of both homogenate and podosome fraction. One unit of β-glucuronidase equals A_{410} of 1.0 per 2 hour incubation.

been reported that activation of the NAD(P)H oxidase is due to a change in the affinity of the enzyme for its substrates.[27] We compared the K_m^{APP} for NADPH in podosomes made from resting and activated cells. Despite a ten-fold difference in the maximum velocity, the K_m^{APP} for NADPH in podosomes from resting cells was the same as that found in podosomes from activated cells. Thus, in guinea pig granulocytes, activation is not due to a change in affinity for NADPH.

TABLE 3

K_m^{APP} FOR REDUCED PYRIDINE NUCLEOTIDES AND pH OPTIMUM FOR PODOSOME AND PHAGO-CYTIC VESICLE ACTIVITIES

	Podosomes	Phagocytic Vesicles
K_m^{APP} NADPH	4.6×10^{-5}M	8.9×10^{-5}M
K_m^{APP} NADH	5.8×10^{-4}M	1.7×10^{-3}M
pH Optimum	7.5	7.5

Kinetics of activation. We have previously shown that there is a lag between the time of interaction of digitonin and the PMN, and the onset of the linear rate of O_2^- production. To show that this delay is related to activation of the O_2^- generating enzyme, podosomes were prepared at various times after the addition of digitonin to PMN. Figure 1 shows the results obtained when this interaction occurs at 25°C. The lag between interaction of digitonin and PMN and the

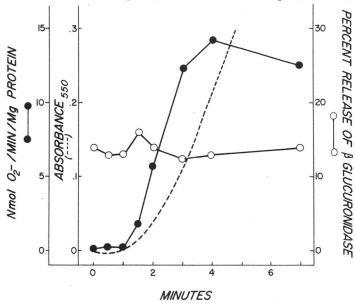

Fig. 1. Relationship between O_2^- production in PMN, NADPH-dependent O_2^- production in podosomes and release of β-glucuronidase in cells activated at 25°C with digitonin. PMN at 2 x 10^7/ml in KRP with glucose were incubated for various lengths of time at 25°C with digitonin (10 μg/ml) and podosomes were made. In addition the supernatant from the final centrifugation was saved. Podosomes were assayed for protein and NADPH-dependent O_2^- production (•) and the supernatants for β-glucuronidase (o). In addition PMN O_2^- production was assayed for continuously at 25°C as described in the Methods section (- - -).

onset of maximum O_2^- production was about three minutes as shown in the dashed line. The specific activity of the podosome NADPH-dependent O_2^- generating enzyme as shown in the solid circles remains low for one minute, then increases over the next two minutes and is maximal at three minutes. Thus, the lag before the onset of linear PMN O_2^- production is due to the time it takes to activate the O_2^- generating enzyme. Figure 1 also shows that at 25°C complete activation occurs prior to release of the granule enzyme β-glucuronidase into the extracellular medium (open circles). The time course for the activation and inactivation of the O_2^- generating enzyme and content of the granule enzyme β-glucuronidase in podosomes from PMN treated with digitonin at 37°C is shown in Figure 2. As has previously been shown for PMN O_2^- production,[10] the enzyme

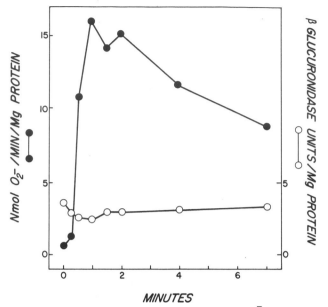

Fig. 2. Relationship between podosome NADPH-dependent O_2^- production and β-glucuronidase content. PMN were incubated in KRP at 37°C for varying lengths of time in the presence of digitonin. Podosomes were made and assayed for NADPH-dependent O_2^- production (●) and β-glucuronidase content (o).

activity increases to a maximum in about 90 seconds (solid circles). As is also true for whole cell O_2^- production the activity decreases after 4 to 7 minutes. During the time of both activation and inactivation there is no change in the granule content of the podosomes as monitored by the constant specific activity of β-glucuronidase (open circles). Similar results are also obtained when lysozyme or myeloperoxidase are used as granule enzyme markers. Thus, activation of the enzyme is independent of translocation of granules to the plasma membrane. Figure 3 shows a comparison of the kinetics of activation of the O_2^- generating enzyme by opsonized particles in the homogenate (squares) and PV (circles). Almost complete activation of the enzyme occurs within one minute. However, the maximum enzyme activity is found in the phagocytic vesicles after four to six minutes. This data is consistent with the interpretation that activation precedes formation of the PV.

Effect of temperature. We have previously reported that temperature affects both the time required for activation and the activity of the O_2^- generating system in PMN (Ref.10). Figure 4 shows the effect of the temperature at which the PMN are incubated on the activation of the podosome NADPH-dependent O_2^- generating enzyme. At 37°C (solid circles) complete activation occurs in $1\frac{1}{2}$ minutes. At 25°C (open circles) complete activation occurs in $3\frac{1}{2}$ minutes. In addition, the maximum specific activity of the enzyme in podosomes produced

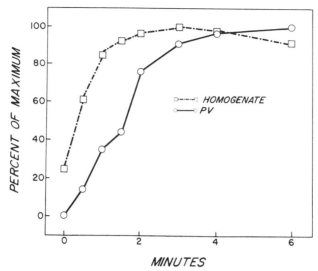

Fig. 3. Kinetics of activation of PV enzyme. Homogenates (□) and PV (o) were made after varying times of incubation of PMN with particles and assayed for NADPH-dependent O_2^- production.

from PMN activated at 37°C is about three times that for podosomes from PMN activated at 25°C. Thus, not only is the rate of activation dependent on temperature, the amount of enzyme activated is also affected by temperature.

Effect of cyanide, glucose and 2-deoxyglucose (2-DOG). The addition of either cyanide (0.5 mM) or glucose (5 mM) to PMN in the presence of digitonin results in an increased rate of O_2^- production without any change in the lag time.[11] The effects of cyanide and glucose are additive and together they double PMN O_2^- production. We determined the specific activity of the podosome NADPH-dependent O_2^- generating enzyme at various times after the addition of digitonin to PMN incubated with and without cyanide and glucose. The rate of activation of the podosomes enzyme is identical in PMN incubated with and without cyanide and glucose. The final specific activity, however, is increased two-fold in cyanide- and glucose-treated PMN. This indicates that the stimulation of PMN O_2^- production by cyanide and glucose is due to the increase in the amount of enzyme activated. Despite the two-fold change in V_{max} for the podosome activity there was no effect of PMN incubation with cyanate and glucose on the K_m^{APP} for either NADPH or NADH. Thus, the same enzyme is activated in the presence of cyanide and glucose.

It has previously been shown that PMN utilize glycolysis as an energy source to undergo phagocytosis.[2] We have shown that PMN incubated in 2-DOG undergo a time and concentration-dependent loss of both intracellular ATP and the ability to generate O_2^-. Cyanate or dinitrophenol when added to 2-DOG-treated

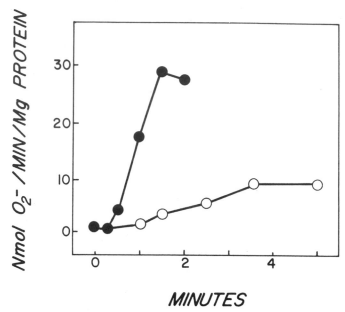

Fig. 4. Effect of temperature on the rate and extent of activation of podosome NADPH-dependent O_2^- production. PMN were incubated in KRP with glucose at 37°C (●) and 25°C (o) in the presence of digitonin for the times indicated. Podosomes were made and protein and NADPH-dependent O_2^- production were measured.

PMN result in a further decrease in digitonin-induced O_2^- generation.[11] To determine whether these changes in O_2^- generating ability were due to changes in the amount of enzyme activated, we examined the specific activity of the podosome enzyme made from cells incubated with these compounds and compared it to their effects on PMN O_2^- production. Table 4 shows that those compounds that stimulate O_2^- production in PMN (cyanide, glucose, dinitrophenol) cause an increase in specific activity of the podosome enzyme. Those compounds that inhibit digitonin-induced O_2^- production in PMN (2-DOG, 2-DOG + cyanide) cause a decrease in specific activity of this enzyme. None of the compounds affected the podosome content of 5'-nucleotidase indicating that the plasma membrane content of these particles was not altered by the additions. The addition of 2-DOG to podosomes had no effect on NADPH-dependent O_2^- production. Thus, these compounds that alter PMN O_2^- production, do so by causing changes in the amount of enzyme activated. In PV incubation of PMN in 5 mM 2-DOG results in a decrease of O_2^- generating enzyme primarily by causing a decrease in PV content.

Effect of detergents. As shown in Table 1, a higher specific activity of the NAD(P)H-dependent O_2^- generating system was seen with PV than with podosomes. In addition, the recovery of total O_2^- generating activity in podosomes was very low. Since podosomes are membrane-bound vesicles, we investigated the possibility

TABLE 4

EFFECTS OF 2-DOG, GLUCOSE, CYANIDE AND DINITROPHENOL ON SUPEROXIDE PRODUCTION BY GRANULOCYTES AMD PODOSOME NADPH-DEPENDENT SUPEROXIDE PRODUCTION[a]

	Granulocytes nmol/min/ 10^6 Cells	% Control	Podosomes nmol/min/mg Protein	% Control
1. -Digitonin	0	0	0.97	5.7
2. -	3.29	100	16.90	100
3. + 2-DOG (10 mM)	0.82	25	3.91	23
4. + Glucose (5 mM)	5.11	155	23.90	141
5. + Dinitrophenol (1 mM)	-	-	23.60	140
6. + Cyanide (0.5 mM)	4.69	143	25.40	150
7. + Glucose + cyanide	7.04	214	41.63	246
8. + 2-DOG + cyanide	0.10	2.9	0.16	0.9

[a]5 ml of PMN at 2×10^7/ml were incubated in KRP for 10 minutes with the additions (3-8) as noted above. 50λ were then removed and assayed for O_2^- production in the presence of the additions and digitonin (10 μg/ml) (except (1) which was assayed in the absence of digitonin). Digitonin (10 μg/ml) was added to samples 2-8 and all the samples incubated for 4 minutes. Podosomes were then made and NADPH-dependent O_2^- production was measured.

that the vesicles might be right-side-out with the active site of the enzyme for NAD(P)H on the inside of the membrane. We examined the effect of Triton X-100 and deoxycholate on NADPH-dependent O_2^- production in podosomes. As can be seen in Figure 5, a two- to three-fold increase in specific activity is observed in the presence of either Triton X-100 or deoxycholate. Similar results are obtained with NADH. That this effect is due to increased activity with respect to the reduced pyridine nucleotide was shown by examining the rate of NADPH oxidation by podosomes. Triton X-100 (0.02%) increased the rate of NADPH oxidation 2.7 fold. If this effect of Triton X-100 was on opening right-side-out vesicles to allow NADPH to enter them, then a prediction would be that there would be no stimulation of inside-out vesicles. PV are inside-out vesicles. Table 5 shows that in PV Triton X-100 fails to stimulate O_2^- production or NADPH oxidation. Deoxycholate was also ineffective in stimulating O_2^- production. Babior and Kipnes[28] have shown that in the human PMN 0.045% Triton X-100 causes a loss of NADPH-dependent O_2^- production and that this is prevented by the addition of flavin adenine dinucleotide (FAD). As can be seen in Table 5 100 μM FAD when added to PV stimulated NADPH-dependent O_2^- production over two-fold and protected against the Triton X-100 loss of activity.

DISCUSSION

The results presented show that O_2^- production by digitonin-stimulated guinea pig PMN is due to the activation of a reduced pyridine nucleotide-dependent enzyme activity. This enzyme utilizes either NADH or NADPH with the same maximal

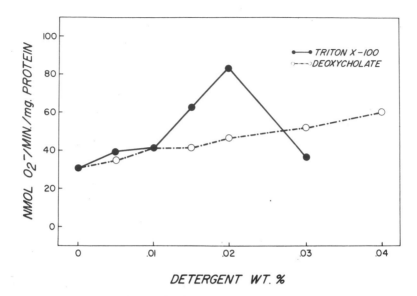

Fig. 5. Effect of detergents on podosome enzyme. Podosomes were assayed for NADPH-dependent O_2^- production in the presence of various concentrations of Triton X-100 (●) and deoxycholate (o).

TABLE 5

EFFECT OF TRITON X-100 ON PHAGOCYTIC VESICLE NADPH-DEPENDENT SUPEROXIDE PRO-DUCTION AND NADPH OXIDATION

| | Percent Control | |
Addition	O_2^- Production	NADPH Oxidation
----	(100)	(100)
0.01% Triton X-100	109	98
0.02% Triton X-100	52	42
0.04% Triton X-100	43	–
+ 100 μM FAD	(100) 236 [a]	
0.01% Triton X-100		
+ 100 μM FAD	85	
0.02% Triton X-100		
+ 100 μM FAD	86	
0.04% Triton X-100		
+ 100 μM FAD	119	

[a]Percent of activity in the absence of FAD

velocity but with a greater affinity for NADPH. This behavior is similar to the O_2^- generating system found in particles obtained from human PMN stimulated by opsonized zymosan.[13]

Podosomes made from PMN by a technique similar to that used for macrophages are enriched for plasma markers and less enriched for granule enzyme markers. The O_2^- generating enzyme is found in podosomes from digitonin-treated cells at an enrichment equal to that obtained for plasma membrane markers. We, therefore, conclude that this enzyme is in the plasma membrane in guinea pig PMN. Digitonin interaction with PMN results in an increase in NADPH-dependent O_2^- production without any change in podosome protein content, SDS gel pattern or specific activity of plasma membrane and granule enzyme markers. Thus, there is no translocation of granules into this fraction with activation. Activation can be separated from degranulation, since at $25^{\circ}C$ complete activation occurs prior to any demonstrable degranulation. The podosomes enzyme has characteristics necessary to account for whole cell O_2^- production. There is very little activity in podosomes made from resting cells, and a ten- to twenty-fold increase in activity when made from digitonin-treated cells. The time course for O_2^- production in PMN correlates with the activation of this enzyme. Compounds such as glucose, 2-DOG, cyanide and dinitrophenol which alter PMN O_2^- produce identical changes in the specific activity of the podosome enzyme made from cells incubated with these effectors. We cannot completely account for PMN O_2^- production by assaying cell-free homogenates. This may be due to the formation of both inside-out and right-side-out vesicles during homogenization or podosome formation. Since the active site for NAD(P)H is probably on the inside of the membrane and since membranes are relatively impermeable to NAD(P)H, the actual enzyme content in the homogenate and podosomes may be greater than measured. The results described using detergents show that the activity of the podosome enzyme can be markedly increased by the addition of low concentration of Triton X-100, whereas no stimulation is seen for the PV enzyme.

Temperature effects both the time necessary for activation and the activity of PMN O_2^- production.[10] As expected, the time course for activation of podosome NAD(P)H-dependent O_2^- production is similarly affected by temperature. Somewhat unexpected, however, was the finding that the specific activity of the podosome O_2^- generating enzyme obtained from cells activated at $25^{\circ}C$ was much lower than that for cells activated at $37^{\circ}C$. Thus, not only is the rate of activation affected by temperature, but the amount of enzyme activated is also changed.

We have presented data to indicate that digitonin-stimulated guinea pig PMN O_2^- production is due to activation of a plasma membrane NAD(P)H-dependent enzyme. Changes in the rate of activation on amount of this enzyme activated account for changes in whole cell O_2^- production. Since complement-opsonized stimulators activate the same O_2^- generating system in PMN,[10] we examined the effect of particulate stimulators on activation of the plasma membrane-bound enzyme and found that the same enzyme is activated and then translocated to PV. The results obtained using Triton X-100 suggest that the active site for NAD(P)H on this enzyme is located on the inner surface of the plasma membrane.

ACKNOWLEDGEMENTS

   This work was supported in part by US Public Health Service grant CA-26506,
and a Research Career Development Award from the National Institutes of Allergy
and Infectious Diseases AI-00311.

REFERENCES

1. Baldridge, C.S. and Gerard, R.W. (1933) Amer. J. Physiol. 103, 235-236.
2. Sbarra, A.J. and Karnovsky, M.L. (1959) J. Biol. Chem. 234, 1355-1362.
3. Babior, B.M., Kipnes, R.S. and Curnutte, J.T. (1973) J. Clin. Invest. 52,
    741-744.
4. Iyer, G.Y.N., Islam, M.F. and Quastel, J.H. (1961) Nature, 192, 535-541.
5. Weiss, S.J., King, G.W. and LoBuglio, A.F. (1977) J. Clin. Invest. 60,
    370-373.
6. Tauber, A.I. and Babior, B.M. (1977) J. Clin. Invest. 60, 374-379.
7. Krinsky, N.I. (1974) Science, 186, 363-365.
8. Rosen, H. and Klebanoff, S.J. (1977) J. Biol. Chem. 252, 4803-4810.
9. Babior, B.M., Curnutte, J.T. and Kipnes, R.S. (1975) J. Clin. Invest. 56,
    1035-1042.
10. Cohen, H.J. and Chovaniec, M.E. (1978) J. Clin. Invest. 61, 1081-1087.
11. Cohen, H.J. and Chovaniec, M.E. (1978) J. Clin. Invest. 61, 1088-1096.
12. Hohn, D.C. and Lehrer, R.I. (1975) J. Clin. Invest. 55, 707-713.
13. Babior, B.M., Curnutte, J.T., and McMurrich, B.J. (1976) J. Clin. Invest.
    58, 989-996.
14. DeWald, B., Baggiolini, M., Curnutte, J.T. and Babior, B.M. (1979) J. Clin.
    Invest. 63, 21-29.
15. Karnovsky, M.L. (1976) in Membrane Structure and Function of Human Blood
    Cells, Vyas, N., ed., American Association of Blood Banks, Washington, D.C.,
    pp.97-100.
16. Badwey, J.A., Curnutte, J.T., Karnovsky, M.L. (1979) N. Engl. J. Med. 300,
    1157-1160.
17. Davies, W.A. and Stossel, T.P. (1977) J. Cell. Biol. 75, 941-955.
18. Stossel, T.P., Pollard, T.D., Mason, R.J. and Vaughan, M.J. (1971) J. Clin.
    Invest. 50, 1745-1757.
19. Mitchell, R.H., Karnovsky, M.J. and Karnovsky, M.L. (1970) Biochem. J. 116,
    207-216.
20. Shugar, D. (1952) Biochim. Biophys. Acta, 8, 302-309.
21. Bretz, U. and Baggiolini, M. (1974) J. Cell. Biol. 63, 251-269.
22. Mitchell, R.H. and Hawthorne, J.N. (1965) Biochem. Biophys. Res. Commun.
    21, 333-338.
23. Lowry, O.H. and Lopez, J.A. (1946) J. Biol. Chem. 162, 421-428.
24. White, A.A. and Zenser, T.V. (1971) Anal. Biochem. 41, 372-396.
25. DePierre, J.W. and Karnovsky, M.L. (1874) J. Biol. Chem. 249, 7111-7120.
26. Rat, T.K. and Forte, J.G. (1974) Biochim. Biophys. Acta, 363, 320-339.
27. Patriarca, P., Cramer, R., Moncalvo, S., Rossi, F. and Romeo, D. Arch.
    Biochem. Biophys. 145, 255-262.
28. Babior, B.M. and Kipnes, R.S. (1977) Blood, 50, 517-524.

SUBCELLULAR LOCALIZATION OF THE ENZYME RESPOSIBLE FOR THE RESPIRATORY BURST
IN RESTING AND PHORBOL MYRISTATE ACETATE ACTIVATED LEUCOCYTES

F. ROSSI[+], P. PATRIARCA[++], G. BERTON[+] AND G. DE NICOLA[+]
+ Istituto di Patologia Generale, Università di Padova, Sede di Verona, Italy;
++ Istituto di Patologia Generale, Università di Trieste, Italy

One of the distinctive properties of polymorphonuclear leucocytes is to undergo a marked activation of their oxidative metabolism upon exposure to appropriate
stimuli. There is a general agreement that the stimulation of the oxidative
metabolism is linked to the activation of an enzymatic system that oxidizes NADPH
with formation of O_2^- and of H_2O_2 (Refs.1-7).

Until 1974 it was almost a general belief that the NADPH oxidizing system
was located intracellularly. The direct evidence supporting this view was obtained in our laboratory when it was demonstrated, with rate zonal sedimentation
experiments, that the oxidase of rabbit granulocytes at rest co-sedimented with
granules containing myeloperoxidase.[8] After 1974 a new trend, based on indirect
evidence, began to emerge which favoured a plasma membrane localization.[9,10]

In the last few years many attempts have been made in different laboratories
to show directly the localization of the oxidase, by using various methods of
fractionation of cell homogenates.[8,11-20] The results of these studies (Table 1)
do not agree. It is worthy of note that the experiments performed in different
laboratories are not similar with respect to the method of cell fractionation,
the functional state of the cells used and the method of assay of the NADPH
oxidase activity. Some authors have fractionated homogenates of cells in the
resting state[8,11,13-16,18] others have used leucocytes stimulated with phago-
cytosable or non-phagocytosable agents.[7,12,16-20] Furthermore, the NADPH oxidative activity has been assayed as superoxide anion production[11,14,17,19,20]
or as NADPH oxidation[14,16,18] or as oxygen consumption.[8,12,15] The disagreement
of the results obtained might depend on these methodological differences.

In order to clarify the matter we have repeated the fractionation experiments
following the experimental design described in Figure 1. Guinea pig granulocytes
in both resting and activated state have been used. The NADPH oxidase activity
has been assayed as oxygen consumption or as O_2^- production.

The stimulatory agent phorbol myristate acetate (PMA) has been used due to
its strong effect on the activation of the oxidative metabolism of granulocytes
(Figure 2). Table 2 shows that the NADPH oxidase activity, measured either as

TABLE 1

SUMMARY OF THE AVAILABLE DATA ON THE SUBCELLULAR LOCALIZATION OF THE ENZYMATIC SYSTEM RESPONSIBLE FOR THE METABOLIC BURST OF NEUTROPHILIC GRANULOCYTES

Authors and References	Animal species	Fractionation technique	Oxidase assay	Stimulatory agent	Subcellular localization
Patriarca et al. (8)	Rabbit	Zonal rate sedimentation	O_2 uptake	None	Azurophilic granules
Baehner (11)	Human	Differential centrifugation and isopycnic equilibration	NAD(P)H–NBT reductase	None	Microsomes or plasmalemmal membrane
Rossi et al (12)	Guinea pig	Discontinuous density gradient	O_2 uptake	Paraffin oil particles	Granules
Segal and Peters (13, 14)	Human	Isopycnic equilibration	$NAD(P)^+$ formed $NAD(P)H$-cyt.c- and NBT-reductase	None	Azurophilic granules Plasma membrane
Iverson et al. (7)	Human	Isopycnic equilibration	$NADP^+$ formation	STZ[a]	Particles of higher density than azurophilic granules
Patriarca et al. (15)	Human	Isopycnic equilibration	O_2 uptake	None	Azurophilic granules
Iverson et al. (16)	Human	Isopycnic equilibration and zonal rate sedimentation	$NADP^+$ formation	None and STZ[a]	Particles of higher density azurophilic granules
Cohen et al. (17)	Guinea pig	Podosomes (differential centrifugation)	O_2^- production	Digitonin	Plasma membrane
Auclair et al. (18)	Human	Differential centrifugation	$NADP^+$ formation	None and STZ[a]	Cytosol, and both heavy and low density particles
Tauber and Goetzl (19)	Human	Discontinuous density gradient	O_2^- production	STZ[a]	Plasma membrane and azurophilic granules
Dewald et al. (20)	Human	Zonal rate sedimentation	O_2^- production	PMA[b]	Plasma membrane

[a] STZ = serum treated zymosan

[b] PMA = phorbol myristate acetate

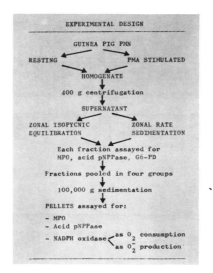

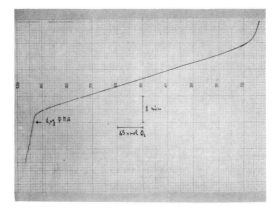

Fig. 2. Polarographic recording of the oxygen consumption by guinea pig granulocytes stimulated by phorbol myristate acetate. Oxygen consumption was monitored by a Clark-type oxygen electrode attached to a plastic vessel. The assay
medium contained: 30×10^6 cells in Krebs Ringer phosphate buffer pH 7.4, 0.5 mM
$CaCl_2$ and 5 mM glucose. Final volume 2 ml. At the arrow 1 µg of phorbol
myristate acetate in dimethyl sulphoxide (1 mg/ml) was added.

oxygen consumption or as superoxide anion production is markedly activated in the
100,00 g pellet of postnuclear supernatants of homogenates from PMA-treated cells,
indicating that the activation state of the enzyme is maintained after cell
homogenization.

TABLE 2

SPECIFIC ACTIVITY OF P-NITROPHENYL PHOSPHATASE (pNPPase), MYELOPEROXIDASE (MPO) AND NADH OXIDASE IN THE 100,000g PELLETS OF THE POSTNUCLEAR SUPERNATANT OF RESTING AND PMA-TREATED LEUCOCYTES

	Resting cells[a]	PMA-treated cells[a]
acid pNPPase (μmoles p-nitrophenol/10 min)	5.5 ± 0.7 (6)	5.7 ± 0.9 (6)
MPO (μmoles tetraguaiacol/min)	1.9 ± 0.2 (6)	2.6 ± 0.9 (4)
NADPH oxidase O_2 consumption (nmoles/min)	61.3 ± 16.5 (5)	693.6 ± 140 (5)
O_2^- production (nmoles/min)	1.2 ± 0.2 (4)	43.2 ± 8.4 (7)

[a]Cells (15×10^6/ml) were suspended in Krebs-Ringer phosphate buffer (KRP) pH 7.4 containing 0.5 mM $CaCl_2$ and 5 mM glucose. After equilibration at 37°C 0.5 μg/ml of PMA dissolved in dimethyl sulphoxide were added. After 2 min of incubation at 37°C under continuous stirring the cells were diluted with cold KRP and centrifuged at 250 g for 7 min. The cell pellets were resuspended in 0.34 M sucrose containing 1 mM $NaHCO_3$ (40×10^7/ml) and were homogenized with a Potter-type homogenizer equipped with a motor driven teflon pestle. For experiments with resting cells the cell suspension was incubated with dimethyl sulphoxide and processed as described above for PMA-treated cells. The homogenates were seven fold diluted with 0.34 M sucrose containing 1 mM $NaHCO_3$ and centrifuged at 400 g for 5 min to sediment nuclei, cell debris and intact cells. Postnuclear supernatants were centrifuged at 100,000 g for 45 min and the pellets were resuspended in 0.34 M sucrose containing 1 mM $NaHCO_3$. Acid pNPPase and MPO were assayed as previously described (Refs.21,22,8). O_2 consumption was assayed with a Clark-type oxygen electrode connected with a plastic vessel under continuous stirring. Final volume 2 ml. The assay medium contained: 0.065 M Na^+/K^+ phosphate buffer pH 5.5, 0.17 M sucrose, 1 mM NADPH, 2 mM NaN_3. T=37°C. O_2^- production was assayed by monitoring the reduction of cytochrome c at 550 nm. The assay medium contained 0.065 M Na^+/K^+ phosphate buffer, 0.17 M sucrose, 0.15 mM NADPH, 2 mM NaN_3, 150 μM cytochrome c and when required, 30 μg of superoxide dismutase. Final volume 1 ml. Temperature = 37°C. The O_2^- production was calculated by the difference between the reduction of cytochrome c in absence and in presence of superoxide dismutase. The values are means ± SEM. The number of experiments is shown in parentheses.

Figure 3 shows the distribution of glucose-6-phosphate dehydrogenase (G6PD), a cytosolic enzyme, of acid p-nitrophenyl phosphatase (acid pNPPase) a marker of the plasma membrane in rabbit granulocytes,[21,22] and of myeloperoxidase (MPO), a marker of the azurophilic granules, among the fractions obtained from isopycnic equilibration of the postnuclear supernatant of PMA-treated cells in a continuous sucrose density gradient. The NADPH oxidative activity could not be assayed with accuracy on single fractions due to dilution of the enzyme in the large volume of each fraction. Therefore selected fractions were pooled in four groups (designated as A,B,C and D in Figure 3) and were centrifuged at 100,000 g for 30 min. The pellets were assayed for NADPH oxidase actvity, and again for MPO and acid pNPPase. Table 3 shows that most of the MPO activity is found in the heavier part of the gradient, whereas acid pNPPase, although more widely distributed, is mainly recovered in the lighter zones of the gradient. The distribution of the NADPH oxidase activity, assayed either as oxygen consumption or as superoxide anion production, clearly differs from that of MPO and resembles that of acid pNPPase.

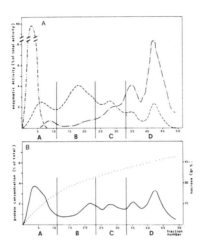

Fig. 3. Isopycnic zonal equilibration of postnuclear supernatants from PMA-treated leucocytes in a sucrose density gradient. The percent distribution of the activities of G6PD (--·--·--··), acid pNPPase (------), MPO (--·--·--·-)(panel A) and of protein (————) and sucrose concentration (·········)(panel B) among the fractions collected from the gradient is shown. Isopycnic density gradeint centrifugation was performed in a A1-14 rotor in a Spinco L-50 ultracentrifuge. 50 ml of postnuclear supernatants were sandwiched between a 20 ml overlayer of 1 mM NaHCO₃ and a 465 ml sucrose gradeint extending linearly with respect to radius with increments of 0.5 cm between density of 1.081 and 1.230. The gradient rested on a 115 ml cushion of density 1.257. Centrifugation was carried out at 35,000 rpm for 90 min. The rotor was filled and emptied at a rate of 20 ml/min while rotating at 2,000 rpm. Sucrose of density 1.304 was used to displace the gradient. 10 ml fractions were collected. On the basis of the distribution of the enzyme activities and of protein concentration, fractions were pooled in four groups (designated A,B,C and D). The means of four experiments are reported. G6PD was assayed as previously described (Ref.8). For the other enzymatic assays see Table 2. Protein concentration was determined as absorbance at 280 nm. Sucrose concentration was determined with an Abbe refractometer.

These results differ from those obtained in our and other laboratories with human and rabbit polymorphs at rest.[8,15,16] However, the recovery of the NADPH oxidase activity in the four zones of the gradient was low with respect to the starting material as shown in Table 3. This inconvenience might well account for the different distribution of the oxidase in PMA-stimulated cells with respect to that previously reported in resting cells. Experiments not reported

TABLE 3

AVERAGE PERCENT DISTRIBUTION OF ACID pNPPase, MPO AND NADPH OXIDASE AFTER ISOPYCNIC EQUILIBRATION OF PMA-TREATED LEUCOCYTES[a]

Gradient zones	Sucrose density range (g/cm^3)	Acid pNPPase	MPO	NADPH oxidase	
				O_2 consumption	O_2^- production
A	1.00 - 1.10	11.9 ± 0.7	3.9 ± 2.2	17.2 ± 4.9	16.3 ± 2.4
B	1.10 - 1.15	37.6 ± 5.9	6.7 ± 2.1	30.2 ± 9.0	32.3 ± 11.6
C	1.15 - 1.19	28.0 ± 4.5	19.2 ± 4.2	33.3 ± 7.5	38.0 ± 9.3
D	1.19 - 1.24	22.5 ± 5.4	70.2 ± 6.9	19.3 ± 2.6	13.4 ± 1.9
Recovery		108 ± 6.1	100 ± 13.3	45 ± 6.8	41 ± 9.5

[a]The four groups of fractions, pooled as described in Figure 1, were diluted with 1 mM NAHCO$_3$ to a final sucrose density of 1.04 and then centrifuged at 100,000 g for 45 min. The pellest were resuspended in 0.34 M sucrose containing 1 mM NaHCO$_3$. For the assays of acid pNPPase, MPO and NADPH oxidase (see Table 2). The results are means of four experiments ± SEM. The recovery was calculated with respect to the 100,000 g pellets of the postnuclear supernatant.

here indicated that the long centrifugation time required to achieve isopycnic equilibration of the particles was responsible for inactivation of the enzyme. Therefore postnuclear supernatants of both resting and PMA-treated guinea pig granulocytes were fractionated by using rate zonal sedimentation technique for 15 minutes and lower sucrose concentrations.

Table 4 shows the data of the percent distribution of acid pNPPase, of MPO and of NADPH oxidase in the pellets of four groups of fractions obtained as already described above for the isopycnic equilibration. It can be seen that 1) the recoveries of the NADPH oxidase activity are considerably improved in these experiments as compared with the isopycnic fractionation; 2) the distribution of MPO and acid pNPPase in subcellular particles of resting and PMA-treated cells is similar, that is most of MPO activity is found in the heavier zones of the gradient and most of the pNPPase activity in the lighter zones; 3) the distribution of NADPH oxidase in subcellular particles of resting and PMA-treated cells is different since in resting cells it follows the distribution of MPO, while in PMA-treated cells the bulk of the oxidase activity is associated with fractions lighter than those containing MPO.

The electron microscopic examination of the pellets of the four groups of fractions obtained with both methods of fractionation was similar in resting and PMA-treated cells. Briefly, groups A and B were mainly composed of empty vesicles of different sizes, while groups C and D were more heterogeneous. The former consisted of vesicles, small granules, mitochondria and amorphous material, the latter of large and medium sized electron dense granules, pieces of membranes and material which may have originated from disrupted nuclei.

TABLE 4

AVERAGE PERCENT DISTRIBUTION OF ACID pNPPase, MPO AND NADPH OXIDASE AFTER RATE ZONAL CENTRIFUGATION OF RESTING AND PMA-TREATED LEUCOCYTES[a]

Gradient zones	Sucrose density range (g/cm^3)	Acid pNPPase	MPO	NADPH oxidase	
				O_2 consumed	O_2^- produced
Resting Cells					
A	1.04 - 1.06	25	5	3	11
B	1.06 - 1.08	35	14	16	19
C	1.08 - 1.12	23	27	34	24
D	1.12 - 1.27	17	54	47	46
Recovery		84	60	89	95
PMA-treated cells					
A	1.04 - 1.06	31	6	26	25
B	1.06 - 1.08	23	8	35	43
C	1.08 - 1.12	16	18	16	18
D	1.12 - 1.27	30	68	23	14
Recovery		87	65	91	70

[a]The fractions were pooled according to the same criteria described for isopycnic equilibration experiments in Figure 3. The values are means of two experiments. The recovery was calculated with respect to the 100,000 g pellets of the post-nuclear supernatant.

Several explanations can be advanced to explain the different subcellular localization of NADPH oxidase in resting and in stimulated leucocytes. Two of these will be briefly mentioned. The most likely is that in resting cells only the oxidase activity of MPO, which is located in the azurophilic granules, is measurable, while the true NADPH oxidase, present in an active form in the plasma membrane or in other intracellular membranes of resting cells, becomes manifest only upon cell activation. The second possibility is that the activated NADPH oxidase activity, found in the fraction consisting of membranes, is contributed by an enzyme associated with the granule membrane which is translocated to the plasma membrane during the process of fusion induced by the stimulating agent.

ACKNOWLEDGEMENTS

This work is supported by grant No.78.02258.04 from the National Research Council of Italy (CNR).

REFERENCES

1. Tyer, G.J.N., Islam, M.F. and Quastel, J.M. (1961), Nature, 192, 535.
2. Rossi, F. and Zatti, M. (1964) Experienta, 20, 21-23.
3. Patriarca, P., Cramer, R., Moncalvo, S., Rossi, F. and Romeo, D. (1971) Arch. Biochem. Biophys. 145, 255-262.

4. Babior, B.M., Kipnes, R.S. and Curnutte, J.T. (1973) J. Clin. Invest. 52, 741.744.
5. Patriarca, P., Dri, P., Kakinuma, K., Tedesco, F. and Rossi, F. (1975) Biochim. Biophys. Acta, 385, 380-386.
6. Hohn, D.C. and Lehrer, R.I. (1975) J. Clin. Invest. 55, 707-713.
7. Iverson, D., De Chatelet, L.R., Spitznagel, J.K. and Wang, P. (1977) J. Clin. Invest. 59, 282-290.
8. Patriarca, P., Cramer, R., Dri, P., Fant, L., Basford, R.E. and Rossi, F. (1973) Biochem. Biophys. Res. Commun. 53, 830-837.
9. Goldstein, I.M., Roos, D., Kaplan, H.B. and Weissmann, G. (1975) J. Clin. Invest. 56, 1155-1163.
10. Roos, D., Homan-Müller, J.W.T. and Weening, R.S. (1976) Biochem. Biophys. Res. Commun. 68, 43-50.
11. Baehner, R.L. (1975) J. Lab. Clin. Med. 86, 785-792.
12. Rossi, F., Patriarca, P., Romeo, D. and Zabucchi, G. (1976) in The Reticu-loendothelial System in Health and Disease: Functions and Characteristics, Reichard, S.M., Escobar, M.R. and Friedman, H., eds., Plenum Press, New York, p.205.
13. Segal, A.W. and Peters, T.J. (1976) Lancet, i, 1363.
14. Segal, A.W. and Peters, T.J. (1977) Clin. Sci. Mol. Med. 52, 429-442.
15. Patriarca, P., Cramer, R. and Dri, P. (1977) in Movement, Metabolism and Bactericidal Mechanisms of Phagocytes, Rossi, F., Patriarca, P. and Romeo, D., eds., Piccin Medical Books, Padua, p.167
16. Iverson, D.B., Wang-Iverson, P., Spitznagel, J.K. and De Chatelet, L.R. (1978) Biochem. J. 176, 175-178.
17. Cohen, H.J., Chovaniec, M.E. and Davies, W. (1978) Fed. Proc. 37, 1276.
18. Auclair, C., Tones, M., Hakim, J. and Troube, H. (1978) Amer. J. Hematol. 4, 113-120.
19. Tauber, A.I. and Goetzl, E.J. (1978) Blood, 52, Suppl. 1, 128.
20. Dewald, B., Baggiolini, M., Curnutte, J.T. and Babior, B.M. (1979) J. Clin. Invest. 63, 21-29.
21. Baggiolini, M., Hirsch, J.G. and De Duve, C. (1969) J. Cell. Biol. 40, 529-541.
22. Baggiolini, M., Hirsch, J.G. and De Duve, C. (1970) J. Cell. Biol. 45, 586-597.

SPIN TRAPPING OF SUPEROXIDE AND HYDROXYL RADICALS PRODUCED BY STIMULATED
HUMAN NEUTROPHILS

MONIKA J. OKOLOW-ZUBKOWSKA AND H. ALLEN O. HILL
Inorganic Chemistry Department, University of Oxford, Oxford OX1 3QR,
England

INTRODUCTION

Neutrophilic polymorphonuclear leukocytes (PMNs) comprise the largest proportion
of leukocytes in the human body and provide the primary defence against bacterial
infection.

Phagocytosis of invading micro-organisms and host debris is associated with a
change in the oxidative metabolism of the cell, the so-called respiratory burst.
It was originally thought that the purpose of the increase in dioxygen consumption
was to provide energy for phagocytosis, but Sbarra and Karnovsky[1] showed that it
was not inhibited by mitochondrial inhibitors such as azide or cyanide. Furthermore
they showed that particle uptake could occur under nitrogen. Iyer et al.[2] found
that stimulated neutrophils produce hydrogen peroxide which was detected in the
medium surrounding the stimulated phagocytes. They postulated that the hydrogen
peroxide was used by the phagocyte as a bactericidal agent and suggested a connection
between the respiratory burst and the microbicidal mechanism of the phagocyte.

More recently several groups[3-6] have shown that stimulated phagocytes reduce
ferricytochrome c[7] or convert nitroblue tetrazolium (NBT) to formazan in the surrounding medium, both processes being inhibited by superoxide dismutase. Stimulated
cells have also been shown to convert methional to ethylene[8], an oxidation thought
to be mediated by hydroxyl radicals.[9] The reduction of NBT and the oxidation of
methional as assays for superoxide and hydroxyl radicals, respectively, are not
specific to these radicals and may lead to mistaken conclusions.[10]

The identity, location and physiological electron donor of the enzyme responsible
for the primary respiratory burst have been controversial issues for a long time.
The origin of the respiratory burst is now believed to be the activation of a flavo-
enzyme which is dormant in resting cells and catalyses the one electron reduction
of dioxygen to superoxide using NADPH as the electron donor.[11] Superoxide can in
turn undergo further reactions to yield hydrogen peroxide and hydroxyl radicals,
these reduced species of dioxygen being used in microbicidal activity.

A more direct assay for superoxide and hydroxyl radicals is the technique of
'spin trapping'[12,13] whereby short lived radical species react with 'spin traps'

to yield relatively long lived adducts which may be characterised by electron paramagnetic spectroscopy (epr). In favourable circumstances it is possible to identify
the trapped radicals from the epr spectra of the adducts. Using this technique
we have previously shown[14] that stimulated human neutrophils produce both superoxide
and hydroxyl radicals when stimulated with phorbol myristate acetate (PMA), a compound which causes the cells to release the contents of their specific granules;
if latex beads opsonised with human IgG were used as a stimulant only hydroxyl
radicals were detected.

In this communication data are presented to relate the radicals produced to
the increase in oxygen consumption of the respiratory burst. The effects of superoxide dismutase, ceruloplasmin and an anion channel blocker have been investigated,
and the location of the trapped radicals studied.

MATERIALS AND METHODS

Venous blood was collected into heparinised plastic syringes from normal adults.
Erythrocytes were sedimented with 1% Dextran 500 (Pharmacia). Neutrophils were
then separated from the leukocyte rich plasma by density centrifugation[15] on ficoll-
hypaque (Pharmacia) for 20 minutes at 400 g. Contaminating erythrocytes were removed by hypotonic lysis and the neutrophils resuspended in RPMI 1640 (Flow Laboratories) at a concentration of 2×10^7 cells/ml containing 97% PMNs.

Oxygen consumption was measured polarographically, before and after stimulation,
with a Clark electrode at 37°C (Yellow Springs Instrument Company). Cells suspended
in RPMI 1640 contained DMPO (100 mmol/l) and diethyltriaminepentaacetic acid (DTPA)
(1 mmol/l) were incubated in the chamber and a steady basal rate of respiration
recorded. The stimulant was then added and oxygen uptake monitored continuously
until a plateau was reached (Figure 1).

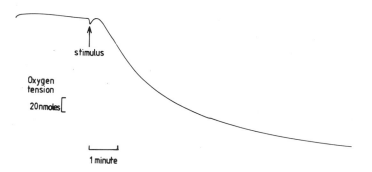

Fig. 1. Polarographic measurement of oxygen consumption by human neutrophils
(3 x 10^7 cells) after the addition of latex particles (100 per cell) opsonised
with human IgG.

Latex particles (Sigma, average diameter 0.8 μ) were opsonised with human IgG (Lister Institute),[16] zymosan was opsonised with human serum.[17] Phorbol myristate acetate (Sigma) was dissolved in dimethyl sulphoxide at a concentration of 1.0 mg/ml and used at a concentration of 10 μg/ml in the reaction mixture.

Cells after stimulation with latex IgG in the presence of DMPO were removed from the cell chamber, and poured into ice cold Hanks balanced salt solution, and centrifuged at 400 g for 5 minutes at 4°C. The supernatant and the pellet were examined by epr (see below). Lysis of the cells was achieved with 2% Triton X-100 (British Drug Houses).

Epr spectra were recorded by means of a Varian E-109 spectrometer with a 100 KHz field modulation in the first derivative mode. Samples of the incubation mixture were withdrawn from the oxygen electrode chamber into a Varian quartz flat cell by a simple syringe technique.[18] To monitor the signal height continuously, the magnetic field sweep was set to zero and the magnetic field adjusted to the resonance position of the peak under study. Signal intensities were calibrated using 4-hydroxy-2,2,6,6, tetramethyl piperidine (Aldrich Chemical Co.). A nitrone spin trap, 5,5 dimethyl-1-pyrroline-N-oxide (DMPO) was used (Figure 2; 1). This has the advantage that the β-hydrogen hyperfine splitting is quite sensitive to the nature of the trapped radical. However a cautionary note should be added that although the hydroxyl adduct of DMPO (Figure 2; II) is relatively long lived with a half life at physiological pH of several minutes in clean systems, the hydroperoxide adduct (Figure 2, III) at pH 7.5 has a half life of less than 45 seconds[19] which may be considerably reduced in a biological milieu; thus only a steady concentration can be detected.

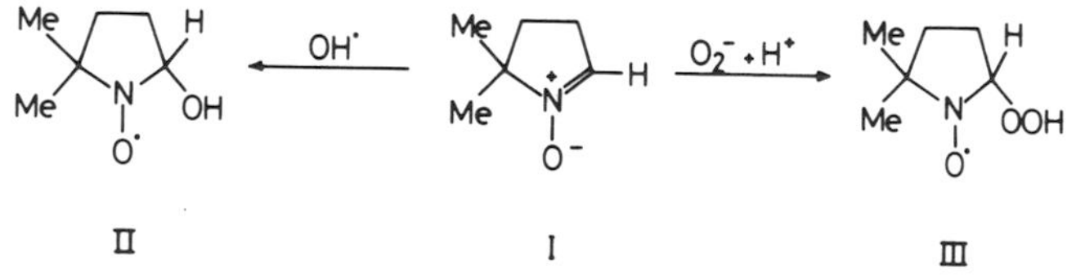

Fig.2. Structure of the spin trap DMPO and its hydroxyl and superoxide radical adducts

DMPO was prepared by the method of Bonnet et al.,[20] diluted with water and further purified.[19] Ceruloplasmin (Sigma) was further purified to an absorbance ratio at 610:280 nm of 0.034 (Ref. 21). Catalase was also obtained from Sigma, bovine erythrocyte superoxide dismutase was a gift from Dr. J.V. Bannister. 1-anilino-8-naphthalene sulphonate (ANS) was obtained from Eastman. All other reagents were 'Analar' grade.

RESULTS

The addition of purified DMPO to the neutrophils had no effect on the oxygen consumption (Table 1), nor did it affect the rate of latex induced oxygen consumption. DTPA, which was present in the reaction mixture, had virtually no effect on oxygen consumption. DTPA is better at sequestering $Fe(III)$ than EDTA (Ref. 22).

TABLE 1

COMPARISON OF THE EFFECTS OF VARIOUS REAGENTS ON THE OXYGEN CONSUMPTION OF HUMAN NEUTROPHILS

Incubation Mixture[a]	Respiratory Burst[b] (nmoles/ml/min)	
Neutrophils + Latex IgG	65.8 ± 1.2	(3.2 ± 0.3)
Neutrophils + Latex IgG + DMPO (100 mM)	65.9 ± 1.0	(3.2 ± 0.3)
Neutrophils + Latex IgG + DMPO (200 mM)	35.7 ± 3.0	(3.0 ± 0.4)
Neutrophils + Latex IgG + DTPA (1 mM)	68.0 ± 3.0	(3.2 ± 0.5)
Neutrophils + Latex IgG + DMPO (100 mM) + DTPA (1 mM)	65.0 ± 3.0	(3.5 ± 0.5)

[a]All samples contained 1×10^7 cells/ml.
[b]Figures in parentheses refer to basal rate of respiration.

The epr spectrum of a sample containing resting, unstimulated neutrophils and the spin trap is shown in Figure 3, a. The absence of any resonances suggests that either no radicals are produced by the unstimulated cells or that any produced do not react with the spin trap. It is also possible that the adduct is rapidly consumed by the cells. As previously shown,[14] the addition of latex IgG to a sample of neutrophils results in an epr spectrum (Figure 3, b) which is consistent with that expected from the hydroxyl radical adduct of DMPO (II), (Refs. 19, 23, 25). Stimulation of cells with zymosan also results in the formation of II, whilst stimulation with PMA results in the formation of a mixture of the hydroxyl (II) and hydroperoxide (III) adducts of DMPO, (Figure 3, c).[23,24]

Addition of superoxide dismutase to the neutrophils prior to stimulation by either latex IgG or PMA in the presence of DMPO has an obvious effect; no radicals

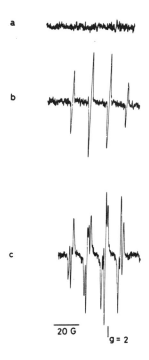

Fig. 3. Epr spectra obtained after incubation at 37°C of human neutrophils (1 x 10⁷ cells/ml) and DMPO (100 mM) with latex particles (1 x 10⁹) or PMA (10 μg/ml). Samples were removed from the thermostated cell and transferred to an epr flat cell after 1 minute. a, Neutrophils and DMPO; b, plus latex IgG; c, plus PMA. (Field 3305 G, frequency 9.5 Ghz, power 30 mw, modulation 1 G time constant 0.128 s, scan rate 0.8 Gs⁻¹, gain 5 x 10⁴).

are observed (Figure 4). In contrast the effect of catalase is far less dramatic, it only causes a small decrease in the amount of hydroxyl radicals trapped (84% of the control signal maximum). Ceruloplasmin at 1.5 and 5 mg/ml reduces the hydroxyl signal height to 63% and 10% of control, respectively, when included in the incubation mixture.

The results of experiments to determine where the radicals are trapped are summarised in Table 2. It can clearly be seen that any radicals that are trapped under these conditions are present in the extracellular medium (Figure 5, upper spectrum). Lysis of the cells reveals that no adducts of DMPO are found intracellularly (Figure 5, lower spectrum), nor is DMPO detected after lysis.

Stimulation of the cells with latex IgG in the absence of DMPO followed by removal of aliquots from the oxygen electrode chamber at time intervals and subsequent

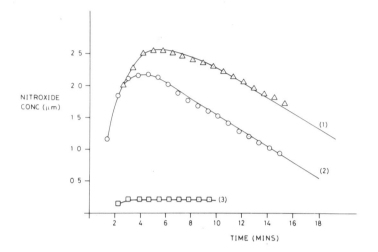

Fig. 4. Plots of the nitroxide concentration against time of the epr spectrum b of Fig. 3. (1) △ Sample as for spectrum b, Fig. 3; (2) ⊙ plus catalase (100 μg/ml); (3) ☐ plus SOD (100 μg/ml). The epr spectra were obtained under the following conditions: field 3381 G, frequency 9.5 GHz, power 10 mw modulation 1 G, time constant 1 s, scan rate 0.2 Gs^{-1}, gain 10^5.

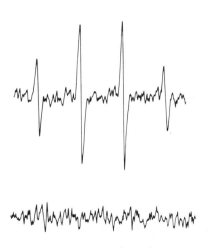

Fig. 5. Epr spectra obtained from stimulated human neutorphils in the presence of DMPO. Reaction stopped after 2 minutes. Upper spectrum obtained from the supernatant; lower spectrum obtained from the lysed cell pellet. Epr conditions were as follows: field 3385 G, frequency 9.5 GHz, power 10 mw, modulation 1 G, time constant 0.5 s, scan rate 0.4 Gs^{-1} gain 8 x 10^4.

TABLE 2

THE AMOUNT OF HYDROXYL ADDUCT OF DMPO DETECTED IN THE SURROUNDING MEDIUM
AT VARIOUS STAGES DURING THE COURSE OF THE RESPIRATORY BURST[a]

Time at which the reaction was stopped (Min)	μmoles of hydroxyl adduct of DMPO
0.5	1.89
2.0	3.23
5.0	1.86
10.0	1.70

[a]Initial concentration of DMPO was 100 mM.

addition of DMPO (100 mmol/1) indicates that the majority of radicals detected
in the medium (Table 3) are produced during the initial stages of increased oxygen
consumption.

TABLE 3

THE RESULT OF ADDING DMPO TO HUMAN NEUTROPHILS STIMULATED WITH LATEX
IgG

Time of addition of DMPO after stimulation of neutrophils by latext IgG[a] (Min)	μmoles of hydroxyl adduct of DMPO
0	3.76
1	1.56
2.5	0.94
10	0.42
20	0.32
30	0.21

[a]Samples contained 1x 10^7 cells/ml; DMPO concentration was 100 nM.

The effect of ANS on both the radicals trapped and the respiratory burst are
shown in Table 4. Addition of this anion channel blocker[26] decreased the amount
of the nitroxide detected but did not have a significant effect on oxygen uptake.[27]

TABLE 4

EFFECT OF THE ANION CHANNEL BLOCKER 1-ANILINO-8-NAPHTHALENE SULPHONATE ON
THE RESPIRATORY BURST AND RADICAL PRODUCTION FROM STIMULATED HUMAN NEUTROPHILS

Addition to neutrophil sample[a]	Respiratory Burst[b] (nmoles/ml/min)		μmoles of hydroxyl adduct of DMPO
Latex IgG	69.5	(3)	1.68
Latex IgG + ANS	67.2	(2)	0.86

[a]Samples contained 1×10^7 cells/ml; ANS concentration was 100 mM
[b]Figures in parentheses refer to average value of number of experiments.

DISCUSSION

The observation of superoxide and hydroxyl radical adducts of DMPO produced by
stimulated human neutrophils is consistent with observations made by somewhat more
indirect methods.[8] The formation of II indicates the presence of hydroxyl radicals
and the inhibition of its formation by superoxide dismutase suggests that these
hydroxyl radicals are formed by a process dependent on superoxide. The lack of
inhibition by catalase would tend to indicate, at first glance, that 'free' hydrogen
peroxide is not the precursor of the hydroxyl radical. However recent investigations[27]
have shown that, in the absence of any haem inhibitors such as azide or cyanide,
only trace amounts of hydrogen peroxide are found in the medium surrounding the
cells. Thus it is quite possible that hydrogen peroxide is not released into the
surrounding cell medium and hence catalase will not be an effective inhibitor.

The superoxide adduct of DMPO has been observed when PMA is used to stimulate
the neutrophils, a steady state concentration of more than 5 μmol/1 is measured.
Stimulation by PMA causes the specific granules to empty their contents and hence
is not a 'complete' stimulation. This may be why superoxide is observed. The
derivation of hydroxyl radicals from superoxide via the Haber-Weiss[28] reaction was
originally postulated but recent work[30] has shown the rate of this reaction to be
very slow. It is possible that the hydroxyl radical may arise via a disproportionation reaction of the superoxide ion in the locus of a reducing metal ion:

$$2O_2^- + Fe^{2+} \longrightarrow OH^{\cdot} + OH^- + O_2 + Fe^{3+}$$

Our results tend to indicate that the majority of radicals we detect are produced
in the first few minutes of the respiratory burst and only a small amount of radicals
is detected in the surrounding medium after the respiratory burst is over (Table 3).

Although we have not detected any radicals within the phagosome this does not
mean that none are formed. The effect of ceruloplasmin in reducing the amount of
radicals trapped is consistent with the observation of Goldstein et al.[21] that

it may act as a scavenger for superoxide ion. It may act as the scavenger in the plasma where the superoxide dismutase levels are very low.

Use of the anion channel blocker ANS indicates that some of the radicals we detected may arise from within the cell and subsequently pass into the extracellular medium through the membranes possibly via anion channels. This has also been suggested on the basis of the inhibition of cytochrome c reduction.[27]

In conclusion, our results indicate that hydroxyl and superoxide radicals are produced during the course of the short respiratory burst, only very small quantities being detected in the medium when the respiratory burst is over. These radicals may arise through diffusion of superoxide along anion channels in the cell membrane. The relationship of these radicals detected to the microbicidal activity of the neutrophil has yet to be established.

ACKNOWLEDGEMENT

M.J.O-Z wishes to thank the Science Research Council for a research studentship. We also thank Drs. M.P. Esnouf, A.W. Segal and S. Chavin and A.R. McEuan and Messrs. A.E.G. Cass and M.R. Green for their help and advice. This is a contribution from the Oxford Enzyme Group of which H.A.O.H. is a member.

REFEREMCES

1. Sbarra, A.J. and Karnovsky, M.L. (1959) J. Biol. Chem. 234, 1355-1362.
2. Iyer, G.Y.N., Islam, F.M. and Quastel, J.H. (1961) Nature, 192, 535-541.
3. Babior, B.M., Kipnes, R.S., and Curnutte, J.T. (1973) J. Clin. Invest. 52, 741-744.
4. Johnstone, R.B., Jr., Keele, B.B., Jnr., Misra, H.P., Leymeyer, J.E., Webb, L.S., Bachner, R.L. and Rajagopalan, K.V. (1975) J. Clin. Invest. 55, 1357-1372.
5. Weening, R.S., Wever, R. and Roos, D. (1975) J. Lab. Clin. Med. 85, 145-252.
6. Drath, D.B. and Karnovsky, M.L. (1975) J. Exp. Med. 141, 257-262.
7. McCord, J.M. and Fridovich, I. (1969) J. Biol. Chem. 244, 6049-6055.
8. Tauber, A.I. and Babior, B.M. (1977) J. Clin. Invest. 60, 374-379.
9. Beauchamp, C. and Fridovich, I. (1970) J. Biol. Chem. 245, 4641-4646.
10. Pryor, W.A. and Tang, R.H. (1978) Biochem. Biophys. Res. Commun. 81, 498-503.
11. Babior, B.M. and Kipnes, R.S. (1977) Blood, 50, 517-524.
12. Janzen, E.G. (1971) Account. Chem. Res. 4, 31-40.
13. Langercrantz, C.J. (1971) J. Phys. Chem. 75, 3466-3475.
14. Green, M.R., Hill, H.A.O., Okolow-Zubkowska, M.J. and Segal, A.W. (1979) FEBS Lett. 100, 23-26.
15. Boyum, A. (1968) Scan. J. Clin. Lab. Invest. 21, 31-50.
16. Singer, J.M. and Plotz, C.M. (1956) Amer. J. Med. 21, 888-892.
17. Björksten, B., Bäck, O., Hägglof, B. and Tärnvik, A. (1979) in Inborn Errors of Immunity and Phagocytosis, Güttler, R., Seakins, J.W.T. and Harkness, R.A., eds., MTP Press, Lancaster, pp. 189-198.
18. Miller, R.W. and Rapp, U. (1973) J. Biol. Chem. 248, 6084-6090.
19. Buettner, G.R. and Oberley, L.W. (1978) Biochem. Biophys. Res. Commun. 83, 69-74.
20. Bonnet, R., Brown, R.F.C., Clark, W.M., Sutherland, I.O. and Todd, A. (1959) J. Chem. Soc., 2094-2102.
21. Goldstein, I.M., Kaplan, H.B., Edelson, H.S. and Weissman, G. (1979) J. Biol. Chem. 254, 4040-4045.

22. Halliwell, B. (1978) FEBS Lett. 92, 321-326.
23. Harbour, J.R., Chow, V. and Bolton, J.R. (1974) Can. J. Chem. 52, 3549-3553.
24. Harbour, J.R. and Bolton, J.R. (1975) Biochem. Biophys. Res. Commun. 64, 803-807.
25. Janzen, E.G., Nutter, D.E., Jr., Davis, E.R., Blackburn, B.J., Poyer, J.L. and McCay, P.B. (1978) Can. J. Chem. 56, 2237-2242.
26. Cabantchik, Z.I. and Rothstein, A. (1972) J. Membrane Biol. 10, 311-330.
27. Gennaro, R. and Rodeo, D. (1979) Biochem. Biophys. Res. Commun. 88, 44-49.
28. Dri, P., Bellavite, P., Berton, G., and Rossi, F. (1979) Mol. Cell. Biochem. 23, 109-122.
29. Haber, F. and Weiss, J. (1934) Proc. Roy. Soc. A, 147, 332-355.
30. Weinstein, J. and Bielski, B.H.S. (1979) J. Amer. Chem. Soc. 101, 58-62.

THE RELEASE OF SUPEROXIDE ANION BY MACROPHAGES AND ITS RELATIONSHIP TO PHAGOCYTIC MICROBICIDAL ACTIVITY

RICHARD B. JOHNSTON, Jr., MICHAEL J. PABST AND MASATAKA SASADA
National Jewish Hospital and Research Center and Departments of Pediatrics and Biochemistry, University of Colorado Health Sciences Center, Denver, Colorado 80206, USA

The process of phagocytosis, when carried out by "professional" phagocytes (neutrophils, monocytes, macrophages, and eosinophils), is associated with a remarkable series of biochemical events involving oxygen. These reactions take place in other cell types, but the magnitude with which they occur and the sub-cellular location at which they take place appear unique to phagocytes. These events are interesting from the standpoint of their biochemistry, but they have assumed particular significance because it is now clear that they are essential to the principal physiologic function of phagocytic cells, the elimination of invading microorganisms. The relationship between oxidative metabolism and microbicidal activity has been securely established for neutrophils and, perhaps, blood monocytes, the circulating precursors of tissue macrophages. Until recently, however, concepts of how tissue macrophages kill microorganisms have been based as much on extrapolation from data with neutrophils as on experiments performed with macrophages.

The oxidative metabolic events associated with phagocytosis by neutrophils have been reviewd in detail.[1-3] In summary, they begin on contact with the micro-organism, as consumption of oxygen by the phagocyte. Oxygen consumption appears to be directed by an enzyme (oxidase) that transfers an electron from NADPH or NADH, or both, to oxygen. By this mechanism, inspired oxygen is converted mostly, if not wholly, to superoxide anion (O_2^-). Superoxide radicals here, as elsewhere in nature, interact with each other in a rapid dismuation reaction, the products of which are the anion of H_2O_2 (O_2^{2-}) and oxygen. The potent oxidant hydroxyl radical ($\cdot OH$) is formed, presumably through the interaction of O_2^- and H_2O_2 in the presence of metal ions. Glucose oxidation through the hexose monophosphate (HMP) shunt is stimulated as a result of $NADP^+$ formation during phagocytosis. This $NADP^+$ is probably derived in part from oxidation of NADPH by the initiating oxidase, in addition to its derivation from activation of the glutathione peroxidase cycle by H_2O_2.

A burst of oxygen-dependent luminescence occurs during phagocytosis. It was originally proposed that this luminescence was caused by singlet oxygen ($^1O_2$),

but no direct data currently exist to substantiate this hypothesis. The presence
of multiple broad bands of light in the emission spectrum of phagocytosis-
associated luminescence indicates that, in the least, the light does not emanate
from $^1O_2$ alone.

Although studies of the oxidative metabolism of macrophages have lagged behind
those with neutrophils, it is quite clear that this cell line also undergoes a
phagocytosis-associated "respiratory burst" (reviewed in Refs.3,4). Early studies
showed increased oxygen consumption and HMP shunt activation during phagocytosis
by guinea pig peritoneal macrophages. These findings have been confirmed by many
investigators with peritoneal and alveolar macrophages from several animal species.
Increased H_2O_2 release during phagocytosis has been demonstrated with these same
cell types. Macrophages from guinea pigs and mice have been shown to generate
O_2^- in response to phagocytosis. Phagocytizing macrophages of various species
produce luminescence, but the signal is weaker than that from neutrophils, and
luminol has been used to amplify the effect. Superoxide dismutase virtually
eliminates the chemiluminescence generated by phagocytizing guinea pig peritoneal
and human alveolar macrophages, indicating the importance of O_2^- as an initiator
of chemiluminescence in these cells.

Human blood monocytes undergo these same manifestations of the phagocytosis-
associated respiratory burst, although the changes are generally less pronounced
than those with neutrophils studied under the same conditions.[3] Soluble substances capable of plasma membrane perturbation, such as phospholipase C and
phorbol myristate acetate (PMA), also elicit these manifestations of the respiratory burst in mononuclear phagocytes, as well as in neutrophils.

The neutrophil is a relatively short-lived cell that always exhibits a strong
respiratory burst when appropriately stimulated. In contrast, the magnitude of
the respiratory burst produced by macrophages appears to vary greatly with the
previous experience of these cells in vivo with inflammation, or more specifically,
with their state of "activation".

Increased release of O_2^- by elicited and activated macrophages. Resistance to
infection by intracellular parasites, such as mycobacteria and listeria, appears
to depend on "activation" of macrophages by lymphocytes.[5] In this process,
which is the basis for cell-mediated immunity, lymphocytes sensitized by previous
experience with a microorganism will, in the presence of that organism, stimulate
macrophages to enhanced killing of the ingested parasites. Macrophages obtained
from animals with enhanced resistance to intracellular microorganisms exhibit an
increase in size, spreading, plasma membrane movements, number of mitochondria
and lysosomal granules, content of hydrolytic enzymes, protease secretion, pino-
cytosis, phagocytosis (of some particles), killing of certain microorganisms,
tumoricidal activity, and other manifestations of a state of activation. Macrophages elicited by intraperitoneal injection of materials that induce inflammation,

e.g., casein, thioglycollate broth, or endotoxin (lipopolysaccharide, LPS), have been reported to exhibit most but not all of these properties.

Some comparisons have been made in the oxidative metabolic activities of normal (resident) and activated macrophages. Macrophages from animals previously infected with bacillus Calmette-Guérin (BCG) or with listeria, or from animals injected with an inflammatory agent like LPS have shown an increased relative to resident cells in activation of the HMP shunt, consumption of oxygen, and release of H_2O_2 during phagocytosis.

In order to gain a more complete understanding of the effect of activation on the phagocytosis-associated respiratory burst, we undertook a systematic analysis of O_2^- release in resident, LPS elicited, and immunologically activated (by BCG infection) macrophages.[6] Murine peritoneal macrophages were used because studies of other aspects of macrophage function are most extensive in this species. Macrophages were harvested by peritoneal lavage and cultured overnight in tissue culture dishes. Superoxide anion was measured as superoxide dismutase-inhibitable reduction of ferricytochrome c, after stimulation of the cells by PMA or by zymosan particles coated with antibody and complement. Results were corrected for the protein content of the dish, determined after completion of the assay.

Under resting (unstimulated) conditions, there was minimal O_2^- release from any of the cell types (Table 1). When incubated with the surface-active agent PMA or with opsonized zymosan particles, however, each cell type consistently released significant quantities of O_2^-. The response with cells elicited chemically or with cells from BCG-infected mice was much greater than that achieved with resident cells using either PMA or aymosan at stimulus. Since elicited and BCG cell types were larger than resident macrophages, the enhanced O_2^- response of these "activated" macrophages was even more pronounced when considered as a function of cell number rather than quantity of protein on the dish.[6]

The enhanced oxidative response of elicited and BCG macrophages was noted over a wide range of concentrations of PMA (0.004-30 µg/ml) and opsonized zymosan (0.1-1.3 mg/ml). Analysis of the time course of stimulated O_2^- release by the four cell types showed that the greater response by elicited or BCG macrophages was present at the earliest timepoint studied (20 min), was most pronounced in the first 30-60 min, and persisted through 120 min (Ref.6).

The capacity to effectively elaborate O_2^- in culture appears to be confined to professional phagocytes. The macrophage-like tumor lines J774.1, PU5-1.8, and P388D1 released O_2^- on exposure to PMA or opsonized zymosan. In contrast, human peripheral blood lymphocytes, fibroblast and endothelial cell lines, and a primary culture of embryonic cells did not. None of the cell types released O_2^- if unstimulated.[6]

TABLE 1

RELEASE OF SUPEROXIDE BY MOUSE PERITONEAL MACROPHAGES ON CONTACT WITH PHORBOL
MYRISTATE ACETATE OR DURING PHAGOCYTOSIS OF OPSONIZED ZYMOSAN

Cell type	Unstimulated cells[a]	Stimulated cells	
		PMA[a]	Opsonized zymosan[a]
Resident	3 ± 10 (13)	49 ± 34 (21)	233 ± 80 (22)
Thioglycollate-elicited	6 ± 10 (15)	493 ± 168 (11)	304 ± 78 (8)
LPS-elicited	0 (12)	570 ± 71 (5)	454 ± 99 (4)
BCG	1 ± 3 (9)	518 ± 187 (10)	305 ± 88 (10)

[a]Nanomoles of cytochrome c reduced during 90 min incubation per mg of protein
on the culture dish. Values are expressed as mean ± SD of averages obtained
with duplicate dishes; the number of experiments is given in parentheses. Only
results with dishes that contained 40-100 µg cell protein were collated. The
LPS used was a crude extract (Difco). (Data from Ref.6).

The enhanced capacity of activated macrophages to release O_2^- (reviewed here),
as well as H_2O_2 (Ref.7), is associated with an increased capacity of these
cells to resist infection and tumors. Thus, the greater oxidative response
of activated macrophages could contribute significantly to their enhanced ability
to carry out the killing functions required to mediate cell-mediated immunity.

Increased release of O_2^- by macrophages exposed in vitro to bacterial products.
The increased production of O_2^- observed with macrophages which have been activated
in vivo can also be reproduced in vitro . Resident mouse peritoneal macrophages,
placed in culture with or without serum and exposed to certain bacterial products,
become "primed" to release increased amounts of O_2^- in response to stimulation
by PMA or opsonized zymosan. We believe that this "priming" is a manifestation
of macrophage activation that has been induced in vitro.

The bacterial products used in our studies were phenol-extracted bacterial
LPS (Ref.8), N-acetylmuramyl-L-alanyl-D-isoglutamine (muramyl dipeptide, or MDP),
and analogs of MDP. Although the MDP we used was synthesized chemically, the
molecule represents a fragment of bacterial cell wall peptidoglycan, and is, in
fact, the minimal fragment which displays adjuvant activity (the capacity to
enhance antibody formation). Its muramic acid moiety and D-amino acid character-
ize it as a product of bacterial origin. MDP is found in all bacteria which
have cell walls, both gram positive and gram negative.

The response of resident mouse peritoneal macrophages to LPS, MDP and some
related compounds is shown in Table 2. In the case of the series of compounds
related to MDP, the ability to prime macrophages to produce increase O_2^- paralleled
the ability of these compounds to act as immune adjuvants[9] or to protect animals
against a lethal bacterial infection[10]: MDP and N-acetylmuramyl-L-alanyl-D-
isoglutamic acid were active, whereas the L-L and D-D stereoisomers of MDP were
inactive.

TABLE 2

PHORBOL MYRISTATE ACETATE STIMULATED SUPEROXIDE RELEASE BY MOUSE PERITONEAL

MACROPHAGES INCUBATED IN VITRO WITH BACTERIAL PRODUCTS OR CHEMOTACTIC FACTORS

Substance added to cultures	O_2^- released[a] (treated/control)
N-Acetylmuramyl-L-alanyl-D-isoglutamine (1 µM)	3.0[b]
N-Acetylmuramyl-L-alanyl-D-isoglutamic acid (1 µM)	2.3[b]
N-Acetylmuramyl-L-alanyl-L-isoglutamine (1 µM)	1.1
N-Acetylmuramyl-D-alanyl-D-isoglutamine (1 µM)	1.2
FMLP (1 µg/ml, 0.5 µM)	1.1
C5a (0.1 µg/ml, 10 nM)	1.2
C5a$_{des Arg}$ (0.1 µg/ml, 10 nM)	0.9
LPS (10 ng/ml)	6.8[b]
Phthalyl LPS (10 ng/ml)	1.1
Trypsin (10-250 µg/ml)	3.9[b]
LPS (1 µg/ml)	3.4[b]
LPS (1 µg/ml) + Diisopropylfluorophosphate (0.1 mM)	3.6[b]

[a]The amount of O_2^- released in 60 min per mg of cell protein was determined by the cytochrome c reduction assay, then compared in the same experiment with the response of control cells which had not been exposed to the priming agent (bacterial product or chemostatic factor). Values are ratios for the means of triplicate assays. Cells were exposed to the agent for 16 hr, except for the last three entries in the table. In those experiments, exposure to trypsin was for 10-30 min, and exposure to either LPS or LPS + DFP was for 4 hr.

[b]Treated cells were significantly different control in a paired t test, P < 0.01.

We also tested the chemotactic peptide N-formyl-L-methionyl-L-leucyl-L-phenylalanine (FMLP), hypothesizing that the formyl group might cause the macrophages to recognize this peptide as a product of bacteria. FMLP is chemotactic for neutrophils at low concnetrations (10^{-9} to 10^{-10}M) (Ref.11); and at higher concentrations (10^{-7}M), stimulates O_2^- production in neutrophils.[12] However, FMLP at concentrations of up to 10^{-5}M failed to prime macrophages for enhanced O_2^- release after stimulation with PMA. Similar results were obtained with highly purified human C5a, the complement-derived chemotactic factor, and with C5a$_{des Arg}$, which is C5a that has been inactivated in regard to anaphylotoxin activity.[13] Both molecules stimulate O_2^- production in neutrophils, C5a being about 50 times more potent than C5a$_{des Arg}$ (Ref.13). Neither C5a nor C5a$_{des Arg}$ at concentrations of 5 x 10^{-8}M to 1 x 10^{-9}M primed mouse macrophages to release more O_2^- when stimulated by PMA. Neither FMLP nor C5a directly stimulated O_2^- release by macrophages at concentrations that caused O_2^- release from neutrophils.

In addition to priming macrophages to produce increased amounts of O_2^-, LPS has many effects on host defense systems. It is an adjuvant, a mitogen, a

216

pyrogen, a toxin, and a stimulator of nonspecific resistance to infection. In
an attempt to determine which of these other properties corresponded with macrophage priming, a chemically modified form of LPS was tested, phthalyl LPS. In
preparing this derivarive, LPS was allowed to react with an excess of phthalic
anhydride, thereby acylating free hydroxyl and amino groups. This modification
"detoxifies" the LPS (Ref.8): Phthalyl LPS is less potent than LPS as a mitogen
by a factor of 10^{-2}, as a toxin by 10^{-4}, and as a pyrogen by 10^{-5}. However,
the adjuvant activity of phthalyl LPS appears to be undiminished.[8,14] Figure 1
shows the concentration dependence of the priming response to phthalyl LPS
compared with the parent LPS molecule. Phthalyl LPS appears to be less effective
than LPS in macrophage priming by a factor of 10^{-3}. Thus, despite the corres-
pndence between adjuvanticity and priming in the MDP series, the results with
phthalyl LPS indicate that adjuvanticity may not be directly linked to macrophage priming.

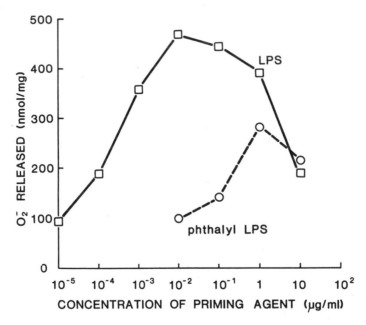

Fig. 1. Macrophage priming by LPS and phthalyl LPS. Resident mouse peritoneal
macrophages were harvested in saline, then allowed to adhere to 35 mm tissue
culture dishes for 2 hr in the presence of Dulbecco's Modified Eagle's Medium
(DMEM), antibiotics, and 10% heat-inactivated fetal calf serum (Ref.6). After
adherence the cells were washed, then incubated for 16 hr in DMEM in the presence of the indicated concentration of LPS or phthalyl LPS. The cells were
then washed and assayed for O_2^- production in response to stimulation by 0.5 µg/ml
of phorbol myristate acetate (Ref.6). Protein was determined by the Lowry method
with bovine serum albumin as standard. The points represent means of six dishes.
The standard error in all cases was less than 25. The response of control cells,
which had not been treated with LPS, was 105 ± 12 nmol/mg (mean ± SEM, n = 6).

Priming requires only minute quantities of the bacterial products and takes place within a matter of hours. For example, treatment of macrophages with only 0.1 ng/ml of LPS for 16 hr doubled their O_2^- response (Figure 2). Using 1 μg/ml of LPS, increased O_2^- release was detected within 30 min of exposure to the LPS, although the optimal response required about 12 hr (Ref.15). As a priming agent, MDP appears to be somewhat less potent than LPS. The first detectable response to MDP occurred after 4 hr of exposure, and the minimum consistently effective concentration of MDP was 0.1 μg/ml (Ref.15). Because of

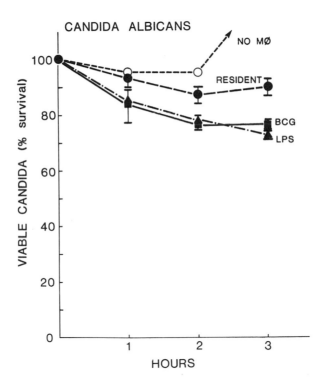

Fig. 2. The killing of C. albicans by mouse peritoneal macrophages. The percantage of surviving organisms (compared to the number of viable organisms at time 0) is plotted as a function of the length of the time of incubation of fungi with macrophages. The points represent the means of 7–16 experiments (16 with resident cells, 9 with LPS-elicited cells, 7 with BCG-activated cells, and 16 with no macrophages (no Mφ)). The bars indicate 1 SEM. (Data from Ref.17).

the rapid response of the cultured macrophages to LPS and MDP, we believe that
priming is not due to the formation of lymphokines by the few (if any) lymphocytes
which might remain adherent to the macrophage culture dishes (The production of
lymphokines usually requires at least 24-48 hr.). It would appear that the bacterial
products act directly on macrophages, as indicated by several lines of experi-
mentation: (i) Deliberate addition of lymphocytes had no effect on macrophage
primimng; (ii) Macrophages from nude mice, which lack mature T cells, responded
normally; (iii) Addition of lymphocytes from a normal strain of mice (C3Heb/FeJ)
failed to correct the defective response of abnormal C3H/HeJ macrophages to
phenol-extracted LPS; and (iv) The macrophage-like cell line J774.1 responded
to LPS and MDP (Ref.15).

Priming of macrophages in vitro may be accomplished by a few minutes' exposure
to proteolytic enzymes. Exposure to 1-500 µg/ml trypsin, chymotrypsin, pronase,
or papain for 5 to 30 min caused macrophages to produce 2-6 times more O_2^- in
response to PMA. The results obtained in 13 experiments with trypsin are shown
in Table 2. The effect of proteolytic enzymes was blocked by inactivating the
enzymes with diisopropylfluorophosphate (DFP), or by preincubating the enzymes
with soybean trypsin inhibitor or dilute serum. This results suggests that
macrophages in an inflammatory site could be activated by proteases from serum
or from neutrophils. However, proteolytic cleavage may not be an obligatory
step in macrophage priming, since the effect of LPS cannot be blocked by DFP
or other esterase inhibitors (Table 2).

The pronounced priming shown here with nanogram quantities of LPS indicates
the importance of considering the LPS content of all components of macrophage
culture systems. Since LPS can induce this aspect of macrophage activation
overnight in vitro, it may be possible through the use of metabolic blockers
to gain information about the molecular events that permit development of the
activated state.

Oxygen metabolites and macrophage microbicidal activity. The assumption that
the increase in phagocytosis-associated oxidative metabolism of LPS-elicited
and BCG-activated macrophages would be associated with increased phagocytic
killing activity was tested using the fungi Candida albicans and Candida
parapsilosis. Candida were added to mouse peritoneal macrophages cultured
overnight.[6] The ratio of macrophages to fungi was 3 or 4 to 1, and 20% heat-
inactivated fetal calf serum was present. As shown in Figure 2, C. albicans
was not easily killed by the macrophages. However, elicited and activated
macrophages killed 2-3 times more fungi that did resident cells, and this
difference was significant at 2 and 3 hr (P < 0.002 at 3 hr, paired t test).

Macrophage killing of candida is dependent at least in part on phagocytic
generation of oxygen metabolites, as shown by the inhibition of candidacidal
activity by superoxide dismutase (SOD), catalase, benzoate, which is believed

to remove ˙OH, and azide, which is believed to scavenge singlet oxygen (Table 3). Data are shown for C. parapsilosis; similar results were obtained with C. albicans, but the differences were less pronounced since fewer fungi were killed under any conditions. None of the scavengers decreased phagocytosis of the candida, which was quantiated by counting the number of fungi ingested by 200 macrophages using smears stained with Wright's stain; and non-decreased macrophage viability, as determined by tryptan blue dye exclusion. Heat-inactivated SOD or catalase had no effect on macrophage candidacidal activity. It is not possible from these data with scavengers to define with certainty which oxygen metabolites are involved in macrophage candidacidal activity. However, the significant inhibition achieved with SOD, catalase, and benzoate favors the involvement of ˙OH in killing fungi, as we have previously shown to be true for bacteria.[16]

TABLE 3

EFFECT OF OXYGEN RADICAL SCAVENGERS ON THE KILLING OF CANDIDA PARAPSILOSIS
BY MOUSE PERITONEAL MACROPHAGES

Macrophage type	Oxygen radical scavenger	Viable candida (% survival)[a]
Resident	None	23.3 ± 8.4 (20)
	Superoxide dismutase, 20 μg/ml	60.6 ± 6.1 (5)[b]
	Superoxide dismutase, 100 μg/ml	71.9 ± 11.5 (9)[c]
	Catalase, 66 μg/ml	30.6 ± 8.9 (12)[b]
	Catalase, 132 μg/ml	40.5 ± 2.0 (4)[c]
	Benzoate, 10 mM	31.4 ± 9.9 (5)
	Benzoate, 20 mM	38.5 ± 4.7 (4)[b]
	Azide, 10 mM	37.4 ± 10.7 (5)[b]
	Azide, 30 mM	53.2 ± 13.1 (4)[b]
LPS-elicited	None	14.8 ± 4.9 (6)
	Superoxide dismutase, 100 μg/ml	70.0 ± 7.2 (5)[c]
	Catalase, 66 μg/ml	24.8 ± 6.5 (5)[b]
BCG-activated	None	19.5 ± 6.9 (6)
	Superoxide dismutase, 100 μg/ml	73.3 ± 11.6 (3)[b]
	Catalase, 66 μg/ml	34.6 ± 8.1 (5)[b]

[a]The percentage of candida surviving after 2.5 hr incubation with macrophages is shown. Values are expressed as mean ± SEM of averages obtained with duplicate dishes; the number of experiments is given in parentheses. (Data from Ref.17).

[b]Results are significant compred to those without scavenger, using paired t test: P < 0.01.

[c]Results are significant compred to those without scavenger, using paired t test: P < 0.001.

Each of the three macrophage types used in these studies (resident, LPS-elicited, and BCG-activated) killed C. parapsilosis much more effectively than C. albicans (compare results of Figure 2 with those of Table 3). This greater killing was not due to increased serum antibody to C. parapsilosis or greater activation of complement by this species, since the difference was similar in the presence of complement-inactivated (heated) serum or no serum.[17] Resident and LPS-elicited macrophages phagocytized C. parapsilosis slightly faster than C. albicans, but the number of organisms ingested was the same for the two species by 60 min of incubation, and all candida of both species were ingested by this time. However, the effect of the two types of candida on macrophage oxidative metabolism was markedly different. C. parapsilosis stimulated much more O_2^- release and oxygen uptake by macrophages than did C. albicans, and the difference was evident within 3 min of exposure.[17] The difference was not modified by pretreatment of the macrophages with cytochalasin B to suppress phagocytosis. Although C. albicans contained higher concentration of the protective enzymes SOD and catalase, this species did not scavenge more reagent H_2O_2 or more O_2^-, generated by xanthine oxidase and acetaldehyde, than did C. parapsilosis; and both species were killed to approximately the same extent by H_2O_2 or the enzymic O_2^--generating system. Therefore, it appears likely that the greater susceptibility of C. parapsilosis to macrophage killing is due, at least in part, to its greater capacity to stimulate the respiratory burst in this phagocytic cell.

SUMMARY

Phagocytizing macrophages undergo a "respiratory burst" in which oxygen is converted to O_2^-, H_2O_2, and other radicals. Activated macrophages, which have enhanced capacity to resist infection and tumors, exhibit a pronounced increase in the production of O_2^- and H_2O_2 in response to stimulation by surface perturbation or phagocytosis. Macrophages can also be primed for this increased oxidative response by incubation in vitro with small amounts of LPS or the adjuvant, muramyl dipeptide. The greater respiratory response of activated macrophages could play a significant role in their accelerated capacity to effect cell-mediated immunity. In agreement with this concept, activated macrophages kill candida species more effectively than do normal macrophages, and this killing can be markedly inhibited by agents that remove oxygen metabolites.

ACKNOWLEDGEMENTS

The work described here was supported by US Public Health Service grant AI 14148 from the National Institutes of Health.

REFERENCES

1. Babior, B.M. (1978) N. Engl. J. Med. 298, 659–668.
2. Klebanoff, S.J., and Clark, R.A. (1978) The Neutrophil: Function and Clinical disorders, North-Holland, Amsterdam.
3. Johnston, R.B., Jr. (1978) Fed. Proc. 37, 2759.
4. Johnston, R.B., Jr., Chadwick, D.A., and Pabst, M.J. in Mononuclear Phagocytes – Functional Aspects, van Furth, R., ed., Martinus Nijhoff, The Hague, in press.

5. Mackaness, G.B. (1970) in Infectious Agents and Host Reactions, Mudd, S., ed., W.B. Saunders Company, Philadelphia, p. 61.
6. Johnston, R.B., Jr., ·Godzik, C.A., and Cohn, Z.A. (1978) J. Exp. Med. 148, 115-127.
7. Nathan, C.F., and Root, R.K. (1977) J. Exp. Med. 146, 1648-1662.
8. McIntire, F.C., Hargie, M.P., Schenck, J.R., Finley, R.A., Sievert, H.W., Rietschel, E.T., and Rosenstreich, D.L. (1976) J. Immunol. 117, 674-678.
9. Chedid, L., Audibert, F., Lefrancier, P., Choay, J., and Lederer, E. (1976) Proc. Nat. Acad. Sci. USA, 73, 2472-2475.
10. Chedid, L., Parant, M., Parant, F., Lefrancier, P., Choay, J., and Lederer, E. (1977) Proc. Nat. Acad. Sci. USA, 74, 2089-2093.
11. Showell, H.J., Freer, R.J., Zigmond, S.H., Schiffmann, E., Aswanikumar, S., Corcoran, B., and Becker, E.L. (1976) J. Exp. Med. 143, 1154-1169.
12. Lehmeyer, J.E., Snyderman, R., and Johnston, R.B., Jr. (1979) Blood, 54, 35-45.
13. Webster, R.O., Hong, S.R., Johnston, R.B., Jr., and Henson, P.M., submitted for publication.
14. Chedid, L., Audibert, F., Bona, C., Damais, C., Parant, F., and Parant, M. (1975) Infec. Immunity 12, 714-721.
15. Pabst, M.J., and Johnston, R.B., Jr. J. Exp. Med., in press.
16. Johnston, R.B., Jr., Keele, B.B., Jr., Misra, H.P., Lehmeyer, J.E., Webb,L.S., Baehner, R.L., and Rajagopalan, K.V. (1975) J. Clin. Invest. 55, 1357-1372.
17. Sasada, M., and Johnston, R.B., Jr., submitted for publication.

ON THE QUESTION OF SINGLET OXYGEN PRODUCTION IN LEUCOCYTES, MACROPHAGES AND THE DISMUTATION OF SUPEROXIDE ANION

CHRISTOPHER S. FOOTE, ROSANGELA B. ABAKERLI, ROGER L. CLOUGH AND FREDERICK C. SHOOK
Department of Chemistry, University of California, Los Angeles, California 90024,
USA

Oxygen toxicity and its damaging effects in living organisms have frequently been associated with superoxide anion (O_2^-) production.[1] There is great interest in the mechanism of the biological toxicity of O_2^-, since it has been shown to induce red blood cell membrane damage,[1c] peroxidize lipids,[2] and kill viruses[3] and cells,[4] for example. Although a great majority of these effects are oxidative in character, superoxide anion is by itself a very poor oxidizing agent in aqueous solution.[5] On the other hand, the dismutation of O_2^- (reaction 1) is fast in protic environments[6] and may give products that could be responsible for the oxidative damage.

One suggestion is that singlet oxygen, known to react with many biological molecules, may be produced during the uncatalyzed dismutation of O_2^- in water. While several reports have provided suggestive evidence for this reaction,[7] several others have produced moderately convincing negative evidence,[8] and others have produced inconclusive results.[9]

$$2O_2^- + 2H^+ \longrightarrow H_2O_2 + O_2 \; (^1\Delta_g \text{ or } ^3\Sigma_g) \tag{1}$$

One problem with all of these studies is that the amount of $^1O_2$ produced (or the upper limit for its production) has not been determined quantitatively. Another problem is that the quenching of any $^1O_2$ produced by O_2^- (reaction 2), which has been shown to have a rate constant of $\sim 10^9 M^{-1} sec^{-1}$ in dipolar aprotic solvents,[9a] could obscure the formation of $^1O_2$ in chemical model systems.

$$O_2^- + {}^1O_2 \longrightarrow {}^3O_2 + O_2^- + 22 \text{ kcal} \tag{2}$$

We have designed a system which avoids both difficulties, and have used it to test for the production of singlet oxygen during the uncatalysed dismutation of superoxide, and in several other systems involving "oxygen toxicity". Cholesterol gives a characteristic product with singlet oxygen, the 5-α-hydroperoxide (1a), which is distinct from the products of radical autoxidation, which include the 7α- and 7β-hydroperoxides (2a) (Ref.10). Since cholesterol is virtually insoluble in water, a system was designed uisng cholesterol (4-^{14}C) supported on polystyrene latex microbeads ((P)-chol) in buffer.[11]

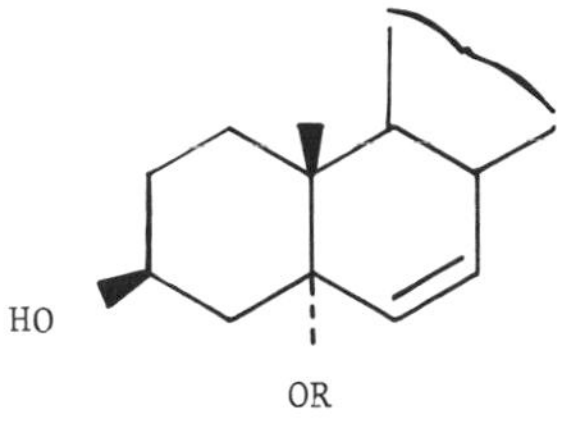

HO

OR

1

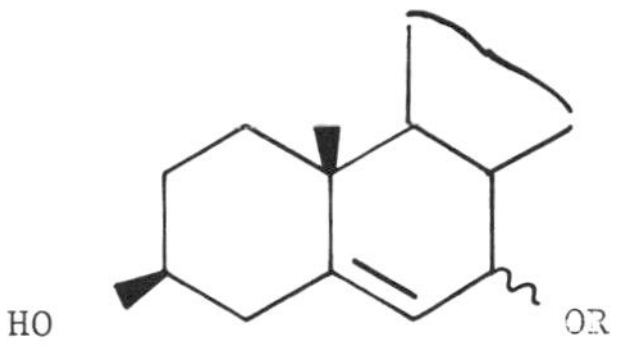

a, R = OH

b, R = H

(^{1}O$_2$ Product)

HO

OR

2

+ complex product mixture

(Radical Products)

To the stirred dispersion, $(CH_3)_4\overset{+}{N}\cdot O_2^-$ (0.1 M) in 25 ml dry DMSO was slowly added through a Teflon capillary with a syringe pump.[12] Following reaction, the organic products were extracted and hydroperoxides were reduced with $(C_6H_5)_3P$. TLC of the mixture was carried out with added known products,[10] and the bands corresponding to cholesterol, the ^{1}O$_2$ product (3β, 5α-diol (1b)), and the 3β, 7α- and 3β, 7β-diols along with other zones (2b) were scraped from the plate, extracted and counted.[13]

To quantiate the technique, the following method was used. To a suspension of (P)-chol, prepared as above, histidine (5 x 10^{-1}M) and methylene blue (2 x 10^{-6}M) were added. The amount of 5α product produced from (P)-chol in this mixture after irradiation for 10 min was determined as above. The amount of histidine in the aqueous layer which had reacted was determined.[14] From the amount of histidine reacted, corrected for its trapping efficiency for ^{1}O$_2$ at the concentration used,[15] the amount of ^{1}O$_2$ generated photochemically under this set of conditions was determined, and thus the trapping efficiency (moles 5α pdt/moles ^{1}O$_2$ produced) of the (P)-chol system could be calculated, and was found to be 2.5 x 10^{-5} under the conditions used. Although this efficiency is low, because of the sensitivity of the ^{14}C radioassay, it is sufficient.

As a final control, two solutions containing (P)-chol and rose bengal were photooxidized under the same conditions. To one of them $(CH_3)_4\overset{+}{N}\cdot O_2^-$/DMSO (0.1 M) was added at 8.8 ml/hr (6 times the rate used in the dismutation experiment); the O$_2^-$ addition caused a decrease of 7 ± 6% in the amount of 5α product formed, so that O$_2^-$ quenching of ^{1}O$_2$ is not significant under the conditions.[16]

The results are shown in Table 1. The fraction of oxygen appearing as ^{1}O$_2$ is listed as a function of pH. The amounts found in the range pH 4-8 are probably not significantly greater than zero. The amount found at pH 10, although small, may be significant, but further work will be necessary to establish this. Thus we conclude that $^1O_2$ is produced in amounts of no more than 0.2% in the O_2^- dismutation under the conditions we have studied, which include correlation for trapping efficiency and O_2^- quenching.

TABLE 1

YIELDS OF SINGLET OXYGEN BASED ON SUPEROXIDE ADDED AT VARIOUS pH, CORRECTED FOR TRAPPING EFFICIENCY

pH	Yield $^1O_2$ (% x 10^4)
4	2.7
6	6.0
7	7.1
8	2.8
10	16

Closely related to the chemistry of O_2^- is the interesting oxidative microbicidal activity of polymorphonuclear leucocytes[19](PMNs) and macrophages.[20] During phagocytosis, molecular oxygen consumption with O_2^- release and subsequent H_2O_2 production are well documented.[21] Either O_2^- or H_2O_2 could be bactericidal by itself. However, the lack of production of O_2^- and H_2O_2 in chronic granulomatous PMNs[21,22] which are deficient in bactericidal activity emphasizes the role of the neutrophil peroxidase (myeloperoxidase), since these cells can show bactericidal activity towards microorganisms that generate H_2O_2 as a metabolic product.[19b]

However, several suggestions for the involvement of $^1O_2$ in the PMN activity have been based on chemiluminescence of cells after phagocytosis,[22] or on the possible formation of $^1O_2$ from dismutation of superoxide[24] or its reactions with other substrates.[2,25] The reported production[26] of $^1O_2$ products in a cell-free system of myeloperoxidase, H_2O_2 and chloride strengthened previous suggestions of the role of $^1O_2$ as the microbicidal agent. However, the $^1O_2$ traps used (furans) lack specificity, as pointed out by many workers.[27] To date no specific test for $^1O_2$ production has been applied to phagocytic neutrophils.

We have also used the same system described above and a related one to test for the production of $^1O_2$ in phagocytosing leucocytes and macrophages. $\circledP$ -^{14}C-cholesterol beads prepared similarly[28] to those described above were incubated with a suspension of freshly isolated human PMNs[29] in 2 ml phosphate buffer (pH 7.4) for 65 min. The oxygen uptake[30] rate at 37°C was 48.6 nmoles/10^7 cells/min. After separation of ingested from uningested particles[31] the cholesterol was extracted and analyzed by a procedure similar to that described above; unlabelled standards (cholesterol, 3β, 5α-diol, 3β, 7α and 7β-diols, 5α, 6αepoxide and 7-ketone) were synthesized by published methods.[32] Autoradiograms of the TLC plates[33] were also carried out. The products formed in one of the many runs are shown in Table 2.

TABLE 2

RADIOACTIVITY DISTRIBUTION ON TLC PLATE FROM INCUBATION WITH POLYMORPHONUCLEAR LEUCOCYTES

Product	$R_c^{a,b}$	R_c^a (Standards)	% of Total Recovered in Extracellular Fraction	Intracellular Fraction
	0.0 -0.04		0.30	0.31
	0.04-0.12		0.04	0.11
	0.12-0.17		0.07	0.19
	0.17-0.24		0.08	0.09
	0.24-0.35		0.03	0.07
	0.35-0.47		0.02	0.06
$3\beta,7\alpha$-diol	0.47-0.53	0.51	0.03	0.07
	0.53-0.57		0.02	0.11
$3\beta,7\beta$-diol	0.57-0.61	0.58	0.13	0.52
$3\beta,5\alpha$-diol	(0.65)	0.63	0.27	0.31
7-ketone	(0.70)	0.68	0.58	0.59
$5\alpha,6\alpha$-epoxide	(0.76)	0.75	2.26	2.10
	0.80-0.92		0.40	0.35
cholesterol	(1.0)	1.0	95.7	95.0
	1.09-1.3		0.08	0.08

$^a R_c = R_f$ of products/R_f cholesterol.
$^b R_c$ of observed spots in autoradiogram are in parentheses. Other R_c values refer to arbitrarily removed zones.

In the fraction isolated from the cells and assumed to be inside the phagocytic vacuoles, 1.0×10^{-8} moles of labelled material were recovered, of which no more than 3.1×10^{-11} moles were $3\beta,5\alpha$-diol. It is very probable that not all of the radioactive material collected in this spot was the 5α product since there is a substantial background of radioactivity where the known products do not occur. Although we attempted to carry out a quantitative study of the upper limits for singlet oxygen production in this system using the quantitation technique described earlier, it was found that the efficiency of the beadlet system depended strongly on concentration. Since the beads are concentrated locally after ingestion, but in an unknown fashion, we decided that it was meaningless to calculate the efficiencies. However, the following statement can be made: although new products are formed, the singlet oxygen products are at best a minor constituent. In Table 2, the upper limit for the 5α product is 0.3%; many runs had up to an order of magnitude less. Thus, it seems very unlikely that singlet oxygen is a major contributor to the oxidation of unsaturated materials in the membranes of phagocytized organisms.

The major product (over 2% of material formed both inside and outside the cells) appears at an R_f value identical to that of the 5α, 6α-epoxide and is

not separated from it on any thin layer system investigated. Although its
identity has not been established completely, we believe that this material is
the epoxide. Smaller amounts of other unknown materials are also formed.
Indeed, the TLC spectrum of the total product mixture is similar to that formed
on reaction of cholesterol with hydrochlorous acid; the product distribution of
this reaction is shown in Table 3. We have not yet established the structure of
these compounds but we believe many of the products with PMNs may well be products of reaction of HOCl with cholesterol.

TABLE 3

RADIOACTIVITY DISTRIBUTION ON TLC PLATE FROM HOCl REACTION[a]

Product	R_c[b,c] (Products)	R_c[b] (Standards)	% of Total Recovered in Reaction	Control
	0.0 -0.19		1.40	0.63
	0.19-0.22		0.65	0.30
3β,7α-diol	(0.25)	0.24	1.53	0.30
	0.27-0.30		1.41	0.52
	0.30-0.34		0.32	0.08
3β,7β-diol	(0.37)	0.37	2.32	0.48
3β,5α-diol	0.39-0.43	0.42	2.07	0.38
	(0.46)		5.14	0.53
7-ketone	0.49-0.54	0.52	1.21	0.39
5α,6α-epoxide	(0.57)	0.57	7.62	2.43
	0.61-0.73		0.76	0.34
	0.73-0.85		1.30	0.62
	0.85-0.96		1.44	1.02
cholesterol	(1.0)	1.0	67.63	90.57
	1.05-1.07		1.99	0.34
	(1.11)		9.36	0.69
	(1.23)		1.46	0.37

[a]Carried out in acetate buffer (pH 4.65); [HOCl] = 1.22 x 10^{-5}M.

[b]R_c=R_f of products/R_f cholesterol.

[c]R_c of observed spots in autoradiogram are in parentheses, other R_c values
refer to arbitrarily removed zones.

A very similar series of experiments was carried out using [4-^{14}C]-cholesterol
dissolved in a mineral oil dispersion.[34] The technique was quantiated by the
same method used for the beadlets. If the efficiency is assumed to be independent
of the local concentration of the dispersion, an upper limit of 2 ± 2% of the
oxygen taken up by the PMNs during the ingestion of the mineral oil droplets was
found to be singlet. Although we did not carry out experiments to test the
concentration dependence with the mineral oil dispersion, it is likely that
localization effects would also affect the efficiency of this system.

The cholesterol-bead technique was also used with rat lung macrophages.[35] In
these systems, it was not possible to meaure oxygen uptake but a variety of products were formed (although in low total conversion) (Table 4), but they are very
different from those formed in the PMNs. Other runs gave larger amounts of

products resembling the free-radical distribution. Again, no more than a fraction of a percent of the 5α-product is formed and $^1O_2$ seems to be absent. In this system, no epoxides seem to be formed; what products there are are similar to those of free radical oxidation.

TABLE 4

RADIOACTIVITY DISTRIBUTION IN TLC PLATES FROM INCUBATION WITH RAT LUNG MACROPHAGES

Product	R_c [a]	R_c [b] (Standards)	% of Total Recovered in Intracellular Fraction	Extracellular Fraction	Control
	0.00-0.04		0.50	0.50	0.60
	0.04-0.08		0.11	0.04	0.07
	0.08-0.16		0.29	0.04	0.10
	0.16-0.24		0.14	0.06	0.11
3β,7α-diol	0.24-0.29	0.26	0.19	0.05	0.09
	0.29-0.35		0.07	0.04	0.06
3β,7β-diol	0.35-0.40	0.38	0.12	0.05	0.10
3β,5α-diol	0.40-0.49	0.45	0.11	0.08	0.14
	0.49-0.53		0.04	0.08	0.08
7-ketone	0.53-0.58	0.55	0.13	0.07	0.13
5α,6α-epoxide	0.58-0.65	0.60	0.30	0.14	0.20
	0.65-0.76		0.12	0.15	0.15
	0.76-0.85		0.40	0.16	0.19
	0.85-0.94		0.53	0.65	0.97
cholesterol	(1.0)	1.0	95.90	97.4	96.64
	1.07-1.14		0.70	0.50	0.30
	1.14-1.29		0.40	0.06	0.10

[a] R_c values refer to arbitrarily removed zones. Value in parentheses is sole spot observed in autoradiogram.

[b] $R_c = R_f$ of standards/R_f of cholesterol.

Since the macrophages do not contain a myeloperoxidase,[20] it would not be expected that hypochlorous acid would be formed and it is significant that products possibly caused by its reaction in the case of the PMNs are absent in macrophages. Thus the macrophage system actually serves as a useful control for the PMNs since the product distribution is different.

The conclusion we draw from these studies is that singlet oxygen does not play a major role in the bactericidal action of either PMNs or macrophages, and is at best a very minor product in the dismutation of superoxide.

ACKNOWLEDGEMENTS

The PMN work was carried out with the collaboration and advice of Dr.Robert Lehrer of the UCLA Department of Medicine; that with macrophages was done with Professor Anne P. Autor and Dr.J.B. Stevens, Department of Pharmacology, University of Iowa. We express gratitude to these collaborators, without whom the work could not have been done. This work was supported by Public Health Service grant No. GM 20080. R.B. Abakerli thanks the Fundacão de Amparo a Pesquisa do Estado de S. Paulo for a fellowship.

228

REFERENCES

1. Fridovich, I. (1972) Account.Chem. Res. 5, 321-326; b) McCord, J.M. and
   Fridovich, I. (1968) J. Biol. Chem. 243, 5733-5760; c) Lynch, R.E. and
   Fridovich, I. (1978) J. Biol. Chem. 253, 1838-1845; Fridovich, I. (1978)
   Science, 201, 875; Fridovich, I. (1979) Adv. Inorg. Biochem. 1, 67.
2. Kellogg, E.W.,III and Fridovich, I. (1975) J. Biol. Chem. 250, 8812-8817.
3. Lavelle, F., Michelson, A.M. and Dimitrejevic, L. (1973) Biochem. Biophys.
   Res. Commun. 55, 350-357.
4. Michelson, A.M. and Buckingham, M.E. (1974) Biochem. Biophys. Res. Commun.
   58, 1079-1086.
5. Fee, J.A. and Valentine, J.S. (1977) in Superoxide and Superoxide Dismutases,
   Michelson, A.M., McCord, J.M. and Fridovich, I., eds., Academic Press,
   New York, pp.19-60.
6. Behar, D., Czapski, G., Rabani, J., Dorfman, L.M. and Schwartz, H.A. (1970)
   J. Phys. Chem. 74, 3209-3213.
7. a) Khan, A.U. (1970) Science, 168, 476-477; b) Bors, W., Saran, M.,
   Lengfelder, E., Spottl, R. and Michel, C. (1974) Curr. Top. Radiat. Res.
   9, 247-309; c) Mayeda, E.A. and Bard, A.J. (1974) J. Amer. Chem. Soc. 96,
   4023-4024; d) Goda, K., Chu, J., Kimura, T. and Schaap, A.P. (1973)
   Biochem. Biophys. Res. Commun. 52, 1300-1306; e) Bus, J.S., Aust, S.D. and
   Gibson, J.E. (1974) Biochem. Biophys. Res. Commun. 58, 749-755.
8. a) Nilsson, R. and Kearns, D.R. (1974) J. Phys. Chem. 78, 1681-1683;
   b) Poupko, R. and Rosenthal, I. (1973) J. Phys. Chem. 77, 1722-1724;
   c) Barlow, G.E., Bisby, R.H. and Cundall, R.B. (1979) Radiat. Phys. Chem.
   13, 73.
9. a) Guiraud, H.J. and Foote, C.S. (1976) J. Amer. Chem. Soc. 98, 1984-1986;
   b) Kobayashi, S. and Ando, W. (1979) Biochem. Biophys. Res. Commun. 88,
   676-681.
10. a) Schenck, G.O., Gollnick, K. and Neumüller, O.-A. (1957) Liebigs Ann.
    603, 46; Nickon, A. and Bagli, J.F. (1959) J. Amer. Chem. Soc. 81, 6330;
    Nickon, A. and Bagli, J.F. (1961)
    b) Schenck, G.O. (1957) Angew. Chem. 69, 579-599; c) Kulig, M.J. and
    Smith, L.L. (1973) J. Org. Chem. 38, 3639-3642; d) Teng, J.I. and Smith, L.L.
    (1973) J. Amer. Chem. Soc. 95, 4060-4061; e) Smith, L.L., Kulig, M.J. and
    Teng, J.I. (1977) Chem. Phys. Lipids, 20, 211-227; f) Smith, L.L. and Kulig,
    M.J. (1975) J. Amer. Chem. Soc. 98, 1027-1029; g) Smith, L.L., Kulig, M.L.,
    Miller, D. and Ansari, G.A.S. (1978) J. Amer. Chem. Soc. 100, 6206-6211;
    h) Smith, L.L., Teng, J.I., Kulig, M.J. and Hill, F.L. (1973) J. Org. Chem.
    38, 1763-1806.
11. A solution of TLC-purified 4-C^{14} cholesterol (50 µl) and 0.5 ml diethyl
    ether were stirred with 0.2 ml of a 10% aq. suspension of polystyrene latex
    beads (Dow Diagnostic, 0.82 µm average diameter) for 30 min; 2-3 ml dist.
    H_2O was added and the ether removed by passing N_2 over the sample. 25 ml of
    the appropriate 0.1 M buffer (pH 4, acetate; pH 6,7 and 8, phosphate; pH 10,
    carbonate) were then added.
12. a) Prepared by the method of McElroy, A.D. and Hashman, J.S. (1964) Inorg.
    Chem. 3, 1798-1799. Caution: explosions have been reported during this
    reaction. Extraction should be carried out in all-glass apparatus; avoid
    paper thimbles. b) Material so prepared had m.p. 101°C uncorr. (Ref.12a,
    m.p. 97°C) and assayed 96-98% O_2^- by O_2 evolution.
13. Using a scintillation counter; 10 ml Biofluor (NEN) were added; the internal
    standard technique was used to correct for quenching.
14. The aqueous layer was centrifuged and the amount of histidine reacted measured
    by the Pauly reagent (Ref.17); unreacted controls were used for comparison.
    Controls also established that the extraction process did not remove histidine,
    and that buffer, histidine photooxidation products, or methylene blue did not
    interfere with the analysis.
15. The fraction of $^1O_2$ trapped is $[A]/(\beta+[A])$ and was 1/8 at the 5×10^{-4}M
    concentration of histidine used (Ref.18). Controls showed that the P -chol
    did not trap a substantial fraction of the $^1O_2$ generated.

16. It is interesting to calculate the steady state concentration of $[O_2^-]$ present. From the rate of addition of O_2^- and the rate of decay at pH 7, calculated from Czapski's relationship (Ref.6), $6.36 \times 10^5 M^{-1} sec^{-1}$, the steady state concentration of O_2^- is calculated to be $3.9 \times 10^{-6}M$, and the fraction of $^1O_2$ quenched (from the O_2^- quenching rate, Ref.9a) to be about 1.6%. The steady state concentration of O_2^- increases with pH because of the slower dismutation of O_2^- (Ref.6); however, even at pH 10, the amount of $^1O_2$ quenching is calculated to be no more than 20% under the conditions of the yield experiment at pH 10.

17. McPherson, H.T. (1946) Biochem. J. 40, 470-475.

18. For definitions of terms and values of constants, see Foote, C.S. (1976) in Free Radicals in Biology, Vol.2, Pryor, W.A., ed., Academic Press, New York, p.94.

19. For a review see a) Karnovsky, M.L. (1973) Fed. Proc. 32, 1527; b) Klebanoff, S.J. (1975) Semin. Hematol. 12, 117-142; c) Sbarra, A.J., Selvaraj, R.S., Paul, B.B., Poskitt, P.F.K., Zgliczynski, J.M., Mitchell, G.W., Jr. and Louis, F. (1976) Int. Rev. Exp. Pathol. 16, 249.

20. Cohn, Z.A. (1970) Semin. Hematol. 7, 107.

21. Babior, B.M., Kipnes, R.S. and Curnutte, J.T. (1973) J. Clin. Invest. 52, 741-742.

22. Curnutte, J.T., Whitten, D.M. and Babior, B.M. (1974) N. Engl. J. Med. 290, 593-597.

23. Allen, R.C., Stjernholm, R.L. and Steele, R.H. (1972) Biochem. Biophys. Res. Commun. 47, 679-684.

24. Stauff, J., Sander, V. and Jaeschke, W. (1973) in Chemiluminescence and Bioluminescence, Cormier, M.J., Hercules, D.M. and Lee, J., eds., Plenum Press, New York, pp.131-141.

25. Khan, A.U. (1976) J. Phys. Chem. 80, 2219.

26. a) Klebanoff, S.J. (1977) J. Biol. Chem. 252, 4803-4810; b) Rosen, H. and Klebanoff, S.J. (1976) Fed. Proc. 35, 1391.

27. a) Held, A.M. and Hurst, J.K. (1978) Biochem. Biophys. Res. Commun. 81, 878-885; b) Harrison, J.E., Watson, B.D. and Schultz, J. (1978) FEBS Lett. 92, 327-332. c) See also a discussion of this topic by Krinsky, N.I. (1979) in Singlet Oxygen, Wasserman, H.H. and Murray, R.W., eds., Academic Press, New York.

28. A benzene solution of $[4-^{14}C]$-cholesterol (0.740 ml; 17.6 μCi/ml) were dried under a nitrogen stream and redissolved in 1 ml methanol. To this solution, 0.5 ml of a 10% polystyrene suspension was added and dried under nitrogen. The residue was resuspended in 1 ml of doubly distilled water and diluted to 2 ml with phosphate buffer (PBS). The suspension was filtered through a capillary, homogenized and 10 ul were checked for radioactivity.

29. The PMNs were isolated as described by Boyum, A. (1968) Scand. J. Clin. Lab. Invest. 21, 77, and resuspended in PBS, pH 7.4. A count of the cells gave 1.0×10^7 cells/ml.

30. Followed in a Gilson Oxygraph with a Clark electrode.

31. The separation of the particles [by layering over a density gradient (Ficoll type 400, density 1.06 g/cm^3) and centrifuging at 2000 rpm for 10 min] did not pellet the cells completely. This separation step was then performed by several saline solution washing - centrifuging steps. Microscopic examination showed that the supernatants did not contain cells. However, some unigested particles remained in the cell fraction due to agglomeration during centrifuging.

32. The 3β,5α-diol was synthesized by the procedure of Kulig anf Smith (Ref.10c); the 7-ketone by that of Nickon, A. and Mendelson, W.L. (1965) J. Amer. Chem. Soc. 87, 3928-3934; the 3β, 7α,β-diols through the reduction of LAH of the 7-ketone and the 5α,6α-epoxide by reacting cholesterol and m-chloroperbenzoic acid at room temperature. All standards were purified by column chromatography.

33. The TLC's were carried out on 0.25 mm silica gel HF 254 plates (Merck). The solvent system was ether:hexane (9:1) for the first two irrigations and benzene ethyl acetate (3:1) for the third one. The plates were stained by

230

spraying with 50% methanol-sulfuric acid and heating. SB54 Kodak films and intensifying screens (Cronex Quanta IIB) were used for autoradiography; the system was kept at -70ºC for one week.

34. Klebanoff, S.J. (1974) J. Biol. Chem. 249, 3724-3728.

35. The macrophages were extracted following the procedure of description II of Stevens, J.B. and Autor, A.P. (1977) Lab. Invest. 37, 470-478. Three cell preparations containing 4.4×10^6, 4.8×10^6 and 1.5×10^6 cells, respectively, were resuspended in 10 ml of Joklik's minimum essential medium and 0.4 ml of a saline suspension of cholesterol-supported microbeads was added to each and incubated for 1 hour at 37ºC with shaking every 10 min. The cells containing ingested particles were separated by layering over a 75% Ficoll, type 400 density medium and centrifuging for 10 min at 2000 rpm, which precipitated the cells. The supernatants were centrifuged further at 10,000 rpm for 20 min to pellet the uningested particles. Both fractions were treated with 1 ml of methanol-triphenylphosphine solution (3.8×10^{-3}M) and the cholesterol was extracted by treating several times with methanol and centrifuging. The combined extracts were dried, redissolved in 0.2 ml of methanol and applied to a TLC plate and worked up as previously described.

THE CYTOCHROME b COMPONENT OF THE MICROBICIDAL OXIDASE SYSTEM OF HUMAN
NEUTROPHILS

ANTHONY W. SEGAL[+] AND OWEN T.G. JONES
[+]Department of Haematology, University College Hospital, London, WC1E 6AU,
England; Department of Biochemistry, University of Bristol, Bristol, BS8 1TD,
England

It has been known since 1933 that neutrophilic polymorphonulcear leukocytes
(neutrophils) demonstrate a marked increase in oxygen consumption when exposed
to a phagocyctic stimulus.[1] The respiratory burst is important for the killing
of certain bacteria[2,3] by neutrophils. This is dramatically demonstrated in
the syndrome of Chronic Granulomatous Disease (CGD).[4-6] Patients with this
disease have an unusual susceptibility to bacterial infection and their neutrophils show no evidence of enhanced oxygen metabolism in response to a variety
of stimuli.[7,8] The syndrome of CGD includes a number of distinct groups of
patients.[9,10] The condition is most commonly diagnosed in boys since the mode
of inheritance appears to be X-linked in about 80% of the subjects and autosomal
recessive in most of the others.[9,11] The diagnosis is confirmed by isolating
leukocytes from the patients' blood and demonstrating defective killing of
bacteria[12] and the absence of the normal burst of respiration and associated
activities. The normal enhancement of oxygen consumption[7], hexose monophosphate shunt activity[7], reduction of cytochrome c[13], or nitro blue tetrazolium,[14]
H_2O_2 generation in the presence of azide[15] and of chemiluminescence[16] are not
seen after treatment with either particulate[7] or soluble stimuli such as phorbol
myristate acetate (PMA)(Ref.8) and cytochalasin E,[17] although a patient has
been described in whose cells the respiratory burst could be activated by soluble
stimuli despite the inability of particles to stimulate the system.[18]

Ever since the observation[19] that the phagocytosis associated respiratory
burst is not decreased by cyanide and azide, the classical inhibitors of cytochrome oxidase, there has been much search[20] for the "oxidase enzyme". The
reducing equivalents appear to originate from glucose. The conversion of
glucose to CO_2 is enhanced with characteristics[19,21] suggesting a relatively
greater increase in the rate of metabolism through the hexose monophosphate
(HMP) shunt. NADH and NADPH have both been proposed as direct substrates of
the oxidase system. A number of investigators have measured NADH (Refs.22-24)
and NADPH (Refs.25-28) "oxidase" activity in a variety of subcellular locations
and the ratios of NADH to NAD and NADPH to NADP have been shown to change after

phagocytosis.[29,30] Of these two substrates, NADPH has been favoured as the natural one because it is produced by the HMP shunt. In the rare patients with concomitant glucose-6-phosphate dehydrogenase deficiency, the activity of the shunt is diminished and a mild variant of CGD is observed.[10,31]

The "enzyme" which has received the most support as candidate for the role of microbicidal oxidase is an NADPH "oxidase"[25-28] which is assayed by superoxide dismutase inhibitable, NADPH-dependent reduction of cytochrome $\underline{c}$ and which is thought to be a membrane bound flavoprotein.[32] The activity of this "enzyme" is enhanced when normal cells are stimulated with zymosan, and this increase in activity is not observed in cells from patients with CGD (Refs.26,27,33). There are, however, a number of serious objection to the acceptance of this NADPH "oxidase" activity as being due to the natural activity of a single enzyme. The "enzyme" is assayed at $25^{\circ}C$ in sucrose solutions, and even under these optimal conditions its activity can only account for a small proportion of the overall oxygen consumption. There are, however, two observations that indicate that if this NADPH "oxidase" is other than an artefact it is of doubtful physiological relevance as an independent microbicidal enzyme. Firstly, unlike zymosan, latex particles fail to activate this superoxide generating "enzyme",[34] although they stimulate the respiratory burst equally well. Secondly, preparations of cells from patients with CGD, do not initially have the additional activity after stimulation that is observed with normal cells; however, they develop this enhanced activity after dialysis [35] or refrigeration overnight at $4^{\circ}C$.[36] This suggests that the NADPH "oxidase" activity could be due to an oxidation product of the cells or zymosan, which is normally produced in normal cells by the respiratory burst, and which develops more slowly in the extracts of CGD cells as a result of prolonged exposure to atmospheric oxygen.

In the absence of a clearly defined oxidase enzyme, we have examined the spectral characteristics of neutrophils in an attempt to identify any redox agents which could be implicated in this oxidase system. Reduced minus oxidised difference spectra of purified homogenised neutrophils demonstrated dramatic spectral changes,[37] the most obvious of which were a broad peak due to myeloperoxidase at 474 nm, and smaller peaks at 559, 530 and 429 nm corresponding to the α, β and Soret bands of a cytochrome $\underline{c}$ (Figure 1). A number of Japanese workers had previously described and similarly interpreted almost identical spectral changes in horse[38,39] and rabbit[40] neutrophils. The shift of the 559 nm absorption band to 557 nm upon the addition of pyridine and alkali (see Figure 2 and Refs.38,41) is characteristic of the cytochromes of the $\underline{b}$-type, which have a protohaem prosthetic group.

In studies on homogenates of animal neutrophils the Japanese workers localised the cytochrome $\underline{b}$ to the neutrophil granules[38] and then, more precisely, to the specific or A granules[38,39] by differential centrifugation. By the use of

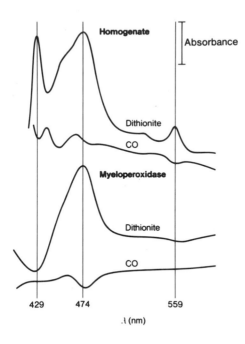

Fig. 1. Dithionite and CO difference spectra of an homogenate
of neutrophils (4 mg protein/
cuvette) from a pool of 20 normal
donors, and of purified human
myeloperoxidase (50 µg protein/
cuvette). Absorbance scale
marker = 0.0056 absorbance unit.

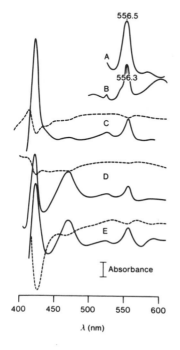

Fig. 2. Dithionite (———) and
CO (-----) difference spectra of
fractions separated from homogenates of human neutrophils by
centrifugation on discontinous
sucrose gradients (Ref.41).
Spectra of the fractions collected
at the interfaces between sucrose
of densities 1.10 and 1.17 g/ml
(A,B and C) and between 1.17 and
1.21 g/ml (D), and of the whole
cell homogenate (E) are shown.
Specimen A was measured in (alkaline) pyridine and B was at 77°K.
Absorbance scale marker = 0.0056
absorbance unit.

analytical subcellular fractionation on continuous sucrose gradients we found
that in human cells this cytochrome b is largely localised to the plasma membrane
(Figure 3 and 4) with an additional small amount in the region of the specific
granules. Further evidence of a plasma membrane location comes from the observations that this cytochrome b is incorporated into the phagocytic vacuole as soon

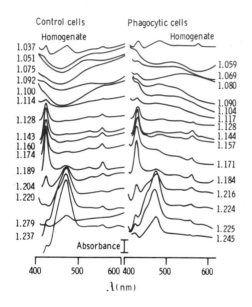

Fig. 3. Dithionite difference
spectra of fractions of homogenates of neutrophils before
(control) and after (phagocytic
cells) the phagocytosis of latex
particles. The organelles have
been separated on continuous
gradients of sucrose and show
the difference in the density
profiles of the cytochrome b
and myeloperoxidase. Absorbance
scale marker = 0.0056 absorbance
unit. The numbers indicate the
density (g/ml) of the respective
fractions.

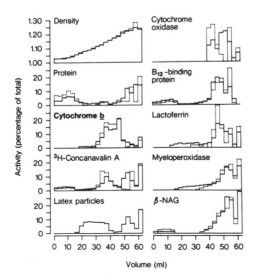

Fig. 4. Analytical subcellular
fractionation of neutrophils
showing the density profiles of
the markers for the various organelles which were separated on
the gradients shown in Fig.3.
³H-Concanavalin A was used as
the marker of the plasma membrane.

as these vacuoles are formed,[42] much earlier than myeloperoxidases (Figure 5), and that dithionite, which only slowly penetrates membranes,[43] readily reduces the b cytochrome in intact cells while only very little of the myeloperoxidase, which is intracellular, becomes reduced (Figure 6). The myeloperoxidase does however become rapidly reduced when the integrity of the membranes is disrupted by the addition of a detergent (Figure 6).

Dithionite difference spectra of phagocytic
vacuoles at various intervals after phagocytosis

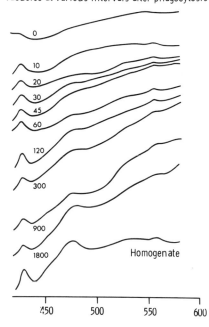

Fig. 5. Dithionite difference spectra of phagocytic vacuoles containing latex particles isolated from human neutrophils by floatation at different times after phagocytosis. The integrated areas under the curves have been corrected for variations in the numbers of phagocytic vacuoles. The figures associated with each spectrum indicate the time in seconds after the addition of latex particles at which phagocytosis was stopped and the vacuoles prepared.

The evidence that this cytochrome b might function as a component of the microbicidal oxidase system comes from studies on its oxidation state after stimulation of the cell, and on the examination of neutrophils from patients with CGD.

We have examined the spectral changes induced in intact neutrophils after stimulation with either an opsonised particulate stimulus, latex particles coated with IgG,[41] or a soluble stimulus, PMA. The latex particles induced the appearance of a broad band of absorption with a peak at about 415 nm with a shoulder at about 440 nm (Figure 7), but did not reveal the spectral changes of a reduced cytochrome c. PMA stimulation, on the other hand, produced the classical spectrum of the cytochrome (see Figure 8 and Ref.44). In addition, this reduction of the cytochrome b was enhanced under anaerobic conditions and reversed by the reintroduction of oxygen, indicating the reversible, oxygen

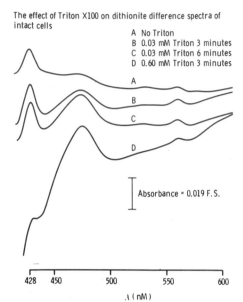

The effect of Triton X100 on dithionite difference spectra of intact cells

A No Triton
B 0.03 mM Triton 3 minutes
C 0.03 mM Triton 6 minutes
D 0.60 mM Triton 3 minutes

A

B

C

D

Absorbance = 0.019 F.S.

428   450   500   550   600

λ (nM)

Fig. 6. Reduction of the cyto-
chrome $\underline{b}$ after the addition of
dithionite to intact human
neutrophils and to cells in the
presence of Triton X–100

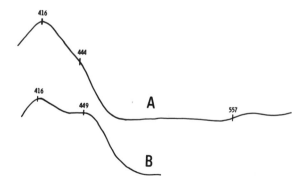

416
444
416
449   A   557
B

Fig. 7. Difference spectra of intact neutrophils after the addition of
opsonised latex particles in the absence (A) and presence (B) of 2 mM KCN
(Ref.37).

dependent, oxidation and reduction of this molecule (Figure 8). Thus, a (soluble) stimulus that activates the respiratory burst also results in the reduction
of the cytochrome b̲.

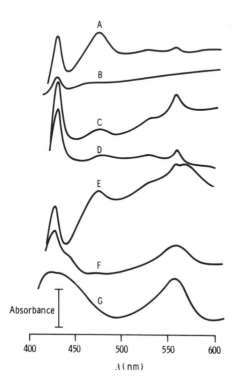

Fig. 8. Effect of phorbol
acetate (PMA) on the spectra of
human neutrophils. A: Cells +
dithionite. B: Aerobic cells +
PMA (after ∿2 min). C: Anaerobic cells + PMA (after ∿2 min).
D: Spectrum of cytochrome b̲ in
purified neutrophil membranes.
E: Sample C after the reintroduction of air (5 min after
stimulation). F: Sample E 15
min later. G: Cells ± latex
particles (after 2 min).

Cells from patients with CGD were examined for spectroscopic evidence of this
cytochrome b̲. In preparations of cells from the first four patients that we
studied, the cytochrome b̲ appeared to be completely missing in two, and in the
other two, including a girl, the spectra were bizarre.[45] Since the quantity of
the cytochrome appeared to be diminished in cells from two mothers, who were
heterzygotes, we suggested that a defect of this cytochrome represented the
basic abnormality in this disease. It was subsequently shown, in studies
which we successfully repeated, that three Danish patients with CGD, including
a brother and sister, had spectroscopic evidence of a normal cytochrome c̲ (Ref.
46). Upon re-examination of our four patients, spectral evidence of this
cytochrome was still missing in the three boys, but was present in the girl.[47]

Unfortunately interpretation of the original spectra had been complicated as a
result of contamination by eosinophil peroxidase, which has major peaks of
absorbance in the same regions as cytochrome b̲ (Ref.52). We have now examined

cells from six English males, all of whom have lacked the cytochrome b, and from two females, who were found to contain it. The demonstration that this cytochrome b was present in some of the patients with CGD, in whom the respiratory burst was lacking, suggested that the cytochrome was not the only component involved in the microbicidal oxidase system. We thus examined the effect of stimulating cells from these patients with PMA under anaerobic conditions to determine whether or not they were able to reduce the cytochrome.[48] Whereas the control cells demonstrated the customary reduction of the cytochrome b, the spectrum of the reduced cytochrome was not produced either by cells from male patients, which appear to lack this molecule, or, what is more important, by cells from female patients in whom spectroscopic evidence of a normal cytochrome is observed after dithionite reduction (Figure 9 and 10). It thus seems that the microbicidal oxidase is a complex electron transport chain in which the cytochrome b is a component. Possibly this chain is not functional in the common X-linked forms of CGD becuase of a primary abnormality of the cytochrome, and inactive in the female patients with the disease, who have the cytochrome, because of the absence of malfunction of a proximal component, or because of defective activation or integration of the system.

In this study we have found it convenient to reduce intact cells, rather than cell homogenates, by the addition of dithionite. This was a technically superior

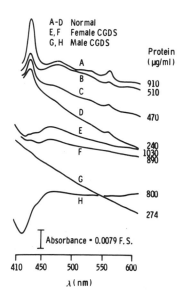

Difference spectra of phorbol stimulated neutrophils

Fig. 9. Dithionite difference spectra of intact neutrophils from normal subjects and male and female patients with CGD.

Dithionite difference spectra of intact neutrophils

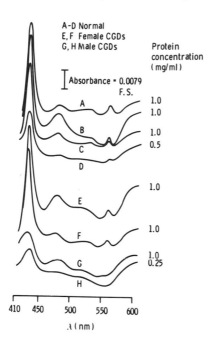

A-D Normal
E,F Female CGDs
G,H Male CGDs

Protein
concentration
(mg/ml)

Absorbance = 0.0079
F.S.

A   1.0
     1.0
B
C   1.0
     0.5
D

E   1.0

F   1.0

G   1.0
     0.25
H

410 450   500   550   600

λ (nm)

Fig. 10. Difference spectra of neutrophils from normal subjects and patients with CGD after stimulation of the cells with phorbol myristate acetate. The letters A-H refer to the same subjects as shown in Fig.9 and the figures refer to the protein concentration of the cells in the cuvettes.

way of demonstrating the spectra of the cytochrome b because the interference by myeloperoxidase and eosinophil peroxidase was greatly diminished as these substances are located within cytoplasmic granules, and are consequently reduced by the dithionite which penetrates membranes poorly.[43] In the two male patients who were found to lack cytochrome b, other changes could be observed because of the absence of the strong absorbance of the b cytochrome in the region of the Soret band. These absorption bands were broader than that of the cytochrome b, with peaks at 431 and 435 nm rather than at 429 nm (Figure 9). These peaks could be due to contamination by haemoglobin, (see M.N. Hammers et al., this volume) or else they could be due to other electron carrier molecules which are usually concealed by the much larger absorption band of the cytochrome b. Similarly,whereas normal cells show the spectrum of the reduced cytochrome after stimulation with PMA, all four samples of CGD cells showed a pattern of spectral changes which were similar to each other and very different from normal. These changes, with troughs at about 422 and 444 nm and broad peaks at about 433 and 463 nm which are obscured by the other spectral changes in normal cells, are very similar to the reduced minus oxidised spectra of flavoproteins.[49]

On the basis of our data the microbicidal oxidase system appears to us to be a complex electron transport chain in which the cytochrome b might play a role.

Many different defects of such a system can be responsible for the syndrome of CGD, a condition that is analogous to the dysfunction of complex electron transport systems in bacterial mutants which may be due to abnormalities of any one of several individual components.[50] Different components of this electron carrier system could have diaphorase activity, each of which has probably been separately assayed by different investgators, thereby accounting for the many different "oxidase enzymes" that have been described in these cells and found to be abnormal in CGD.[22,25-28,51]

ACKNOWLEDGEMENTS

We thank the Wellcome Trust and the Medical Research Council for support.

REFERENCES

1. Baldridge, C.W. and Gerard, R.W. (1933) Amer. J. Physiol. 103, 235-236.
2. Selvaraj, R.J. and Sbarra, A.J. (1966) Nature, 211, 1272-1276.
3. Mandell, G.L. (1974) Infect. Immunity, 9, 337-341.
4. Berendes, H., Bridges, R.A. and Good, R.A. (1957) Minnesota Med. 40, 309-312.
5. Johnston, R.B. and Newman, S.L. (1977) Pediat. Clin. Amer. 24, 365-376.
6. Holmes, B. and Good, R.A. (1971) Immunobiology, pp.55-61.
7. Holmes, B., Page, A.R. and Good, R.A. (1967) J. Clin. Invest. 46, 1422-1432.
8. Repine, J.E., White, J.G., Clawson, C.C. and Holmes, B.M.(1974) J. Clin. Invest. 54, 83-90.
9. Windhorst, D.B., Page, A.R., Holmes, B., Quie, P.G. and Good, R.A. (1968) J. Clin. Invest. 47, 1026-1034.
10. Gray, G.R., Klebanoff, S.J., Stamatoyannopoulos, G., Austin, T., Naiman, S.C., Yoshida, A., Kliman, M.R. and Robinson, G.C.F. (1973) Lancet, 2, 530-534.
11. Quie, P.G., Kaplan, E.L., Page, A.R., Gruskay, F.L. and Malawista, S.E. (1968) N. Engl. J. Med. 278, 976-980.
12. Quie, P.G., WHite, J.G., Holmes, B. and Good, R.A. (1967) J. Clin. Invest. 46, 668-679.
13. Curnutte, J.T., Whitten, D.M. and Babior, B.M.(1974) N. Eng. J. Med. 290, 593-597.
14. Baehner, R.L. and Nathan, D.G. (1968) N. Engl. J. Med. 278, 971-976.
15. Homan-Müller, J.W.T., Weening, R.S. and Roos, D. (1975) J. Lab. Clin. Med. 85, 198-207.
16. Stjernholm, R.L., Allen, R.C., Steele, R.H., Waring, W.W. and Harris, J.A. (1973) Infect. Immunity, 7, 313-314.
17. Nakagawara, A., Kakinuma, K., Shin, H., Miyazaki, S. and Minakami, S. (1976) Clin. Chim. Acta, 70, 133-137.
18. Harvath, L. and Andersen, B.R. (1979) N. Engl. J. Med. 300, 1130-1135.
19. Sbarra, A.J. and Karnovsky, M.L. (1959) J. Biol. Chem. 234, 1355-1362.
20. Badwey, J.A., Curnutte, J.T. and Karnovsky, M.L. (1979) N. Engl. J. Med. 300, 1157-1159.
21. Stjernholm, R.L. and Manak, R.C. (1970) J. Reticuloendothelial Soc. 8, 550-560.
22. Cagan, R.H. and Karnovsky, M.L. (1964) Nature, 204, 255-257.
23. Baehner, R.L., Gilman, N. and Karnovsky, M.L. (1970) J. Clin. Invest. 49, 692-700.
24. Briggs, R.T., Drath, D.B., Karnovsky, M.L. and Karnovsky, M.J. (1975) J. Cell. Biol. 67, 566-586.
25. Iyer, G.Y.N. and Quastel, J.H. (1963) Can. J. Biochem. Physiol. 41, 427-434.
26. Hohn, D.C. and Lehrer, R.I. (1975) J. Clin. Invest. 55, 707-713.
27. DeChatelet, L.R., McPhail, L.C., Mullikin, D. and McCall, C.E. (1975) J.Clin. Invest. 55, 714-721.

28. Gabig, T.G., Kipnes, R.S. and Babior, B.M. (1978) J. Biol. Chem. 253, 6663–6665.
29. Selvaraj, R.J. and Sbarra, A.J. (1967) Biochim. Biophys. Acta, 141, 243–249.
30. Aellig, A., Maillard, M., Phavorin, A. and Frei, J- (1977) Enzyme, 22, 207–212.
31. Baehner, R.L., Johnston, R.B. and Nathan, D.G. (1972) J. Reticuloendothelial Soc. 12, 150–169.
32. Babior, B.M. and Kipnes, R.S. (1977) Blood, 50, 517–524.
33. Curnutte, J.T., Kipnes, R.S. and Babior, B.M. (1975) N. Engl. J. Med. 293, 628–632.
34. Tauber, A.I. and Curnutte, J.T. (1978) Blood, 52, Suppl. 1, 128.
35. DeChatelet, L.R., Shirley, P.S., Bass, D.A. and McCall, C.E. (1979) Clin. Res. 27, 343A.
36. Johnston, R.B., personal communication.
37. Segal, A.W. and Jones, O.T.G. (1978) Nature, 276, 515–517.
38. Hattori, H. (1961) Nagoya J. Med. Sci. 23, 362–378.
39. Ohta, H., Takahashi, H., Hattori, H., Yamada, H. and Takikawa, K. (1966) Acta Haemotol. Japan, 29, 799–808.
40. Shinagawa, Y., Shinagawa, Y., Tanaka, C. and Teraoka, A. (1966) J. Electron Microsc. 15, 81–85.
41. Segal, A.W. and Jones, O.T.G. (1979) Biochem. J. 180, 33–44.
42. Segal, A.W. and Jones, O.T.G., submitted for publication.
43. Bendall, D.A., Davenport, H.E. and Hill, R. (1971) Methods Enzymol. 23, 23A, 327–344.
44. Segal, A.W. and Jones, O.T.G. (1979) Biochem. Biophys. Res. Commun. 88, 130–134.
45. Segal, A.W., Jones, O.T.G., Webster, D. and Allison, A.C. (1978) Lancet, 2, 446–449.
46. Borregaard, N., Staehr-Johansen, K., Taudorff, E. and Wandall, J.H. (1979) Lancet, 1, 949–951.
47. Segal, A.W. and Jones, O.T.G. (1979) Lancet, 1, 1036–1037.
48. Segal, A.W. and Jones, O.T.G., submitted for publication.
49. Mahler, H.R. and Cordes, E.H. (1966) Biological Chemistry, p.207.
50. Haddock, B.A. and Jones, C.W. (1977) Bacteriol. Rev. 41, 47–99.
51. Segal, A.W. and Peters, T.J. (1976) Lancet, 1, 1363–1365.

CYTOCHROME b AND THE SUPEROXIDE-GENERATING SYSTEM IN HUMAN NEUTROPHILS

MIC N. HAMERS[+], RON WEVER[++], MARGRIET L.J. VAN SCHALK[+++], RON S. WEENING[+] AND DIRK ROOS[+]
+ Central Laboratory of the Netherlands Red Cross Blood Transfusion Service and Labortaory of Clinical and Experimental Immunology, University of Amsterdam;
++ B.C.P. Jansen Institute, Labortaory for Biochemistry, University of Amsterdam;
+++ Pediatric Clinic, Binnengasthuis, University of Amsterdam, Amsterdam, The Netherlands

ABSTRACT

The role of a novel cytochrome b in the plasma membrane of neutrophilic granulocytes (Ref.3) was investigated in relation to the superoxide-generating system in these cells. The absence in cytochrome b in spectra of neutrophils from patients with chronic granulomatous disease (CGD) (Ref.4) who lack the ability to generate superoxide, could not be confirmed in our studies. One female CGD patient was found to have a normal amount of cytochrome b in her neutrophils; cell preparations from four male patients showed clear contamination with eosinophilic peroxidase, which obscures the presence of cytochrome b. Although minor shoulders were found at the position of cytochrome b, no definite conclusion could be drawn about the amount of cytochrome b in the neutrophils of these male CGD patients.

The amount of cytochrome b in normal neutrophils was estimated from the reduced minus oxidized ferrohemochrome spectra and found to be $1.87 \pm 0.21 \times 10^{-17}$ moles/ cell (n=4). Cytochrome b was extracted from the neutrophils by treatment with Brij 35. With this cytochrome b preparation, the absorbance coefficients of the reduced minus oxidized forms were established as 106 $mM^{-1} cm^{-1}$ at 425 nm and 21.7 $mM^{-1} cm^{-1}$ at 558 nm. Cytochrome b did not bind CO; therefore, cytochrome b is not an 0-type cytochrome as was suggested by Segal and Jones (Refs.3,18).

Finally, the superoxide-generating system (NADPH oxidase) was separated from cytochrome b without loss of activity. From these observation, we conclude that cytochrome b is not involved in the superoxide-generating system, neither as an 0-type cytochrome nor as part of an electron-carrying system.

INTRODUCTION

Neutrophilic granulocytes (neutrophils), when exposed to bacteria or other partculate matter, show a series of changes in oxidative metabolism which together are termed the 'respiratory burst'. The 10-15 fold increase in oxygen consumption by phagocytosing neutrophils is not inhibited by cyanide or azide, distinguishing it from mitochondrial respiration. Instead, the oxygen consumed is used to produce hydrogen peroxide, as a substrate for the myeloperoxidase-mediated bactericidal system.

Patients with chronic granulomatous disease (CGD) suffer from recurrent bacterial infections due to a defect in this hydrogen-peroxide-generating system (for review, see Refs.1,2). The nature of the molecular defect in chronic granulomatous disease is the subject of considerable controversy.[1,2] Recently, a novel b-type cytochrome, found in phagocytic vacuoles of human neutrophils,[3] was reported to be absent in four different patients with chronic granulomatous disease.[4] This suggests that deficiency of a cytochrome b is the molecular defect in these cells. Moreover, CO-difference spectra of whole homogenates from normal neutrophils suggested that cytochrome b reacts with CO and thus also with O_2. This could imply a role of cytochrome b as a terminal oxidase, comparable to the O-type cytochromes in bacteria.[4,5]

We have studied the possible role of cytochrome b in the superoxide-generating system from normal and CGD neutrophils. The conclusion of our studies is that cytochrome b is not involved in this system.

MATERIALS AND METHODS

Venous blood was collected and neutrophils were prepared as described in Ref. 6. After isolation, the cells were counted electronically with a Coulter counter and differentiated morphologically. Oxygen consumption,[6] hydrogen peroxide production,[7] killing capacity towards Staphylococcus aureus[8] and activity of hexose monophosphate shunt enzymes[9] were determined according to previously published methods. The diagnosis of chronic granulomatous disease was based on the criteria of Johnston and Newman.[10]

Sodium dithionite reduced minus oxidized spectra were recorded on an Aminco-Chance Dual Wavelength (DW 2) spectrophotometer operated in the split-beam mode (slit 2.0 nm, scan speed 2 nm/s). To avoid contamination with hemoglobin, which would interfere with the cytochrome b determination, the cells were first lysed in water and subsequently centrifuged (100,000 x g, 15 min, 4°C) and the pellets resuspended in phosphate-buffered saline (pH 7.4). Ferrohemochromes from these preparations were obtained in pyridine (2.1 M)-NaOH (0.075 M), and spectra were recorded as described above. Cytochrome b was extracted from normal neutrophils as follows: the cells were homogenized in a potassium phosphate-buffered solution (100 mM, pH 7.0) of Brij 35 (2% w/v), in a Potter with a tight pestle. The homogenate was subsequently centrifuged (350 x g, 10 min, 4°C) and the resulting pellet was discarded. The remaining supernatant was once more centrifuged (2000 x g, 10 min 4°C). The second supernatant was concentrated on an Amicon UM-20 filter and dialysed overnight against the extraction buffer without detergent.

In the fractionation studies, a slightly different method was used. First, the cells were stimulated with serum-treated zymosan (10 mg zymosan per 10^8 cells in 2 ml phosphate buffer (5 mM), pH 7.4) for 2 min in the presence of 1 mM sodium azide. Immediately after stimulation, the cells were homogenized. The homogenate was diluted 5 times with potassium phopshate buffer (100 mM, pH 7.0) containing Brij 35 (2% w/v), and mixed vigorously. Subsequently, the centrifugation steps were performed. The 350 x g fraction, 2000 x g fraction and the second supernatant were analysed for protein with a modified Lowry procedure[11] with bovine serum albumin as a standard. Cytochrome c and myeloperoxidase were quantitated with the spectral method described above; an absorbance coefficient of 75 mM^{-1} cm^{-1} at 472 nm was used to quantitate myeloperoxidase.[12] NADPH oxidase was assayed as the superoxide dismutase-(SOD)-inhibitable cytochrome c reduction under the following conditions: 50 mM potassium phopshate buffer, pH 7.0; 1 mM sodium azide; 1 mM NADPH; 150 µM cytochrome c; 0-0.25 mg cellular protein per ml, 25^oC. The reduction of cytochrome c was recorded at 550 nm on a Cary-219 spectrophotometer for about 2 min, followed by the addition of superoxide dismutase (final concentration of 15 µM) and another recording for about 2 min. The difference between both rates was taken as the superoxide-generating NADPH oxidase activity.

RESULTS

Figure 1(trace A) shows the reduced minus oxidized difference spectrum of a neutrophil homogenate before the 100,000 x g centrifugation step described above. The maximum at 472 nm corresponds to myeloperoxidase,[12] whereas the maxima at 429 nm and 558 nm as well as the characteristic troughs at 540 nm and at 574 nm are due to contamination of the preparation with hemoglobin. This is illustrated in Figure 1(trace B), which represents the difference spectrum of the supernatant after the centrifugation step; the presence of hemoglobin is obvious. The difference spectrum of the resuspended pellet is shown in Figure 1(trace C). The maximum at 472 nm corresponds to myeloperoxidase, and the maxima at 427 nm and 558 nm to cytochrome b (Ref.3). The recovery of myeloperoxidase in this experiment was 103%. Although the trough at 574 nm is still observed to some extent, the trough at 540 nm disappeared and changed into an isosbestic point. This experiment demonstrates that the presence of hemoglobin obscures the small cytochrome b peaks. Thus, to show the presence of cytochrome b, it is essential that the preparations are free of hemoglobin.

During our studies, we observed that another type of contamination with hemoglobin occurs if the collected blood is allowed to stand for a few hours. In that case, the hemoglobin could no longer be removed from the homogenate even by extensive washing with distilled water.

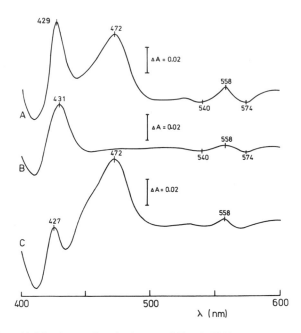

Fig. 1. Sodium dithionite-reduced minus oxidized difference spectra of a granulocyte preparation. A. Before the 100,000 x g centrifugation step. B. Supernatant after the 100,000 x g centrifugation step. C. Resuspended pellet after the 100,000 x g centrifugation step.

Figure 2(trace A) show the difference spectrum of a normal neutrophil preparation with less than 1% eosinophils. The difference spectrum of a granulocyte preparation from a bronchial carcinoma patient with eosinophilia (88% eosinophils, 10% neutrophils and 2% monocytes plus lymphocytes) is shown in Figure 2(trace B). The absorbance maxima at 448 nm, 562 nm and 595 nm correspond to eosinophilic peroxidase and are comparable to the maxima from the spectra published for eosinophilic peroxidase from the rat.[13]

Figure 2(trace C) shows the sum of the spectra represented by traces A and B which leads to a spectrum equivalent to a granulocyte preparation with 13% eosinophils. The maximum of eosinophilic peroxidase is clearly shifted to 453 nm, while the cytochrome b peak is transformed to a minor shoulder at 430 nm. Thus it is conceivable that contamination of neutrophils with eosinophils results in a difference spectrum in which cytochrome b appears to be abnormal or even absent, while the shifted peak of eosinophilic peroxidase could be misinterpreted as compound III, the inactive form of myeloperoxidase.[4]

The difference spectrum of a granulocyte preparation from a male CGD patient is shown in Figure 2(trace D). Since contamination with eosinophils is evident (morphological differentiation gave about 17% eosinophils), a definite con-

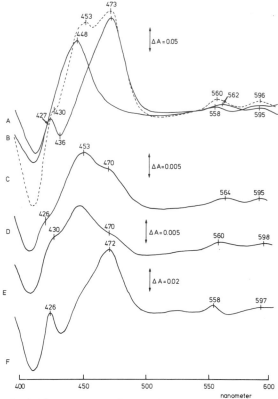

Fig. 2. Sodium dithionite-reduced minus oxidized difference spectra of granulocyte preparations. A. Neutrophils (<1% eosinophils); B. Eosinophils (88% eosinophils, 10% neutrophils); C. Theoretical spectrum of neutrophils plus 13% eosinophils (obtained by addition of spectra A and B); D. Neutrophils from a male CGD patient (17% eosinophils); E. Neutrophils from the healthy brother of the patient in D; F. Neutrophils from a female CGD patient (<1% eosinophils).

clusion about the absence or presence of cytochrome b or the meaning of the shoulder at 426 nm cannot be drawn. More or less the same spectrum was obtained from a granulocyte preparation from the healthy brother of this patient (Figure 2(trace E). The difference spectrum of a granulocyte preparation from a female CGD patient is shown in Figure 2(trace F). This preparation contained less than 1% eosinophils. A normal myeloperoxidase as well as normal cytochrome b peak were observed.

The reduced minus oxidized difference spectrum of the ferrohemochromes from purified myeloperoxidase are shown in Figure 3(trace A). The maxima at 436 nm and 586 nm correspond to the ferrohemochromes of myelperoxidase and that at 476 nm to a small amount of residual native myeloperoxidase. Figure 3(trace B) shows the difference spectrum of the ferrohemochromes from normal neutrophils with less than 1% eosinophils. In addition to the maxima mentioned above, maxima at 418 nm, 522 nm and 554 nm are present, corresponding to the ferrohemochromes of protoporphyrin-IX (ref.14). The ferrohemochromes from isolated cytochrome b showed maxima at 419 nm, 523 nm and 554 nm (not shown). With the

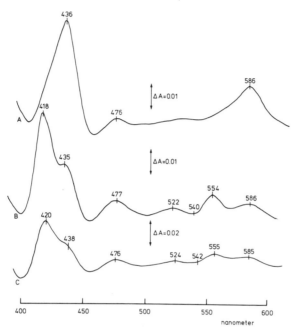

Fig. 3. Sodium dithionite-reduced minus oxidized difference spectra of ferrohemochromes. A. Purified myeloperoxidase; B. Neutrophils (<1% eosinophils); C. Neutrophils from the female CGD patient (<1% eosinophils; see also Figure 2, trace F).

use of an absorbance coefficient of 20.7 $mM^{-1}cm^{-1}$ for the absorbance difference between 554 nm and 540 nm (Ref.14), the amount of cytochrome b can be estimated from the ferrohemochrome spectra. The presence of hemoglobin, however, could interfere because it too, contains protoporphyrin-IX. Even the presence of eosinophlic peroxidase would interfere, because its ferrohemochromes show maxima very close to the maxima of the protoporphyrin-IX-derived hemochromes. Catalase could also interfere because this enzyme contains protoporphyrin-IX. Although catalase is not reducible with dithionite and thus not observed in the difference spectrum, its ferrohemochromes are dithionite-reducible and contribute to the total protoporphyrin-IX content of the cells. However, catalase is a cytosolic enzyme and only a very small amount was detectable in the particulate fraction. The catalase-derived protoporphyrin-IX amounted to about 2% of the total protoporhyrin-IX content observed in the particulate fraction.

Since hemoglobin and eosinophilic peroxidase were absent from our preparation and since catalase contributes only 2%, we were able to estimate the amount of cytochrome b in this and three pther preparations (see Table 1). For comparison with the ferrohemochrome spectrum of normal neutrophils, the corresponding spectrum of the neutrophils from the female CGD patient (see Figure 2(trace F)) is shown in Figure 3(trace C). The amount of cytochrome b calculated from this spectrum was within the range of the controls (see Table 1).

TABLE 1

AMOUNT OF CYTOCHROME b IN HUMAN NEUTROPHILS

Granulocyte preparation from	Cytochrome b^a (moles/cell)
Control 1	2.18×10^{-17}
Control 2	1.75×10^{-17}
Control 3	1.78×10^{-17}
Control 4	1.76×10^{-17}
Female patient with chronic granulomatous disease	1.88×10^{-17}

^{a}Estimated from the ferrohemochrome spectra.

The dithionite reduced minus oxidized difference spectrum of a cytochrome b preparation is shown in Figure 4(trace A). The absorbance maximum at 472 nm corresponds to a small amount of myeloperoxidase. In Figure 4(trace B), the dithionite-reduced, CO-treated minus oxidized spectrum of the same preparation is shown. No shift of the cytochrome b peaks was observed, indicating that cytochrome c does not react with CO. The peak heights are somewhat diminished due to a very slow denaturation of cytochrome b by dithionite during the interval between both recordings.

The reduced minus oxidized spectrum and the ferrohemochrome spectrum were used to estimate the amount of cytochrome c and the absorbance coefficients. We found 106 $mM^{-1}cm^{-1}$ at 425 nm and 21.7 $mM^{-1}cm^{-1}$ at 558 nm.

Finally, the results of the fractionation experiments with zymosan-stimulated neutrophils are shown in Table 2. It is evident that, when the homogenization is increased, the localization of the markers in the difference fraction is gradually shifted. The NADPH-oxidase activity shows up in the 2000 x g fraction first, followed by myeloperoxidase, whereas cytochrome b gradually dissolves into the supernatant. Thus, cytochrome b shows a different fractionation pattern than NADPH oxidase.

DISCUSSION

From our studies with the CGD neutrophils, it can be concluded that the presence of eosinophilic granulocytes in the neutrophilic granulocyte preparation interferes

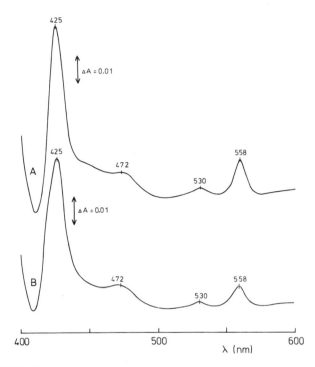

Fig. 4. Difference spectra of a cytochrome c preparation from human neutrophils. A. Dithionite-reduced minus oxidized; B. Dithionite-reduced, CO-treated minus oxidized.

TABLE 2

FRACTIONATION OF ZYMOSAN-STIMULATED HUMAN NEUTROPHILS BY DIFFERENTIAL CENTRI-FUGATION[a]

	Relative specific activity or amount of marker[b]									
Marker	Homogenization 4 strokes			Homogenization 6 strokes			Homogenization 8 strokes			
	350xg	2000xg	super-natant	350xg	2000xg	super-natant	350xg	2000xg	super-natant	recovery
NADPH oxidase	1.15	1.44	0.73	0.86	1.90	0.82	0.44	3.16	1.21	107%
Cyto-chrome b[c]	2.24	0.41	0.35	1.70	0.29	0.52	0.78	0.54	1.62	89%
Myelo-peroxidase	2.58	0.40	0.16	1.89	0.57	0.20	0.78	1.93	1.06	96%

[a]In Brij 35 (see MATERIALS AND METHODS)

[b]Percent activity or % amount divided by % protein.

[c]The amount of cytochrome b was calculated with an absorbance coefficient of 21.7 $mM^{-1}cm^{-1}$ at 558 nm (see RESULTS).

with the determination of cytochrome b. Unfortunately, we were unable to find
male patients without large quantities of eosinophils in their granulocyte preparations (four male patients were tested) or to reduce the eosinophil contamination
to acceptable low percentages (<3%). Thus, we were unable to clearly demonstrate
the presence of cytochrome b in male CGD patients. However, since cytochrome b
is present in normal amounts in the neutrophils of the female CGD patients
(Table 1), and since it is not clear whether female patients suffer from an
autosomal form of CGD or represent extreme lyonization of the X-linked form[15],
we conclude that cytochrome b is probably not the molecular defect in chronic
granulomatous disease. Moreover, the presence of cytochrome b was demonstrated
recently in a sister and a brother suffering from CGD, as well as in an unrelated
male patient.[16]

The fact that cytochrome b failed to react with CO rules out the possibility
that this cytochrome is an O-type cytochrome functioning as a terminal oxidase.[17]
Contrasting results have been obtained by Segal and Jones,[18] possibly due to
contamination of their preparations with hemoglobin. Cytochrome b could still
play a role as an intermediate electron carrier in an oxidation chain. However,
this possibility can be ruled out, since we separated cytochrome b from the NADPH
oxidase without affecting the recovery of the oxidase activity.

We conclude that cytochrome b is not involved in the generation of superoxide
via the SOD-inhibitable NADPH/cytochrome c reductase from human neutrophils.
Nevertheless, theoretically, cytochrome b could still be involved as an electron
carrier in an oxidation chain that directly produces hydrogen peroxide, but not
superoxide. In that case, we have to postulate an NAD(P)H-specific flavoprotein,
coupled to cytochrome b and followed by a compound X with the properties of a
terminal oxidase catalyzing a two-electron reduction. To our knowledge, no
enzyme activity found in neutrophils fulfils this prerequisites of such a
compund X.

REFERENCES

1. Cheson, B.D., Curnutte, J.T. and Babior, B.M. (1977) Progr. Clin. Immunol.
    3, 1-65.
2. Klebanoff, S.J. and Clark, R.A. (1978) in The Neutrophil: Function and
    Clinical Disorders, North-Holland, Amsterdam.
3. Segal, A.W. and Jones, O.T.G. (1978) Nature, 276, 515-517.
4. Segal, A.W., Jones, O.T.G., Webster, D. and Allison, A.C. (1978) Lancet,
    ii, 446-449.
5. Segal, A.W. and Allison, A.C. (1979) in Oxygen Free Radicals and Tissue
    Damage, Ciba Foundation Symposium 65, Excerpta Medica, Amsterdam, pp.205-224.
6. Weening, R.S., Roos, D. and Loos, J.A. (1974) J. Lab. Clin. Med. 83, 570-576.
7. Homan-Müller, J.W.T., Weening, R.S. and Roos, D. (1975) J. Lab. Clin. Med.
    85, 198-207.
8. Johnston, R.B., Keele, B.B., Misra, H.P., Lehmeyer, J.E., Webb, L.S.,
    Baehner, R.L. and Rajagopolan, K.V. (1975) J. Clin. Invest. 55, 1357-1372.
9. Weening, R.S., Roos, D., Weemaes, C.M.R., Homan-Müller, J.W.T. and van
    Schaik, M.L.J. (1976) J. Lab. Clin. Med. 88, 757-768.

10. Johnston, R.B. and Newman, S.L. (1977) Pediat. Clin. N. Amer. 24, 365-376.
11. Cadman, E., Bostwick, J.R. and Eichberg, J. (1979) Anal. Biochem. 96, 21-23.
12. Bos, A., Wever, R. and Roos, D. (1978) Biochim. Biophys. Acta, 525, 37-44.
13. Archer, G.T., Air, G., Jackas, M. and Morell, D.B. (1965) Biochim. Biophys. Acta, 99, 96-101.
14. Fuhrhop, J.-H. and Smith, K.M. (1975) in Porphyrins and Metalloporphyrins Smith, K.M., ed., Elsevier, Amsterdam, pp.804-807.
15. Mills, E.L., Rholl, K.S. and Quie, P.G. (1979) Clin. Res. 27, 516A.
16. Borregaard, N., Johansen, K.S., Taudoff, E. and Wandall, J.H. (1979) Lancet, i, 949-951.
17. Lemberg, R. and Barret, J. (1973) The Cytochromes, Academic Press, New York, pp.225-233.
18. Segal, A.W. and Jones, O.T.G. (1979) Biochem. J. 182, 181-188.

RADIOACTIVE LABELLING OF SUPEROXIDE DISMUTASES

A. BARET[+], P. SCHIAVI[+] AND A.M. MICHELSON[++]
+ CERB, HIA Sainte-Anne, Toulon Naval, France;
++ Institut de Biologie Physico-Chimique, 13 rue P. et M. Curie, 75005 Paris, France.

INTRODUCTION

The radioactive labelling of superoxide dismutases (SODs) constitutes an important step in the study of various problems connected with these enzymes. Indeed, the preparation of highly radioactive tracers, which have preserved the immunological properties of the native enzyme is essential for the elaboration of specific and sensitive radioimmunoassays of these proteins.[1-3] Moreover, pharmacokinetic and metabolic studies of SODs[4] also require, the preparation of radioactive compounds which have preserved certain characteristics. These conditions, stability of the radioactive label, high specific radioactivity, integrity of immuniological properties and, if possible, retention of enzymic activity, are essential for the validity of results given by the use of these tracers.

Superoxide dismutases constitute a class of metalloenzymes. Therefore, two ways of radioactive labelling may be considered, either the iodination of the protein moiety, or the incorporation of radioactive metal such as ^{57}Co or ^{60}Co into the apoenzyme.[5,6] We discuss here these two possibilities, applied to the labelling of rat, human and bovine Cu SODs, human Mn SOD (and bacterial Fe SOD).

MATERIALS AND METHODS

Enzymes. Rat, human and bovine Cu SODs were prepared according to the technique of McCord and Fridovich.[7] After purification, the specific activity of bovine enzyme was 3300 Riboflavin-NBT units/mg.[8] Human Mn SOD was prepared from liver according to the technique of McCord et al.[9]

^{125}I-labelling of SODs. Preliminary studies showed that the standard labelling procedure according to Greenwood et al.[10] did not give highly radioactive SODs. Therefore, two techniques were used, a modified chloramine T method[11] for the rat and bovine Cu SODs using very small quantities of oxidizing agent, and coupling of the proteins with N-succinimidyl 3-(4-hydroxy,5-(^{125}I)-iodophenyl proprionate (^{125}I-NSHPP), according to Bolton and Hunter[12] in the case of human Cu and Mn SODs.

The chloramine T labelling is effected with 2 µg of protein in 2 µl of 0.5 M phosphate buffer pH 7.5, 200 µCi of Na ^{125}I (I M S 30, The Radiochemical Centre

Amersham) in 2 µl of NaOH. Chloramine T (30 µg/ml in 0.5 M phosphate buffer pH 7.5, is added stepwise until about 40 - 50 % of the radioactivity is protein bound.

The yield of iodination is measured by precipitation with 1 ml of 10% tri-chloroacetic acid of an aliquot part of iodinated enzyme diluted in 1 ml of 0.05 M phosphate buffer, pH 7.5 containing 1 mg of bovine serum albumin (BSA).

The necessary quantity of chloramine T is generally about 1.5 µg and the reaction is stopped by addition of sodium metabisulfite, the quantity of which is twice that of the chloramine T used. The labelling medium is diluted with 1 ml of diluent buffer (0.1 M phosphate, pH 7.5 containing 0.5% BSA). The product if purified on Sephadex G-100 (Pharmacia) or Ultrogel AcA 44 (Industrie Biologique Francaise) with the same buffer as eluent.

A low specific radioactivity labelling was effected to measure the specific enzymatic activity of the iodinated enzyme. The labelling was made with 2 mg of bovine Cu SOD using the same molar ratios of the different constituents of the reaction, but at higher concentrations (final volume 1 or 2 ml) and the iodination was effected with cold NaI in presence of 1 µCi of Na ^{125}I. The yield of iodi-nation was measured as previously described. The enzymatic activity was estima-ted directly in the labelling medium and in the different fractions from Ultrogel AcA 44 chromatography.

The labelling of human Cu SOD and Mn SOD by coupling with N-succinimidyl 3-(4-hydroxy,5-(^{125}I) iodophenyl) propionate was effected according to the technique of Bolton and Hunter.[12] Briefly, a benzene-0.5% dimethylformamide solution containing 200 µCi of ^{125}I-NSHPP (The Radiochemcial Centre, Amersham) is evaporated to dryness in a teflon microtube with a conical bottom. 10 µg of human Mn SOD or 5 µg of human Cu SOD in 0.1 M borate buffer pH 8.5 are added. The labelling medium is incubated for 1 hour at 0 to 4°C with gentle intermittent agitation, and the reaction is stopped by addition of 1 ml of borate buffer containing 0.2 M glycocol. The tracer is purified by Ultrogel AcA 44 chroma-tography (0.9 x 60 cm) with 0.1 M phosphate, pH 7.5 containing 0.25% gelatin as eluent buffer.

^{57}Co labelling of apoenzymes. 400 µCi of ^{57}Co^{2+} (carrier free, The Radio-chemical Centre, Amersham) are added to 40 µg of apo human Mn SOD in 15 µl of 0.1 M acetate pH 3.6, or to 15 µg of apo human Cu SOD in 200 µl of 0.1 M acetate, pH 3.6. The incorporation of the cobalt into the SODs is effected by slow neutralisation (18 h) to pH 8 with ammonia vapour under reduced pressure. The separation of the excess of ^{57}Co from the labelled enzyme is effected by Ultrogel AcA 44 chromatography with 0.1 M phosphate, pH 7.5 containing 0.5% BSA as the eluent buffer.

Validity of tracers. The enzymatic activity of the tracers was measured according to the riboflavin NBT technique.[8] Their immunoreactivity was tested by the measurement of their fixation on an excess of specific anti-human, rat or bovine Cu SOD or anti-human Mn SOD antisera which were obtained in rabbits.[1-3] 10000 cpm of highly radioactive enzyme in 400 µl of 0.1 M phosphate buffer, pH 7.5 containing 0.5% BSA was incubated during 24 h at 4°C, with either 100 µl of 100-fold diluted antiserum (specific fixation) or 100 µl buffer (non specific fixation).

For the Cu SODs, the separation of antibody-enzyme complex from free fractions was effected by polyethylene glycol.[1]

For the Mn SOD, a second antibody technique[3] was used. 100 µl of normal rabbit serum and 15 µl of a goat anti-rabbit gammaglobulin antiserum are added to each tube. After incubation for 24 h at 4°C, the precipitates are separated by centrifugation and counted in a gamma spectrometer. The results are expressed as percentage of radioactivity specifically bound to antibody.

RESULTS

Labelling of superoxide dismutases by the chloramine T method. Table 1 compares enzymatic activities, as measured by the riboflavin-NBT method, in the chloramine T iodination mixture before and after the action of the oxidizing reagent. There is only about 25% loss of activity of the bovine rat enzymes, even though the yield of iodination is about 85%. On the other hand, the incorporation of ^{125}I into human Cu SOD is very low (15%) and there is higher loss of enzymatic activity (38%). With human Mn SOD, high loss of activity (48%) and low incorporation (26.5%) of ^{125}I are observed: exclusion chromatography shows that all this radioactivity is associated with aggregates.

TABLE 1

LABELLING OF SUPEROXIDE DISMUTASES BY THE CHLORAMINE T TECHNIQUE

	Yield of iodination (%)	Enzymatic activity Riboflavin-NBT Unit/ml		
		Before labelling	After labelling	% variation
Bovine Cu SOD	84.4	232	170	−26.7
Rat Cu SOD	86.5	390	305	−22
Human Cu SOD	15	375	235	−38
Human Mn SOD	26.4	200	104	−48

Figure 1 shows the chromatographic patterns of human and bovine Cu SODs in the chloramine T labelling medium. Besides free Na [125]I (peak III), a single major peak is obtained with the human preparation but two peaks are observed with the bovine enzyme. One of these peaks (peak 1) has an elution volume identical to that obtained with the labelling mixture of human Cu SOD, and the other major peak (peak II) has an elution volume identical to that of native enzyme. The third peak (peak III) corresponds to the saline retention volume column. Similar results are obtained with rat enzyme.

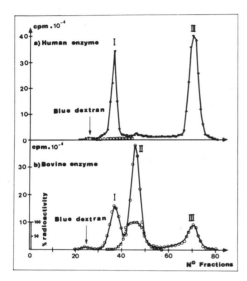

Fig. 1. Chromatography on Ultogel AcA 44 of chloramine T labelling media of human (a) and bovine (b) Cu SODs. (——) cpm; (□—□) % radioactivity bound to antibody.

Examination of elution volumes (Figure 1) shows that peak 1 corresponds to compounds the molecular weight of which is twice that of peak II compounds. This aggregation, which is minor for rat and bovine enzymes, but complete for human SOD, could correspond to a dimerisation of the protein, due to formation of disulfide bridges between two cysteine residues. Indeed, the addition of DMSO, a protecting agent of cysteine residues[13] in human Cu SOD labelling medium, prevents the formation of the dimer, but there is no iodination of the Cu SOD in these conditions. These difficulties for human Cu SOD with the chloramine T technique could be due to absence of tyrosine in this enzyme.[14]

Immunoreactivity studies (Figure 1) effected by radioimmunoassay show that compounds of peak I are not bound to specific anti-Cu SOD antibodies. On the other hand, compounds of peak II show a very high immunoreactivity (>95% in the fraction at the maximum of the peak).

Figure 2 shows the result obtained with the low specific radioactivity labelling performed with 2 mg of bovine enzyme. All the enzymatic activity corresponds to the major peak of radioactivity and optical density at 280 nm, and there is a complete superposition between 280 nm optical density, radio-activity and enzymatic actuvity. These results validate bovine ^{125}I-Cu SOD as radioactive tracer. This chromatographic patterns shows also that no dimeric

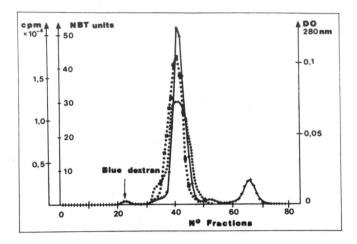

Fig. 2. Chromatography of low specific radioactivity ^{125}I bovine Cu SOD on Ultrogel AcA 44. (———) cpm; (•••) Riboflavin-NBT units; (✗✗✗✗) and DO. optical density.

form is observed with the described labelling conditions. Indeed, the concentrations of all reagents are here higher than in preparation of highly radio-active Cu SOD for which some dimer is obtained. Moreover, no loss of enzymatic activity is observed, compared to that obtained with the highly radioactive protocol. Loss of enzymatic activity thus seems to be correlated with formation of the dimer. Formation of dimer is inversely proportional to concentration of the reagents.

Coupling of superoxide dismutases with ^{125}I-NSHPP. This method was used because of the impossibility to obtain active iodinated human Cu SOD and Mn SOD with chloramine. Figure 3 shows the chromatographic patterns of the labelling mixtures of the two enzymes. For each of them, two peaks are obtained, one corresponding to the coupling product of the excess ^{125}I-NSHPP with glycocol, the other corresponding to labelled enzyme. The two iodinated enzymes present a very high immunoreactivity (>95%). The specific radioactivity of ^{125}I human Cu SOD is 12 µCi/µg (corresponding to the coupling of 0.3 moles of ^{125}I-NSHPP per mole of human Cu SOD), and that of ^{125}I human Mn SOD is 18 µCi/µg (1 mole of ^{125}I-NSHPP per mole of human Mn SOD).

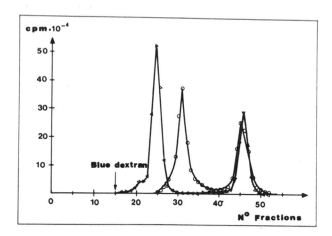

Fig. 3. Chromatography on Ultogel AcA 44 of human Cu SOD and Mn SOD labelled with ^{125}I NSHPP. ($\triangledown$—$\triangledown$) Mn SOD; (O—O) Cu SOD.

Labelling of apoenzymes with ^{57}Co. The labelling of apo human Cu SOD and apo human Mn SOD with carrier free ^{57}Co^{2+} gives the respective tracers after purification and separation of free cobalt ions by Ultrogel AcA 44 chromatography. The specific radioactivity of ^{57}Co Cu SOD is 7.3 µCi/µg (corresponding to the incorporation of 1.9 atoms of cobalt per molecule of protein), and that of ^{57}Co Mn SOD is 4 µCi/µg (2.1 atoms of cobalt per molecule of Mn SOD). (The same procedures were used with the apoenzyme from bacterial Fe SOD with ^{57}Co or ^{60}Co).

Comparison between ^{125}I and ^{57}Co human Cu SOD tracers. Radioimmunoassay[1] using as tracer the iodinated human Cu SOD and a specific rabbit anti-human Cu SOD antiserum indicates that the cross reaction between the native Cu SOD and apo Cu SOD is not complete (Figure 4).

The introduction of ^{57}Co into the apoenzyme does not restore an immunological similitude, since there is no superposition of radioimmuniological standard curves which are obtained with each of the two tracers (Figure 5). The two curves are not parallel and the slope of the curve obtained with ^{57}Co apo Cu SOD is lower than that obtained with the iodinated tracer. On the other hand, the radioimmuniological system ^{125}I Mn SOD – rabbit anti-human Mn SOD[3] shows an immunological similarity of apoenzyme and native enzyme (Figure 6). Moreover, the standard curves which are obtained with the two different tracers (^{125}I or ^{57}Co) are identical (Figure 7). The two labelled molecules therefore appear to be immunologically equivalent.

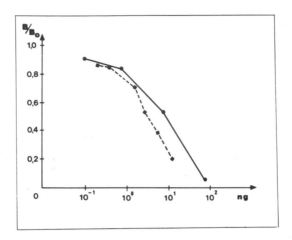

Fig. 4. Radioimmunoassay of human Cu SOD with [125]I Cu SOD as tracer.
Comparison between standard curves obtained with Cu SOD (-----) and apo
Cu SOD (———). Final dilution of anti Cu SOD: 1/20,000. B/B_0 is the
ratio of radioactivity precipitated in the presence of the different cold
SODs compared with that precipitated in their absence.

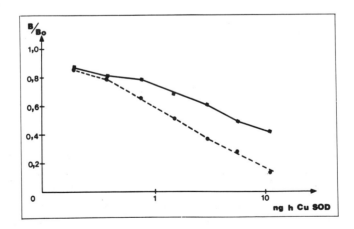

Fig. 5. Radioimmunoassay of human Cu SOD. Comparison of standard curves
obtained with [125]I Cu SOD (----) and [57]Co Cu SOD (———).

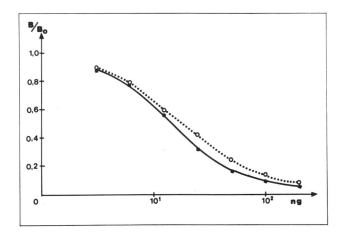

Fig. 6. Radioimmunoassay of human Mn SOD with ^{125}I Mn SOD as tracer.
Comparison between standard curves obtained with Mn SOD (O····O) and apo
Mn SOD (●——●). Final dilution of anti Mn SOD : 1/12500.

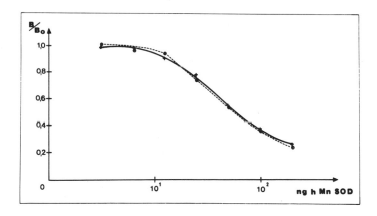

Fig. 7. Radioimmunoassay of human Mn SOD. Comparison of standard curves
obtained with ^{125}I Mn SOD (●---●) and ^{57}Co Mn SOD (+——+).

Stability of labelled superoxide dismutases. Rat and bovine Cu SODs,
iodinated by the chloramine T technique, are stable for at least 4 days at
4°C, with no decrease in immunoreactivity. Moreover, chromatography after
storage for 21 days at -20°C shows only slight degradation of the iodinated
protein. Thus it is possible to use these tracers during 4 weeks without
further purification.

The human Mn SOD and Cu SOD labelled according to Bolton and Hunter, are quickly degraded. A chromatographic purification is needed before each use, and they cannot be stored for more than 3 weeks.

^{57}Co enzymes ate stable for at least five months at -20°C. On the other hand, the apoenzyme itself rapidly undergoes an irreversible aggregation after preparation which renders impossible any incorporation of ^{57}Co.

DISCUSSION

The greatest difficulty which has been encountered in the preparation of highly radioactive SODs is their sensitivity to the action of oxidizing reagents, with a probable dimerisation of Cu SODs (complete for the human enzyme), and, with aggregation for human Mn SOD. These reactions are correlated with a loss of enzymatic activity.

The chloramine T iodination method gives rat and bovine iodinated Cu SODs which are highly radioactive (about 51 µCi/µg, corresponding to 0.8 atom of iodine per molecule), and fully active enzymatically. These tracers permit the elaboration of radioimmunoassays of rat, bovine and human Cu SODs[1] with a sensitivity of 20 pg for the bovine enzyme, 80 pg for the rat enzyme and 800 pg for the human enzyme. Moreover, this technique can be applied to the preparation of larger quantities (up to 1 mCi) of purified bovine ^{131}I Cu SOD which can be used for dynamic scintigraphic studies in animals.

For human Cu and Mn SODs, coupling with ^{125}I-NSHPP or the incorporation of ^{57}Co into the apoenzyme also gives suitable tracers for radioimmunoassay. However, human Cu SOD labelled with ^{57}Co exhibits immunological properties which are not strictly identical to those of the native enzyme, perhaps because of the absence of Zn which plays a structural role in the molecule. On the other hand, the two tracers ^{125}I and ^{57}Co human Mn SOD do not exhibit any immunological differences.

REFERENCES

1. Baret, A., Michel, P., Imbert, M.R., Morcellet, J.L. and Michelson, A.M. (1979) Biochem. Biophys. Res. Commun. **88**, 337-346.
2. Kelly, K., Barefoot, C., Semon, A. and Petkau, A. (1978) Arch. Biochem. Biophys. 190, 531-538.
3. Baret, A., Schiavi, P., Puget, K. and Michelson, A.M., in preparation.
4. Huber, W. amd Saifer, M.G.P. (1977) in Superoxide and Superoxide Dismutases, Michelson, A.M., McCord, J.M. and Fridovich, I., eds., Academic Press, New York, pp. 517-536.
5. Rotilio, G. and Calabrese, L. (1977) in Superoxide and Superoxide Dismutases, Michelson, A.M., McCord, J.M. and Fridovich, I., eds., Academic Press, New York, pp. 193-198.
6. Brock, C.J. and Harris, J.I. (1977) Biochem. Soc. Trans. 5, 1537-1539.
7. McCord, J.M. and Fridovich, I. (1969) J. Biol. Chem. 244, 6049-6055.
8. Beauchamp, C. and Fridovich, I. (1971) Anal. Biochem. 44, 276-287.
9. McCord, J.M., Boyle, J.A., Day, E.D., Jr, Rizzolo, L.J. and Salin, M.L. (1977) in Superoxide and Superoxide Dismutases, Michelson, A.M., McCord, J.M. and Fridovich, I., eds., Academic Press, New York, pp. 129-138.

10. Greenwood, F.C., Hunter, W.M. and Glover, J.S. (1963) Biochem. J. 89, 114-123.
11. Roth, J. (1975) Methods Enzymol. 37B, 223-233.
12. Bolton, A.E. and Hunter, W.M. (1972) J. Endocrinol. 55, xxx-xxi.
13. Stagg, B.H., Temperley, J.M., Rochman, H, and Morley, J.S. (1970) Nature, 228, 58-59.
14. Bannister, W.H., Anastasi, A. and Bannister, J.V. (1977) Superoxide and Superoxide Dismutases, Michelson, A.M., McCord, J.M. and Fridovich, I., eds., Academic Press, New York, pp. 107-128.

DETERMINATION OF HUMAN COPPER CONTAINING SUPEROXIDE DISMUTASE IN BIOLOGICAL
FLUIDS WITH A RADIOIMMUNOASSAY

E. HOLME, L. BANKEL, P-A. LUNDBERG AND J. WALDENSTRÖM
Department of Clinical Chemistry, University of Gothenburg, Sahlgren's
Hospital, S-413 45 Gothenburg, Sweden

INTRODUCTION

The concentration of superoxide anion is controlled by superoxide dismutase.
The concentration of human copper-containing superoxide dismutase (Cu SOD) (E.C.
1.15.1.1) has been studied in erythrocytes and other tissues, but no studies
have been done on the extracellular concentrations of Cu SOD. The methods available have not been sensitive enough to measure low concentrations. We have
therefore developed a radioimmunoassay which makes it possible to determine the
concentration of Cu SOD in extracellular fluids.

MATERIAL AND METHODS

Analytical grade reagents were used without further purification. N-succinimidyl 3- (4-hydroxy, 5-(^{125}I) iodophenyl) propionate was obtained from The Radio-
chemical Centre, Amersham, England; bovine albumin fraction V from Miles Laboratories Ltd., Slough, England; swine anti-rabbit immunoglobulins from DAKO-Immuno-
globulins Ltd., Copenhagen, Denmark; Freund's complete adjuvant from Behringwerke
AG, Marburg, W. Germany; and gelatin from Difco Laboratories, Detroit, Mich.,
USA.

Superoxide dismutase was purified from fresh human blood according to McCord
and Fridovich.[1] The purity of the preparation was checked by agarose gel electrophoresis. The protein concentration of the pure preparation was calculated from
the absorbance at 265 nm ($E_{cm}^{1\%}$ = 4.97) according to Briggs and Fee.[2] The enzyme
activity was determined using the xanthine-xanthine oxidase method;[1] the specific activity of the purified enzyme preparation was 3000 U/mg. The copper content of the enzyme was determined by atomic absorption spectroscopy and was
found to be 1.7 mol/mol of protein, which is the same value as reported by Briggs
and Fee.[2]

Antisera to copper superoxide dismutase. Antisera were obtained by immunisa-
tion of rabbits with 0.8 mg Cu SOD emulsified with Freund's complete adjuvant.
Six to eight injections were given subcutaneously in the back of the rabbit.
The procedure was repeated four times at three-week intervals.

^{125}Iodine labelling. Copper superoxide dismutase was iodinated with N-succi-nimidyl 3-(4-hydroxy,5-(^{125}I)iodophenyl propionate according to Bolton and Hunter.[3] The iodinated Cu SOD was separated from excess reagent by gel filtration on Sepha-dex G-100 Superfine in 0.05 mol/1 sodium phosphate buffer, pH 7.5 with 2.5 g/1 gelatin. The peak fraction of the labelled Cu SOD (specific activity 422 kBq/ug) was diluted 100 times with 10 mmol/1 sodium phosphate buffer, pH 7.5, containing 0.15 mol/1 sodium chloride, 15 mmol/1 sodium azide and 10 g/1 bovine albumin (buffer A) and stored frozen at -20°C.

Radioimmunoassay. One hundred microliter of standards (0-100 µg/1) or sample diluted in buffer A was mixed with 100 µl of the labelled diluted Cu SOD. Rabbit anti-Cu SOD serum (5 µl) and 120 µl of normal rabbit serum were added to 40 ml of buffer A containing 50 mmol/1 EDTA. The diluted antiserum (100 µl) was added to the tubes. The incubations were carried out for 5 days at 4°C before the addi-tion of 10 µl of swine anti-rabbit immunoglobulins. The incubations were then continued for another 24 h at 4°C. Buffer A (500 µl) was then added before centri-fugation of the tubes at 2000 g for 30 min at 4°C. The content of the tubes were then decantated and the precipitates were counted in a LKB Wallac 1270-Rackgamma II. (LKB, Bromma, Sweden). All analyses were done in duplicate.

Samples. Blood samples for Cu SOD determinations in erythrocytes were taken in Vacutainer^R tubes containing heparin. The red blood cells were washed once with 0.15 mol/1 NaCl and the hemoglobin concentration was determined with a ferri-cyanide method. The samples were stored at -20°C. Before use the hemolysates were diluted 1:10,000. The diluted sample (100 µl) was used in the assay. The serum samples were taken in the morning from fasting patients. The samples were centrifuged twice. The sera were stored at -20°C. Prior to assay the sera were diluted 1:2.5. Freshly voided urine samples were centrifuged before they were frozen at -20°C. The samples were diluted 1:5 before assay. The amniotic and cerebrospinal fluids were centrifuged and then frozen at -20°C. A ten-fold dilu-tion of these fluids were used in the assay.

RESULTS

One major radiolabelled protein product is obtained after iodination, as can be seen in Fig. 1. Fig. 2 shows the binding of radiolabelled Cu SOD using diffe-rent amounts of antiserum in the assay with no addition and with 10 ng of pure Cu SOD. The time course of precipitation with secondary antibody is shown in Fig. 3. A standard curve is shown in Fig. 4. Ten percent displacement of the tracer is obtained at a concentration of 1.5 µg/1 (0.15 ng in the assay). This is considered to be the sensitivity limit of the assay. The effect on the stan-dard curve of the addition of 10 µl of undiluted serum is shown in Fig. 4. Serial dilutions of the extracellular fluids give linear curves, as can be seen in Fig. 5. The coefficient of variation in one serie was 3.0% at a level of 15 µg/1 and

7.4% at a level of 42 µg/l of Cu SOD (20 determinations were made at each level).
The coefficient of variation between six different series was 5.2 and 3.8% for
the two concentrations. The distribution of concentrations of Cu SOD in erythrocytes expressed as µg/g hemoglobin in 39 men and women admitted to the hospital
for elective surgery is shown in Fig. 6. The mean value for 18 men is 560 µg/g
Hb with a SD of 44 µg/g Hb. The mean value of 21 women is 592 µg/g Hb with a
SD of 70 µg/g Hb. The distribution of serum concentration values is shown in
Fig.7. The sera are randomly selected samples sent to the laboratory for cholesterol determinations. The samples from pregnant women are from a screening program for neural tube defects, and were all obtained in the 15th to 16th week of
gestation. A few minutes stasis of the arm before taking the sample from the
cubital vein results in a marked elevation of the Cu SOD concentration. In two
normal persons the concentration increased from 21 to 64 and from 27 to 40 µg/l,
respectively. Fig. 8 shows the distribution of concentration of Cu SOD in urine
from the same patients as were used for the determinations in hemolysates. All
urines were negative for albumin and hemoglobin as judged by a dipslide test.
Copper superoxide dismutase content in amniotic fluid aspirated in the 16th week
of gestation are shown in Fig. 9. The amniotic fluids are mainly from women over
the age of 35, screened for chromosomal aberrations of the fetuses. Fig. 10
shows the Cu SOD level in cerebrospinal fluid from consecutive samples sent to
the laboratory.

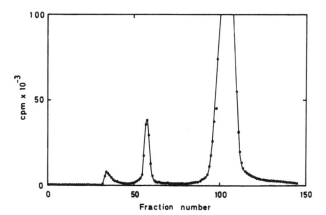

Fig. 1. Purification of (^{125}I)-Cu SOD on a column (1.6 x 80 cm) of Sephadex
G-100 Superfine after iodination according to Bolton and Hunter (Ref. 3). The
column was equilibrated in 0.05 mol/l sodium phosphate buffer, pH 7.5 containing
2.5 g/l gelatin. The fraction volume was 1.75 ml. A part of each fraction
(10 µl) was counted in a gamma counter.

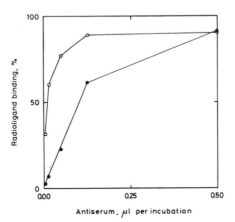

Fig. 2. Binding of radioligand by rabbit antiserum to human Cu SOD with 0 ng Cu SOD (O) and 10 ng Cu SOD (●) added to the tubes.

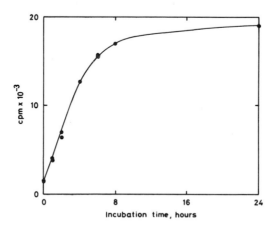

Fig. 3. Time course of precipitation with secondary antibody. For experimental details see MATERIAL AND METHODS.

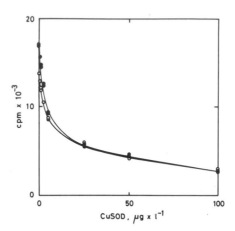

Fig. 4. Standard curve for determination of Cu SOD. One hundred microliter of the standards (concentration range 1.0 to 100 μg/l) was added in the assay. Standards (●); standards with the addition of 10 μl undiluted serum (3.2 nm Cu SOD) (O).

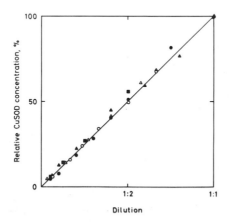

Fig. 5. Cu SOD determination after serial dilution of biological fluids. Diluted samples (100 μl) were added in the assay. Serum (primary dilution 1:1) (●); hemolysate (primary dilution 1:3.300) (O); urine (primary dilution 1:2) (▲); amniotic fluid (primary dilution 1:1.3) (△); cerobrospinal fluid (primary dilution 1:1.3) (■).

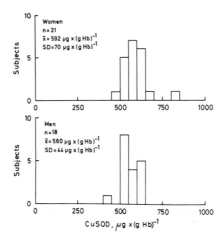

Fig. 6. Cu SOD concentration in hemolysates from 39 men and women admitted to hospital for elective surgery.

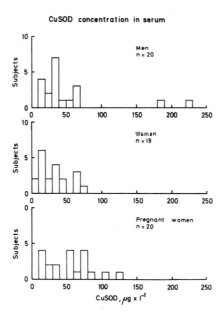

Fig. 7. Cu SOD concentration in serum specimen sent to the laboratory for cholesterol analyses (women and men). Sera from pregnant women were from a screening program for neural tube defects.

CuSOD concentration in urine

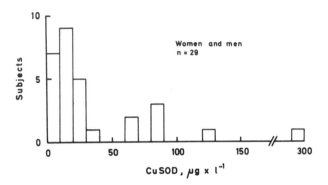

Fig. 8. Cu SOD concentration in urine from 29 men and women admitted to hospital for elective surgery.

CuSOD concentration in amniotic fluid

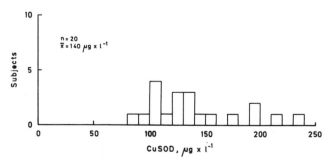

Fig. 9. Cu SOD concentration in amniotic fluid from pregnant women mainly over the age of 35, screened for chromosomal aberrations of the fetuses.

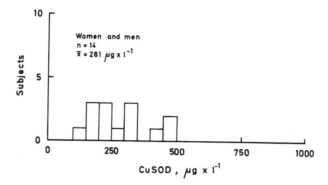

Ref. 10. Cu SOD concentration in cerebrospinal fluid from consecutive samples sent to the laboratory.

DISCUSSION

Since it is reported that copper containing superoxide dismutase contains no tyrosine[2,4] we chose to use the method of Bolton and Hunter[3] for iodination. This procedure gave a stable product with a good recovery. The antiserum obtained is not of a very high titre but it is specific and insensitive to unspecific protein interaction. The assay is sensitive enough to measure Cu SOD concentrations in extracellular fluids with a good precision.

According to our measurements the concentration of Cu SOD in erythrocytes is about 180 mg/1 and in serum about 35 µg/1. This means that the intracellular concentration is about 5000 times higher than the serum concentration. Consequently precaution must be taken to avoid factors that can cause leakage of intracellular Cu SOD when determining the concentration in extracellular fluids. The most important of these factors are hemolysis and stasis. All cellular materials must be carefully removed before analysis.

Our results concerning the mass concentration and distribution of Cu SOD in erythrocytes are in good agreement with those reported by Baret et al.[5] In serum we found a very low concentration of Cu SOD and a much wider distribution of values than in red blood cells.

We believe that in serum the Cu SOD is of little biological importance since serum contains other substances with superoxide scavenging activity. Ceruloplasmin may be an important superoxide scavenger in serum, by virtue of its concentration.[6] McCord[7] has reported a value of 700 µg/1 of Cu SOD in serum. It is not clear, however, if this value really represents mass concentration and not an estimation based upon activity measurements. If the latter is the case interference from other substances must be considered. McCord[7] has also given a value of 350 µg/l for the cerebrospinal fluid Cu SOD concentration. This value is in good agreement with our results ($\bar{x}$ = 281 µg/l), although it must be borne in mind that our material consists of patients with neurological disorders. This might be the reason for our rather wide distribution of values. The better agreement between our cerebrospinal fluid concentration and those reported by McCord[7] as compared to the large discrepancy between our and McCord's[7] serum concentration might be due to the fact that cerebrospinal fluid does contain less of interfering substances than serum. Also in amniotic fluid the Cu SOD concentration ($\bar{x}$ = 140 µg/l) is considerably higher than in serum. Possibly the comparatively high concentration in these two fluids is of biological importance.

In urine the Cu SOD concentration was below 40 µg/l but we found patients with concentration up to 300 µg/l. The low concentration found by us in urine as well as in serum fit well with tracer studies performed _in vivo_ on rats.[8] These show that the half life of intravenously given Cu SOD is 6 min and that most of the tracer is found in the kidneys within 15 min. In a similar experiment performed on dogs it was shown that the Cu SOD accumulated in the cortex of the kidneys.[8] Only a few percent of Cu SOD were recovered from the urine.

ACKNOWLEDGEMENT

This work was supported by a grant from Göteborgs Läkaresällskap.

REFERENCES

1. McCord, J.M. and Fridovich, I. (1969) J. Biol. Chem. 244, 6049-6055.
2. Briggs, R.G. and Fee, J.A. (1978) Biochim. Biophys. Acta, 537, 86-99.
3. Bolton, A.E. and Hunter, W.M. (1973) Biochem. J. 133, 529-539.
4. Bannister, W.H., Dalgleish, D.G., Bannister, J.V. and Wood, E.J. (1972) Int. J. Biochem. 3, 560-568.
5. Baret, A., Michel, P., Imbert, M.R. Morcellet, J.L. and Michelson, A.M. (1979) Biochem. Biophys. Res. Commun. 88, 337-345.
6. Goldstein, I.M., Kaplan, H.B., Edelson, H.S. and Weissmann, G. (1979) J. Biol. Chem. 254, 4040-4045.
7. McCord, J.M. (1974) Science, 185, 529-531.
8. Huber, W. and Saifer, M.G.P. (1977) in Superoxide and Superoxide Dismutases, Michelson, A.M., McCord, J.M. and Fridovich, I. eds., Academic Press, New York, pp 514-536.

LEVELS OF SUPEROXIDE DISMUTASE IN NORMAL AND HEPATIC DISEASE AS RELATED TO
LEVELS OF CERULOPLAMIN AND COPPER IN ERYTHROCYTES AND PLASMA

ARNALDO CASSINI AND VINCENZO ALBERGONI
Instituto di Biologia Animale, Centro C.N.R. Emocianine ed Altre
Metalloproteine, Universita di Padova, Padova, Italy

Plasma and erythrocyte copper content as well as ceruloplasmin and superoxide
dismutase (SOD) content have been determined by many authors.[2,3,4,6] To our knowledge, these parameters have not been correlated with each other in normal individuals nor in patients with diseases that modify them.

In the course of our studies on copper metabolism we have tried to define the
correlations between copper proteins (ceruloplamin and SOD) in the blood, not only
by comparing their mean values in homogenous groups, but also by studying statistically their ratios in single individuals. The aim was to know whether variations
of one parameter are followed by eventual variations of the others and how the
variation occurs.

One hundred and seventeen blood donors and 47 patients affected by hepatic diseases both of infectious (hepatitis, 23 subjects) and degenerative (cirrhosis,
24 subjects) nature have been examined.

Analyses have been carried out on total blood, on plasma and red cells. Ceruloplasmin amount was determined according to Ravin,[6] copper by atomic absorption, and
SOD by measuring the NBT reduction in the xanthine-xanthine oxidase system.[1] The
experimental values have been treated statistically: differences between groups
have been evaluated by Student's t, and the associations between different parameters have been investigated through the correlation coefficients between each
variable and the others. The critical value for statistical significance has been
assumed to be the 5 percent level.

The values concerning patients affected by hepatitis and cirrhosis were homogenous; moreover, Student's t clearly shows that the examined samples cannot be
considered typical of two distinct populations (Table 1). For this reason the
data reported here have been treated as belonging to one single group.

In Table 2 the mean values and the standard deviations of six parameters studied
are reported. The values obtained from normal subjects are in good agreement with
those reported in the literature, and do not require any comment; red cells, hemoglobin, erythrocyte SOD and copper values of patients affected by hepatic diseases
are lower, and the decrease is significant, as indicated by Student's t values.

TABLE 1

MEAN VALUES AND STATISTICAL SIGNIFICANCE OF MEASURED PARAMETERS IN SUBJECTS
AFFECTED BY HEPATITIS AND HEPATIC CIRRHOSIS

| | Hepatitis | | Hepatic cirrhosis | | | |
	Mean	S.D.	Mean	S.D.	t	p[a]
Hemoglobin g/100 ml of blood	14.09 ±	1.06	13.28 ±	1.66	-1.831	ns
Superoxide dismutase mg/100 ml of blood	5.19 ±	2.35	4.36 ±	2.25	-1.075	ns
Erythrocyte copper μg/100 ml of blood	44.83 ±	15.22	38.80 ±	11.87	-1.296	ns
Ceruloplasmin mg/100 ml of plasma	36.42 ±	9.71	38.17 ±	15.00	0.083	ns
Plasma copper μg/100 ml of plasma	122.83 ±	28.92	112.53 ±	32.99	-1.014	ns

[a]ns = not significant

TABLE 2

MEAN VALUES AND STATISTICAL SIGNIFICANCE OF MEASURED PARAMETERS IN NORMAL
SUBJECTS AND PATIENTS WITH HEPATIC DISEASE

| | Normal subjects | | Hepatic diseases | | | |
	Mean	S.D.	Mean	S.D.	t	p[a]
Red cells millions/mm^3 of blood	4.90 ±	0.39	4.45 ±	0.71	-5.121	<0.001
Hemoglobin g/100 ml of blood	14.89 ±	0.98	13.87 ±	1.42	-5.281	<0.001
Superoxide dismutase mg/100 ml of blood	6.25 ±	1.54	4.87 ±	2.19	-4.590	<0.001
Erythrocyte copper μg/100 ml of blood	79.94 ±	30.51	42.45 ±	14.14	-7.312	<0.001
Ceruloplasmin mg/100 ml of plasma	31.90 ±	8.97	37.60 ±	13.10	3.202	<0.002
Plasma copper μg/100 ml of plasma	113.07 ±	29.46	120.85 ±	38.68	1.393	ns

[a]ns = not significant

On the contrary the ceruloplasmin content is higher, and the increase is also significant. No significant difference is found for plasma copper.

The percent of the erythrocyte copper linked to SOD is equal to 35 ± 22 in normal subjects and 46 ± 30 in patients with hepatic disease. The increase is statistically significant ($\underline{t}$ = -2.60 and $\underline{P}$ <0.01).

The correlation coefficients between all the studied parameters are summarized in Table 3, where the statistical significance is also indicated. As shown, SOD is positively correlated with erythrocyte copper and negatively with ceruloplasmin in normal subjects only. The other parameters (ceruloplasmin - plasma copper; hemoglobin - red cells) are correlated in both normal subjects and patients. No significant correlation is found for the parameters indicated in part B of Table 3. We remark also that no correlation is found between the SOD content and the age of the subjects.

TABLE 3

CORRELATION COEFFICIENTS BETWEEN ALL THE MEASURED PARAMETERS

	Normal subjects		Hepatic diseases	
	$\underline{r}$	$\underline{P}^a$	$\underline{r}$	$\underline{P}^a$
(A) Superoxide dismutase - Erythocyte copper	0.232	<0.02	0.064	ns
Superoxide dismutase - Ceruloplasmin	-0.236	<0.02	0.077	ns
Ceruloplasmin - Plasma copper	0.270	<0.01	0.688	<0.001
Hemoglobin - Red cells	0.662	<0.001	0.642	<0.001
(B) Superoxide dismutase - Hemoglobin	-0.011	ns	0.246	ns
Superoxide dismutase - Red cells	-0.048	ns	0.184	ns
Superoxide dismutase - Plasma copper	0.050	ns	0.133	ns
Ceruloplamin - Erythocyte copper	-0.076	ns	0.163	ns
Ceruloplasmin - Hemoglobin	-0.149	ns	0.030	ns
Ceruloplasmin - Red cells	-0.145	ns	-0.006	ns
Erythocyte copper - Plasma copper	0.147	ns	0.252	ns
Erythocyte copper - Hemoglobin	0.105	ns	-0.048	ns
Erythocyte copper - Red cells	0.074	ns	0.122	ns
Plasma copper - Hemoglobin	0.007	ns	-0.123	ns
Plasma copper - Red cells	0.004	ns	-0.006	ns

^{a}ns = not significant

Figure 1 shows how the amount of SOD of normal subjects as compared to that of the patients differs according to the parameters used in the ratio. This is due to the variations of reference parameters as reported in Table 2. Here, in particular, is it seen that the greatest variation concerns the erythrocyte copper

value which in hepatic disease is about 50% lower than in normal subjects, and
justifies the reverse ratio shown in Figure 1.

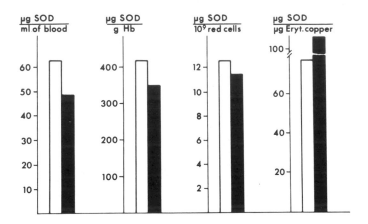

Fig. 1. Amount of superoxide dismutase of normal subjects (□) as compared
to that of the patients (■) in various ways.

So volume seems to be the reference parameters most valid to evaluate the effective variations of the SOD content in the blood and especially the correlations
between different parameters. For this last purpose in fact it is necessary to
have a common reference unit for all the measured values outside the parameters
used in the correlation.

From the reported data the following conclusions can be drawn:

(1) A significant decrease of erythrocyte SOD has been found in the hepatic di-
     seases considered in this study.

(2) In normal subjects the main blood copper proteins, SOD and ceruloplasmin,
     are negatively correlated; the physiological significance of this correlation
     is unknown.

(3) When a decrease of the erythrocyte copper occurs, the SOD content decreases
     also but to a lesser extent. This indicates that the metal not bound to the
     enzyme is more sensitive to the overall variations.

REFERENCES

1.   Beauchamp, C. and Fridovich, I (1977) Anal. Biochem. 44, 276–287.
2.   Cartwright, G.E. and Wintrobe, M.M. (1964) Amer. J. Clin. Nutr. 14, 224–229.
3.   Concetti, A., Massei, P., Rotilio, G., Brunori, M. and Elizier, A.R. (1976)
     J. Lab. Clin. Med. 87, 1057–1064.

4. Michelson, A.M., Puget, K. and Durosay, P. (1977) in Superoxide and Superoxide Dismutases, Michelson, A.M., McCord, J.M. and Fridovich, I., eds., Academic Press, New York, pp. 467-499.
5. Ravin, H.A. (1961) J. Lab. Clin. Med. 58, 161-168.
6. Yunice, A.A., Lindeman, R.D., Czerwinski, A. and Clark, M. (1974) J. Gerontol. 29, 227-281.

EFFECT OF HYDROGEN PEROXIDE ON LYMPHOCYTES

ALESSANDRO FINAZZI AGRO,[+*] PIO CONTI,[+] M.G. CIFONE,[+] P.U. ANGELETTI[+] AND
GUISEPPE ROTILIO[++]
+ Libero Istituto Universitario di Medicina e Chirurgia, L'Aquila, Italy;
++Istituto di Chimica Biologica, Università degli Studi di Roma, Roma, Italy

Lymphocytes are quiescent circulating cells which can be committed to various
specialized roles in the course of inflammatory processes and of the immune
response, via activation of their biosymthetic machinery (blastization, Ref.1).
Despite much experimental effort so far no clear evidence has been obtained
about the mechanism of lymphocyte activation in vivo.[2] In vitro various physical
and chemical agents are able to induce blastization of lymphocytes. This relative
lack of specificity of activating agents makes the solution of the problem very
difficult. Furthermore, heterogeneous cell populations have often been used and
this may be responsible for some contradictory results.

Among the various conditions that produce lymphocyte blastization, oxidation
of some cellular targets seems to play a definite role. An oxidant as simple
as periodate is able to induce lymphocyte activation which has been attributed
to the oxidation of glycoprotein side chains of the cell surface.[3] Similarly
galactose oxidase, a hydrogen peroxide-producing copper-enzyme, has been shown
to be capable of lymphocyte stimulation.[4] On the other hand, hydrogen peroxide
itself, and many other oxidizing agents, can activate guanylate cyclase in spleen
lymphocytes.[5-8] Since cyclic GMP is known to play a key role in the machinery
responsible for cell division, we decided to investigate whether hydrogen peroxide might directly activate lymphocytes; also in view of the fact that hydrogen
peroxide is produced by polymorphonuclear leucocytes and macrophages[9] in critical
steps of host defence against infections.

A preliminary experiment dealt with the determination in circulating lymphocytes of some enzymes concerned with hydrogen peroxide metabolism. Lymphocytes
were prepared from blood of healthy volunteers by the method of Boyum.[16]
Adhering cells were removed by incubating for 120 min at 37°C in plastic Petri
dishes and washing out non-adhering cells by isotonic saline.

[*]Present address: Istituto di Chimica Biologica, Facoltà di Farmacia,
Università di Roma, 00185 Roma, Italy.

A shown in Table 1, lymphocytes contain both H_2O_2-producing and H_2O_2-scavenging enzymes. They contain superoxide dismutase, catalase and gluthathione peroxidase in the range found in erythrocytes.[10] Monoamineoxidase (FAD-dependent) is present, while diamine oxidase (Cu-dependent) is absent in lymphocytes.

TABLE 1

H_2O_2-RELATED ENZYMES IN PURIFIED HUMAN PERIPHERAL LYMPHOCYTES

Enzyme	Concentration in Lymphocytes		Activity μmoles product/mg protein
	μmoles/mg	μg/mg	
Superoxide dismutase[a]	10	0.3	
Catalase[b]	14	3.4	
Glutathione peroxidase[c]			7 ± 2
Monoamine oxidase[d]			0.13 ± 0.01
Diamine oxidase[e]			Nil

[a]Determined according to Misra and Fridovich (Ref.11)

[b]Determined according to Lück (Ref.12)

[c]Determined according to Paglia and Valentine (Ref.13)

[d]Determined according to McEwen and Cohen (Ref.14)

[e]Determined according to Okuyama and Kobayashi (Ref.15)

An effect of hydrogen peroxide on lymphocytes has already been described by Sagone et al.[17] who reported that relatively low (about 1 mM) concentrations of H_2O_2 prevented the cell blastization induced by plant lectins. This sensitivity does not seem to fit with the fairly high concentrations of catalase and glutathione peroxidase present into the cells. Therefore we explored the effects of far lower peroxide concentrations than used by Sagone et al.[17]

Figure 1 shows that even 0.5 μM H_2O_2 is inhibitory for incorporation of radioactive thymidine by lymphocytes.

Figure 1 also shows that lower hydrogen peroxide concentrations may instead stimulate incorporation of the labelled precursor. The stimulation is similar to that induced by phytochemagglutinin and is prevented by the presence of catalase, but not of superoxide dismutase, in the incubation medium before the addition of peroxide. Appropriate controls showed that catalase itself had no effect whatsoever on lymphocytes. Thus the peroxide effect appears to be quite specific. Preliminary experiments have also shown that the inhibition of intracellular catalase with specific catalase inhibitors renders the lympho-cytes much more sensitive to externally added hydrogen peroxide, supporting the hypothesis that the level of H_2O_2-scavenging enzymes inside lymphocytes may be poised to control the stimulatory threshold of the peroxide.

Since some authors[19,20] have demonstrated the effect of periodate and galactose oxidase to take place only in the presence of adhering cells we determined the stimulation of purified lymphocytes (i.e. freed of adhering cells)

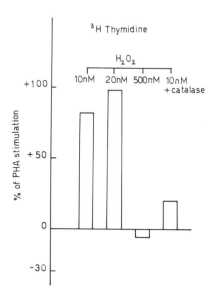

Fig. 1. Diphasic hydrogen peroxide effect on ^{3}H-thymidine incorporation by human circulating lymphocytes. Lymphocytes (0.7-1.0×10^6 cells/ml) were suspended in ROMI 1640 medium containing 10% fetal calf serum (heat inactivated) and supplemented with 100 U/ml penicillin and 100 mg/ml streptomycin. Phytohemagglutinin (PHA) or H_2O_2 were added where required. After 48 h incubation at 37°C, 2 µCi/ml of ^{3}H-thymidine were added. Twelve hours later the acid insoluble material was thoroughly washed and the radioactivity determined. The very low peroxide concentrations used were always checked by the scopoletin-peroxidase method (Ref.18).

by hydrogen peroxide. Figure 2 shows that leucine incorporation into proteins is also stimulated by hydrogen peroxide, though to a lesser extent with respect to PHA, but purified lymphocytes appear to be insensitive to H_2O_2 stimulation. This indicates that adhering cells are the real target of peroxide as well as of periodate and galactose oxidase. This conclusion reconciles more easily with the high activity of H_2O_2-removing enzymes in lymphocytes. In fact this situation is favourable to render lymphocytes insensitive to critically low concentrations of hydrogen peroxide (as suggested by the catalase inhibition experiment), which are instead effective on the adhering cells.

These cells will in turn respond to oxidative stresses by producing substances that are actually able to stimulate lymphocytes. No attempt has been

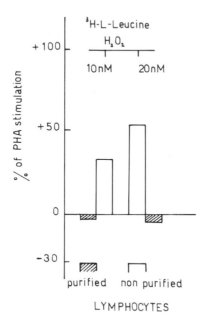

Fig. 2. Stimulation of ^{3}H-leucine incorporation by H_2O_2 in purified and non-purified lymphocytes. Lymphocytes were freed of adhering cells as described in the text. Incubations were performed as in Fig.1 except that 1 μCi/ml ^{3}H-leucine was added after 16 h incubation.

made to isolate these factors, however a significant increase of intracellular cGMP in non-purified lymphocytes preparations after a short exposure to H_2O_2 has been observed. The pertinent data are reported in Table 2.

From these results it seems that hydrogen peroxide increases the activity of guanylate cyclase in intact cells only. Also higher peroxide concentrations, which are toxic to lymphocytes at longer incubation times, increase the guanylate cyclase activity of intact cells. Cell viability was always checked by the trypan blue method[22] and was found to be greater than 90% even after 60 h incubation at the highest peroxide concentration studied.

The data reported above lead to consideration of hydrogen peroxide as a potential messenger molecule which is secreted by specialized cells (for instance granulocytes) and reinforce the concept that this molecule, which so far had been mainly considered a harmful by-product of oxygen reduction, may be a useful metabolite. In fact it has recently been shown to partake in the mechanism of action of insulin.[23,24]

TABLE 2

GUANYLATE CYCLASE ACTIVITY AFTER PREINCUBATION OF LYMPHOCYTES WITH HYDROGEN PEROXIDE[a]

	pmoles cGMP/mg protein per min	
	Intact lymphocytes	Lysed lymphocytes
Addition on Preincubation		
None	20	52
H_2O_2 (50 nM)	46	46
H_2O_2 (5 μM)	42	–

[a]Intact and Triton X-100 treated lymphocytes were incubated at $30^{\circ}C$ for 30 min in the presence or absence of H_2O_2. Then guanylate cyclase activity was measured in the whole cell homogenate according to Kimura and Murad (Ref.21). Enzyme activity was expressed as pmoles of cyclic GMP formed/mg protein per min. Each value was calculated on the basis of 3 enzyme determinations.

ACKNOWLEDGEMENT

This work was supported by C.N.R. grant No. 78.02103.04 to A.F.A.

REFERENCES

1. Rosenthal, A.S. (1975) Immune Recognition, Academic Press, New York.
2. O'Brien, R.L. and Parker, J.W. (1976) Cell, 7, 13-20.
3. Novogrodsky, A. and Katchalski, E. (1971) FEBS Lett. 12, 297-300.
4. Novogrodsky, A. and Katchalski, E. (1973) Proc. Nat. Acad. Sci. USA, 70, 1824-1827.
5. White, A.A., Crawford, K.M., Patt, C.S., and Lad, P.J. (1976) J. Biol. Chem. 251, 7304-7312.
6. Mittal, C.K. and Murad, F. (1977) Proc. Nat. Acad. Sci. USA. 74, 4360-4364.
7. Haddox, M.K., Stephenson, J.H., Moser, M.E. and Goldberg, N.D. (1978) J. Biol. Chem. 253, 3143-3152.
8. Graff, G., Stephenson, J.H., Glass, D.B., Haddox, M.K. and Goldberg, N.D. (1978) J. Biol. Chem. 253, 7304-7312.
9. Klebanoff, S.J. (1975) Semin. Hematol. 12, 117-142.
10. Maral, J., Puget, K. and Michelson, A.M. (1977) Biochem. Biophys. Res. Commun. 77, 1525-1535.
11. Misra, H.P. and Fridovich, I. (1972) J. Biol. Chem. 247, 188-192.
12. Lück, H. (1974) in Methods of Enzymatic Analysis, Bermeyer, H.U. ed., Verlag Chemie/Academic Press, New York, p.886.
13. Paglia, E.D. and Valentine, W.N. (1967) J. Lab. Med. 70, 158-169.
14. McEwen, C.M. and Cohen, J.D. (1963) J. Lab. Clin. Med. 62, 766-776.
15. Okuyama, T. and Kobayashi, Y. (1961) Arch. Biochem. Biophys. 95, 242-250.
16. Boyum, A. (1968) Scand. J. Clin. Lab. Invest. 21, Suppl.97, 77-88.
17. Sagone, A.L., Kamps, S. and Campbell, R. (1978) Photochem. Photobiol. 28, 909-915.
18. Root, R.K., Metcalf, J., Oshino, N. and Chance, B. (1975) J. Clin. Invest. 55, 945-955.
19. Novogrodsky, A., Stenzel, K.H. and Rubin, A.L. (1977) J. Immunol. 118, 852-857.
20. Charoensiri, A. and Katoh, A.K. (1979) Immunol Commun. 8, 407-418.
21. Kimura, H. and Murad, F. (1974) J. Biol. Chem. 249, 6910-6916.
22. Tennant, J.R. (1964) Transplantation, 2, 685-694.
23. Livingstone, J.N., Gurny, P.A. and Lockwood, D.H. (1977) J. Biol. Chem. 252, 560-562.
24. Cascieri, M.A., Mumford, R.A. and Katzen, H.M. (1979) Arch. Biochem. Biophys. 195, 30-44.

SUPEROXIDE DISMUTASE ACTIVITY IN HUMAN MALIGNANT LYMPHOMA

R.D. ISSELS[+] W. WILMANNS[+] AND U. WESER[++]
[+]Ludwig-Maximilians-Universität München, Klinikum GroBhadern, Medizinische
Klinik III, D-8000 München 70, W. Germany; [++]Anorganische Biochemie,
Physiologisch-Chemisches Institut Universität Tübingen, D-7400 Tübingen,
W. Germany

ABSTRACT

Superoxide dismutase activity has been identified in peripheral blood lymphocytes of patients suffering from chronic lymphocytic leukemia. The combination
of the formazan derivative colour formation induced by reaction of O_2^- with nitroblue
tetrazolium and a suitable analytical polyacrylamide disc electrophoresis was used.
In lymphocytes of healthy donors CuZn-SOD as well as Mn-SOD could be identified,
whereas Mn-SOD is diminished or lacking in lymphoma lymphocytes. These findings
are discussed in relation to the well known radiosensitivity of malignant lymphoma.

INTRODUCTION

There are two different forms of superoxide dismutase in eukaryotic cells. One
of these which is found in the cytosol and intermembrane space of mitochondria
contains copper and zinc. The other a manganese containing SOD is found in the
matrix of mitochondria.[14] Diminished amounts of the manganese containing SOD have
been found in different malignant tumors as recently reviewed.[10] This report details
preliminary studies on SOD activity using disc polyacrylamide gel electrophoresis
in normal and lymphoma lymphocytes and its possible relationship to the radiation
sensitivity will be discussed.

MATERIALS AND METHODS

Nitroblue tetrazolium chloride, acrylamide and riboflavin were from Serva, Heidelberg. All other chemicals were from Merck, Darmstadt. Bovine erythrocyte CuZn-
SOD was prepared according to[15] and used as a standard for SOD activity.

Patients. Peripheral blood was obtained from 4 patients suffering from chronic
lymphocytic leukemia (CLL) and from 5 healthy persons as controls. The WBC count
of the CLL patients was more than 50,000 cells/μl indicating the leukemic phase
of the lymphoma. In the differential count more than 90% of the cells were lymphocytes. At the time of investigation the patients were not treated by irradiation
or chemotherapy.

Isolation of cells. Lymphocytes were separated from heparinized venous blood

by centrifugation in a Ficoll-Hypaque gradient.[4] May-Grünwald-Giemsa stain showed
a 95% lymphocyte purity. All cell preparations were washed twice with cold phosphate-
buffered saline (PBS). Contaminating red blood cells were removed by lysis for
3 min in 0.16 M tris-ammonium-chloride solution at room temperature. The cell pellet was suspended and washed twice in 0.9% saline. Cells were counted and the
final concentration was adjusted to 5 x 10^6 lymphocytes/ml in 0.036 M-KPO_4. Cell
suspension were disrupted by sonication at $0^{\circ}C$ with a Bronson sonifier at a power
setting of 110 W. The samples were sonically dispersed for a total elapsed time
of 3 min in bursts of 20 sec to reduce heat effects. For most experiments the rup-
tured cell suspension was centrifuged at 27,000 g for 15 min and the clear supernate was used in the assay procedures. The protein concentration in the sonic
extracts was determined by the method of Lowry et al.[9] with bovine serum albumin
as a standard.

Assay for superoxide dismutase. The reduction of nitroblue tetrazolium by superoxide to develop the formazan derivative colour and its inhibition by SOD was measured using a suitable and reproducible polyacrylamide gel electrophoretic system.[1,3]
The gels were prepared in glass tubes of approximately 4 mm diameter and 100 mm
length. The gels contained 7.5% acrylamide and were prepared by the method of
Davis.[6] They were subjected to electrophoresis for 1 h at approximately 180 V and
1.5 mA per gel prior to the application of the samples. Samples were placed on
the gels in 20% glycerol and electrophoresis was performed at 180 to 220 V (2 mA/
tube) until the bromphenol blue marker dye had swept through most of the gel.
Protein was stained by immersion of the gels in 1% amido black in 7% acetic acid
for 10 min followed by destaining in 7% acetic acid.

Superoxide dismutase was localized by soaking the gels in 2.45 x 10^{-3} M nitroblue tetrazolium for 20 min in the dark, followed by immersion for 15 min in a
solution containing 0.028 M tetra-methylethylenediamine, 2.8 x 10^{-5} M riboflavin,
and 0.036 M KPO_4 at pH 7.8. The gels were placed on an aluminum plate and illuminated for 15 min in an aluminum foil lined box fitted with a 15 W fluorescent lamp.
After illumination the gels were photographed and kept in KPO_4 in complete darkness
to avoid uncontrolled colour changes.

RESULTS

Figure 1 demonstrates the activity of SOD in sonicates of lymphcyte preparations
from different healthy donors. All polyacrylamide gels received equivalent amounts
of sonicate and were stained for SOD activity in the presence and absence of cyanide.
Cyanide inhibits the CuZn-SODs but does not affect the manganese enzymes, thereby
permitting an identification of the bands on the gels. Figure 1 shows that gels
stained for protein show multiple protein bands (gel 1) but all sonicates give
only two distinct bands of activity as representatively shown in gel 2. In the
presence of cyanide the slower moving band is cyanide-insensitive (gel 3). As

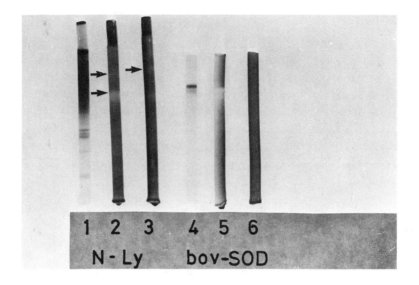

Fig. 1. Polyacrylamide gel electrophoresis of sonicates of lymphocyte preparations from healthy donors. N-Ly: normal lymphocytes; bov-SOD: bovine erythrocyte
CuZn-SOD. 1, protein stain; 2, activity stain; 3, as (2) + 2 mM cyanide; 4,
proton stain; 5, activity stain; 6, as (5) + 2 mM cyanide.

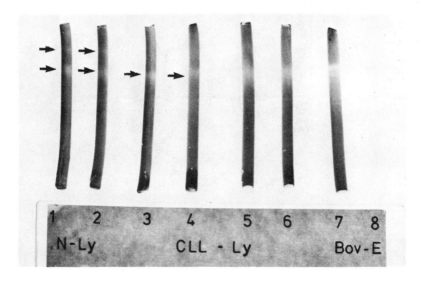

Fig. 2. Superoxide dismutase activity of sonicates of lymphocyte preparations
from chronic lymphocyte leukemia patients. N-Ly; normal lymphocytes; CLL-Ly;
chrome lymphocytic leukemia lymphocytes; Bov-E: bovine erythrocyte SOD.

expected cyanide inhibited the activity of the bovine SOD (gel 6). Figure 2 shows
the SOD activity in lymphocyte preparations of patients suffering from chronic
lymphocytic leukemia (CLL). In marked contrast to the two bands of activity of
normal lymphocytes (gels 1 and 2) equivalent amounts of CLL lymphocyte sonicate
show only a faster moving band (gels 3 and 4), which coincides with the band produced by the bovine SOD (gels 7 and 8). Even by overloading the gells with a 10
times higher amount of CLL sonicate (gels 5 and 6) there is only one detectable
band of activity in the position of the bovine enzyme. In the presence of 2 mM
cyanide this band of activity is inhibited (Figure 3, gels 3 and 4), and is identified in relation to normal lymphocytes (gels 1 and 2) as CuZn-SOD.

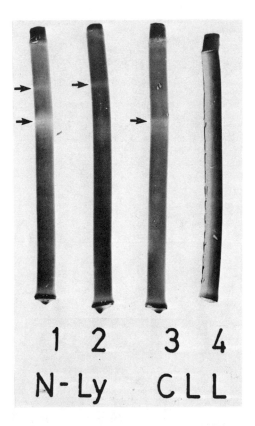

Fig. 3. Superoxide dismutase activity of sonicates of lymphocyte preparations
from chrome lymphocytic leukemia patients. N-Ly: normal lymphocytes; CLL: chronic
lymphocytic leukemia lymphocytes. 1 and 3, activity stain; 2 and 4, as (1) and
(3) + 2 mM cyanide.

DISCUSSION

Normal lymphocytes show two distinct kinds of SOD acitivity, whereas CLL lymphocytes exhibit only one band of activity in the gel electrophoresis due to CuZn-
SOD. Other workers have measured SOD activity by direct assay in L 1210 leukemia
cells and found no activity due to Mn-SOD in contrast to normal thymus lymphocytes.[11]
Furthermore, using polyacrylamide gel electrophoresis to indicate the presence of
the enzyme, there was no Mn-SOD in L 1210 leukemia cells.[16]

The leukemic lymphocyte population of CLL has been identified by immunological
markers as an altered B-type lymphocyte clone.[2] In the peripheral blood of CLL
patients there also exists a residual fraction of lymphocytes (70-80%) in normals.
The increased radiation sensitivity of CLL lymphocytes compared with normal lymphocytes is well known.[8] Our in vitro studies on CLL lymphocytes showed an augmented
T-lymphocyte fraction after irradiation.[13]

Radio-protection by SOD in whole-body irradiated mice[12] and high radiation sensitivity in human tissues with a low SOD content[7] suggest that SOD content might be
associated with radiation sensitivity. Whether this relation depends on the total
SOD content in cells or on levels of both SOD forms, CuZn- and Mn-SOD, is not yet
known. However, the loss of the cyanide insensitive Mn-SOD in CLL lymphocytes
might be related to the high radiation sensitivity of B leukemic lymphocytes.

REFERENCES

1. Beauchamp, C. and Fridovich, I. (1971) Anal. Biochem. 44, 276-287.
2. Bentwich, Z. and Kunkel, H.G. (1973) Transplant. Rev. 16, 29.
3. Bohnenkamp, W. and Weser, U. (1975) Hoppe-Seylers Z. Physiol. Chem. 356, 747-
    754.
4. Bøyum, A. (1968) Scand. J. Clin. Lab. Invest. 21, Suppl. 97.
5. Catovsky, D., Miliani, E., Okos, A. and Galton, D.A.G. (1974) Lancet,2, 751.
6. Davis, B.J. (1964) Ann. N.Y. Acad. Sci. 121, 404-427.
7. Hartz, J.W., Funakoshi, S. and Deutsch, H.F. (1973) Clin. Chim. Acta, 46,
    125-132.
8. Johnson, R.R. (1967) Cancer Res. 27, 39-43.
9. Lowry, O.H., Rosebrough, M.J., Farr, A.L. and Randall, A.J. (1951) J. Biol.
    Chem. 193, 265-275.
10. Oberley, L.W. and Buettner, G.R. (1979) Cancer Res. 39, 1141-1149.
11. Oberley, L.W., Luttenegger, D.W., Buettner, G.R., Sahu, S.K. and Leuthauser,
    H.C., in preparation.
12. Petkau, A., Chelack, W.S., Kelly, K., Barefoot, C. and Monasterski, L. (1975)
    Biochem. Biophys. Res. Commun. 65, 886-893.
13. Theml, H., Begemann, H. and Issels, R. (1974) Haematologia, 8, 43-60.
14. Weisiger, R. and Fridovich, I. (1973) J. Biol. Chem. 248, 3582-3592.
15. Weser, U., Bunnenberg, E., Cammack, R.,Djerassi, C., Flohé, L., Thomas, G.
    and Voelter, W. (1971) Biochim. Biophys. Acta, 243, 203-213.
16. Yamanaka, N., Ota, K. and Usumi, K. (1978) in Biochemical and Medical Aspects
    of Active Oxygen, Hayaishi, O. and Asada, K., eds., University Park Press,
    Baltimore, pp. 183-190.

STUDIES ON SUPEROXIDE DISMUTASE FUNCTION AND PROPERTIES IN DIFFERENT TISSUES OF
ANIMALS OF VARIOUS AGES

UZI REISS, AHUVA DOVRAT AND DAVID GERSHON
Department of Bilogy, Technion-Israel Institute of Technology, Haifa, Israel

INTRODUCTION

We have shown in recent years that as a function of age there is a progressive
accumulation of defective protein molecules in cells of several tissues of eukary-
otes (for a review see Ref.1). Essentially our work has demonstrated the appearance with age of a proportion of enzyme molecules which are either entirely
devoid of catalytic activity or are partially inactive. This has been determined
by measuring either the specific catalytic activity per unit of enzyme antigen
or per milligram purified enzyme.

Peroxidative damage has long been implicated as one of the principal causes
of age-related damage to cells (see e.g. Refs.2,3). Work of several laboratories,
including our own,[4] has indicated that such damage to several cellular components
accumulates in various cell types in ageing organisms. Moreover, the administration of vitamin E, an antioxidant, reduced the accumulation of the peroxidative
damage in cells of ageing nematodes and concomittantly increased the mean life
span of the treated organisms by approximately 30 per cent.[4]

A substantial body of evidence has accumulated which shows that superoxide
dismutase (SOD) constitutes the primary defence against the toxic effect of
superoxide in aerobic organisms. Because of its protective role it was deemed
important to investigate the activity and properties of SOD in a variety of
tissues of ageing organisms. Cytoplasmic SOD is also a relatively small protein
which has been well characterized. It is thus suitable for studies on the nature
of the modifications which lead to enzyme inactivation in ageing animals.

MATERIALS AND METHODS

Preparation of crude enzyme. Rat (WF) and mouse (C57 B1/6J) liver, heart and
brain were homogenized in three volumes of water. The homogenate was centrifuged
at 13,000 x g for 15 minutes and the supernatants were used for enzyme assay.
Lenses of WF rats were extracted and surgically separated into cortex and nucleus.
Care was taken that all work with the lenses was carried out in solutions pre-
saturated with N_2. Nuclei and cortex preparation from 40 lenses were homogenized
in 5 ml each of 0.05 M potassium buffer pH 7.0. The homogenate was spun at

100,000 x g. SOD activity was assayed in the supernatant. SOD activity was
also assayed after extraction of the 100,000 x g supernatant with chloroform and
ethanol whenever immunotitration was used in the lens system. This extraction was
necessary as it was found that lens proteins interfered with SOD-antibody interaction.

Enzyme assay and purification. Enzyme activity was determined according to
the pyrogallol method of Marklund and Marklund.[5] Purification of SOD from rat
liver was carried out according to the method of Weisiger and Fridovich.[6]

Antiserum preparation, immunoprecipitation, Ouchterlony precipitin test and
polyacrylamide gel electrophoresis were carried out as previously described.[7,8]
SOD activity on polyacrylamide gels was detected according to the method of
Beauchamp and Fridovich.[9]

RESULTS AND DISCUSSION

The specific activity of SOD in homogenates of various rat organs was compared.
Results of this comparison are presented in Table 1. It can be seens that the
specific activity declines considerably as a function of age in liver and in lens
nucleus but there is no decline in lens cortex, brain and heart. Substantial
variation in specific activity was found among the organs. In the lens there is
a considerable difference in activity between the cortex and the nucleus. Both
have SOD activity that is much lower than that of the other three organs.

TABLE 1

SPECIFIC ACTIVITY OF CYTOPLASMIC SUPEROXIDE DISMUTASE IN BRAIN, HEART, LIVER AND
LENS OF RATS OF VARIOUS AGES[a]

Age (months)	Specific Activity (Units/mg protein)				
	Liver	Brain	Heart	Lens Cortex	Lens Nucleus
5-8	20	8	6	0.57[b]	0.182
12	18	8	5	–	–
20	14	8	5	–	–
27	10	–	–	0.51	0.076
32	9	8	5	–	–

[a]The contribution of mitochondrial SOD in our supernatant preparations was < 5%
in liver, heart and brain and could not be detected in the lens.

[b]The specific activity of SOD in whole lens homogenates of 2 day old animals
has been found to be 0.91 U/mg.

It was necessary to verify that the same isozymic forms of SOD existed in all
organs and at all ages of the rat. For this purpose we initially compared the
electrophoretic pattern of the enzyme from all sources. No differences could be
discerned in the enzyme preparations derived from liver, heart and brain. Also,
no age-related change in electrophoretic pattern could be observed in any of the
organs.[7]

The enzyme was subsequently purified to apparent homogeniety from livers of young (6 months) and old (27 months) rats. The extent of recovery and degree of purification from both age sources were essentially indistinguishable. However, the specific activity per mg purified "old" SOD was 40% of that of "young" enzyme.[7] At least seven such preparations have yielded "old" enzyme with 40-50% of the specific activity of "young" SOD. The fact that the yield and recovery of SOD from young and old animals was similar eliminated the possibility that during purification from old livers non-specific inactivation of SOD took place. This finding meant that inactive or partially active molecules of SOD existed in livers of old animals.

This interpretation was bolstered by the use of immunotitration with monospecific anti-SOD antibody.[7] Approximately twice as much antiserum was required to precipitate the same amount of SOD activity from old than from young livers. Ouchterlony precipitin test showed complete cross reactivity of "young" and "old" SOD.[7] These results attested to the fact that antigenically cross-reacting but catalytically defective molecules (CRM) accumulated in ageing animals. They also further established the fact that no new isozymic forms accumulated in ageing animals.

Ouchterlony tests also demonstrated that liver, brain and heart SOD were antigenically identical.[8] We have recently shown the same for rat liver and lens SOD (Figure 1). In addition, SOD from young and old lenses was shown to be antigenically identical. They also did not differ in electrophoretic pattern. Since there is practically no protein turnover in the lens cortex and nucleus, the macromolecular components of this organ are essentially of the same age as that of its cells, most of which are as old as the whole rat. Photooxidation is an important contributing factor to lens damage which is expressed as increase opacity that may result in impairment of vision. From Table 1 it can be seen that SOD activity in the lens is very low as compared to that of other organs. This may presumably be attributed to the relatively low oxygen supply to the lens. On the other hand, damaged molecules cannot be replaced in this organ and thus damage accumulates in it throughout the life span of the organism. By the use of immunotitration techniques it has been possible to show for SOD that significant amounts of CRM accumulate in both the lens nucleus and cortex of 6 month old rats (Figure 2). Young liver preparation (11 units/mg protein) served as standard for comparison of SOD precipitation. Liver, lens nucleus and lens cortex preparations were adjusted to the same activity before the addition of antiserum. It is obvious from Figure 2 that much larger quantities of antiserum were required to precipitate the same amount of SOD activity from lens nucleus and cortex than from liver. Lens nucleus contains slightly higher amounts of CRM than cortex. This seems consistent with the facts that nucleus cells are older than cortex cells and that the inactivation of enzyme molecules

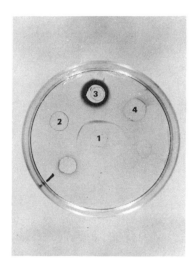

Fig. 1. Immunodiffusion of monospecific antiserum prepared against purified rat liver SOD from young animals.
1) antiserum; 2) young rat liver homogenate; 3) lens cortex homogenate (6 month old rats; 4) lens nucleus homogenate

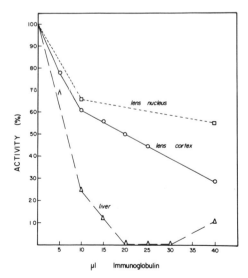

Fig. 2. Immunoprecipitation with monospecific antibody of SOD from homogenates of liver of young rats and lens nucleus and cortex of 6 month old rats. Initial enzyme activities of all preparations were adjusted to the same level.

is a progressive process. This finding is also in accordance with the trend observed in the data shown in Table 1 that SOD activity in the lens declines as a function of age, and that it is considerably lower in the nucleus than in the cortex.

It is important to understand the nature of the modification which leads to inactivation of SOD as a function of age. Indirect evidence has previosuly led us to postulate that age-dependent modifications in various enzymes were post-translational. Since the rate of protein turnover is extremely low in the lens, the decline in SOD activity without substantial loss in SOD antigen unequivocally means that the enzyme is altered by post-translational modifications. Another indication that SOD age changes are due to post-translational modifications has recently been found in our laboratory.[10] Activity of "old" liver SOD can be almost completely restored to levels of "young" enzyme by preparations from young rat liver (Figire 3). This restoration is NADH dependent and is much less effective with a similar preparation from old liver. Such restoration indicates that the modifications in the "old" SOD are post-translational rather than alterations in the primary structure. It is not known as yet what is the exact chemical nature of the modification. We have shown that it does not in-volve charge differences as the "old" form could not be separated from "young" enzyme by isolelectric focusing in polyacrylamide gels.[11] Under the same con-ditions two forms of aldolase which differ from each other by the deamidation of one asparagine residue per subunit could be separated.

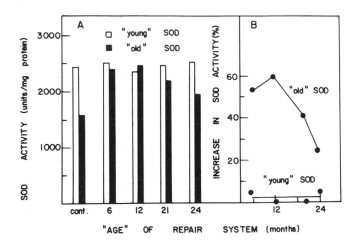

Fig. 3. Effect of "repair" liver preparations derived from animals of various ages on the activity of "young" and "old" SOD. A: SOD activity before and after exposure to the "repair" preparation. B: The result depicted in A expressed as % increase in activity over control.

Another interesting point has emerged from our comparative studies of age-related changes in SOD of various tissues. In order to maintain constant levels of SOD activity in brain and heart, the "old" system carries about double the amount of molecules per cell. This fact was deduced from our results which showed that in brains and hearts of old animals the amount of SOD CRM increased considerably while the level of activity remained constant throughout the life span.[8] The same phenomenon was not observed in the liver. This raises an intriguing possibility that in tissues in which SOD protection is essential, or where auxiliary protective mechanisms are not available, the system compensates for the defective molecules by raising the number of SOD molecules per cell. The nature of a "sensing" mechanism that might be involved in such adaptation can only be speculative at present. It may probably be related to the adaptive system of SOD in lungs of rats exposed to hyperoxice conditions.[12,13]

In conclusion, it can be said that SOD, a ubiquitous protective enzyme, shows an accumulation of altered molecules in cells ageing organisms (compare M. Somville and J. Remacle, this volume). This phenomenon may be expressed as either reduced absolute activity per tissue or maintenance of constant levels of activity by carrying larger numbers of SOD molecules per cell as a compensatory mechanism. The former process results in reduced protection against super-oxide related damage in old cells and tissues; the latter probably constitutes a metabolic stress on the cellular system. Both types of adverse effects have been indirectly implicated in ageing organisms.

ACKNOWLEDGEMENTS

The work reported here was supported by grants from the United States-Israel Binational Science Foundation and U.S. PHS grant AG 00459-06.

REFERENCES

1. Gershon, D. (1979) Mech. Age. Dev. 9, 189.
2. Barber, A.A. and Bernheim, F. (1967) in Advances in Gerontological Research, Vol.2,Strehler, B.L., ed., Academic Press, New York, pp.355-403.
3. Hougaard, N. (1968) Physiol. Rev. 48, 311-373.
4. Epstein, J. and Gershon, D. (1972) Mech. Age. Dev. 1, 257.
5. Marklund, S. and Marklund, G. (1974) Eur. J. Biochem. 47, 469-474.
6. Weisiger, R.A. and Fridovich, I. (1973) J. Biol. Chem. 248, 3582-3592.
7. Reiss, U. and Gershon, D. (1976) Eur. J. Biochem. 63, 617-623.
8. Reiss, U. and Gershon, D. (1976) Bioch. Biophys. Res. Commun. 73, 255-262.
9. Beauchamp, C.O. and Fridovich, I. (1971) Anal. Bioch. 44, 276-278.
10. Reiss, U. and Gershon, D. (1978) in Liver and Ageing, Kitani, K., ed., pp.55-61.
11. Goren, P., Reznick, A.Z., Reiss, U. and Gershon, D. (1977) FEBS Lett. 84, 83-86.
12. Crapo, J.D. and Tierney, D.F. (1974) Amer. J. Physiol. 226, 1401-1407.
13. Crapo, J.D. and McCord, J.M. (1976) Amer. J. Physiol. 231, 1196-1203.

COMPARATIVE STUDY OF MITOCHONDRIAL AND CYTOPLASMIC SUPEROXIDE DISMUTASE DURING
AGEING OF HUMAN FIBROBLASTS IN CULTURE

M. SOMVILLE AND J. REMACLE
Biochimie Cellulaire, Facultés Universitaires, Namur, Belgium

Human cells have a limited capacity for division when cultured in vitro
and show at the end of the culture degenerative process similar to in vivo
ageing.[1,2] Numerous hypotheses are now proposed to explain the degenerative
processes observed during ageing.[3,4] One involves the free radicals which are
continually produced in aerobic cells.[5,6] The radicals may attack cellular
components like polyunsaturated lipids and initiate a chain reaction detrimental
to the cell.[7,8]

Cells contain several enzymes whose function is to protect the cells against
reactive molecules. One of these enzymes is superoxide dismutase (SOD) which
catalyses the dismutation of O_2^- radicals and produces H_2O_2 and O_2.[6] SOD is a
very interesting enzyme to study during the ageing of the cell since it is
involved in the elimination of free radicals and also because two different SOD
enzymes are present in two different subcellular compartments. In this work,
we examined whether SOD was modified during the "in vitro" ageing of the WI-38
fibroblasts, whether the modification affected both the cytoplasmic and mitochondrial enzymes, and whether the detected modification resulted or not from
the presence of a cytoplasmic factor.

Our results show that SOD activity is more important in mitochindria than
in the cytoplasm and that both activities slightly increase during the ageing
of cells, going from 1.09 to 1.44 unit per mg protein in a supernatant fraction,
and from 0.80 to 1.04 unit per mg protein in a fraction containing particles.
When free radicals were measured, an increase from 10.5 to 15.3 µmoles per mg
protein was observed between the cells at the 30th and the 50th population
doubling level. Thermolability curves were then performed in order to detect
heat labile enzyme. Old cells contained a cytoplasmic heat labile enzyme which
represents 35% of the enzyme activity. No other modification could be detected
in the mitochondrial fraction from old cells or in any fraction from young cells.
We also observed that antibody raised against purified SOD from bovine erythrocytes cross-reacted with the human SOD. Using this antibody to make immunoprecipitation curves against cytoplasmic SOD from young and old fibroblasts, we
could demonstrate the presence of a cross-reacting material in the cytoplasm of

the old cells; in fact 25% of the SOD from old cells could react with the antibody without displaying an enzymatic activity.

In another experiment, we inactivated SOD from old cells with the antibody and incubated the mixture with a supernatant from young cells; no alteration could be detected in the SOD. When the opposite experiment was performed,
the alteration which was present in the SOD from the old cell disappeared. We
conclude from these experiments that a modified enzyme is present in the cytoplasm of old cells (but not in their mitochondria), that the modified enzyme is
not produced by a special substance present in old cytoplasm but rather because
of the absence of a stabilizing factor which prevents the alteration of the
enzyme. The properties of this factor are now under study but, whatever its
nature, the alteration of SOD during ageing is a translational effect
peculiar to the cytoplasmic enzyme and linked to a change in the physiological
environment.

REFERENCES

1. Hayflick, L. (1976) N. Engl. J. Med. 295, 1302-1308.
2. Cristofalo, V.J. (1972) in Advances in Gerontological Research, Vol.4.
   Strechler, B.L., ed., Academic Press, New York, 45-79.
3. Orgel, L.E. (1973) Nature, 243, 441-445.
4. Holliday, R., Huschtscha, L.I., Tarrant, G.M. and Kirkwood, T.B.L. (1977)
   Science, 198, 366-372.
5. Harman, D. (1956) J. Gerontol. 11, 298-300.
6. Fridovich, I. (1975) Annu. Rev. Biochem. 44, 147-159.
7. Tappel, A.L. (1973) Fed. Proc. 32, 1870-1874.
8. Sheldrake, A.R. (1974) Nature, 250, 381-385.

A STUDY ON HUMAN SUPEROXIDE DISMUTASE IN HUMAN TISSUE: AMOUNT AND
ISOENZYME PATTERN, IN CONNECTION TO LIPOFUCSIN

J.F. KOSTER, R.G. SLEE AND TH.J.C. VAN BERKEL
Department of Biochemistry I, Medical Faculty, Erasmus University
Rotterdam, P.O. Box 1738, 3000 DR Rotterdam, The Netherlands

INTRODUCTION

Superoxide anion is proposed to be involved in the mechanism of lipid peroxidation. This lipid peroxidation is thought to be involved in the process of aging
and turnover of intracellular components.[1-4] It is furthermore stated that formation
of aging pigments (lipofucsin) is the result of lipid peroxidation. These pigments
accumulate as function of age in the heart, while for liver contradictionary data
are reported.[5-7] It is known that univalent reduced oxygen superoxide anion (O_2^-)
is formed in the mitochondria,[8,9] as well in the cytosol[1,10] during consumption
of oxygen. Besides the beneficial role of O_2^- in the killing of bacteria, as occurs
in leucocytes, the cell has to protect itself against the cytotoxicity of O_2^-, which
is exerted by the enzyme superoxide dismutase (SOD). For mammalian species two
forms of SOD are known: a copper/zinc-containing enzyme present in the cytoplasm
and a manganese-containing enzyme predominantly present in the mitochondria but
also found in the cytoplasma.[11] Reiss and Gershon[12,13] have suggested that the
age-dependent lowering of cytosolic SOD might mediate the aging phenomena in the
different cells. If this suggestion is correct, there might be a relation between
the amount of SOD present and the formation of aging pigment. In order to establish
the validity of this hypothesis we investigated the isoenzyme pattern and the amount
of SOD together with the amount of lipofucsin in different tissues. We had the
opportunity to perform these experiments on human tissues, which were taken within
three hours after death and immediately frozen in liquid nitrogen.

MATERIALS AND METHODS

Human liver, heart, skeletal muscle and brain were obtained post mortem within
3 h after death. The tissues were frozen immediately in liquid N_2 and stored at
-70°C before use. Heparinized human blood was obtained by venepuncture. Erythrocytes were washed thoroughly with buffered saline to get rid of contaminating white
blood cells. All preparations were kept frozen at -70°C. The tissues were homogenized in 50 mM phosphate buffer (pH 7.0) + 5 mM EDTA with the polytron. The homogenates were sonicated 2 times for 30 sec at 21 kHz in a MS 100 W ultrasonic

desintegrator and subsequently centrifuged for 30 min at 13,000 g. The supernatants were used for the activity determinations, for electrophoresis and for isoelectro-focusing experiments. Superoxide dismutase was tested with a method based on the inhibition of the enzyme on the oxidation of epinephrine to adrenochrome as described by McCord and Fridovich.[14]

The mitochondrial enzyme was measured in the presence of 1 mM CN^- and the cyto-solic content was obtained by subtracting the CN^--insensitive part from the total. A standard curve was made with purified bovine blood SOD (Sigma Type 1, 2900 U/mg protein, Lot 38c-8190). In order to check if this procedure could be applied, the amount of SOD in human erythrocytes was measured. A value of 463 µg/g Hb was found which is in good agreement with the literature.[15]

Malonyl-CoA decarboxylase was measured in a mixture of phosphate buffer 0.2 M (pH 7.0), L-carnitine 4 mM, rotenone 30 µM and $[1,3-^{14}C]$malonyl-CoA 0.3 mM (final concentrations). The liberated $^{14}CO_2$ was captured with hyamine on paper and counted (H.R. Scholte, personal communication).

Lipofucsin was extracted from tissues, which were dried overnight at $110^{\circ}C$ with chloroform-methanol (2:1). The fluorescence was measured with excitation wavelength 360 nm and emission wavelength 430 nm. Quinine hydrobromide was used as standard. The amount of fluorescence was estimated as the equivalence of quinine per mg dry weight. Iron content was measured according Van Eijk et al.[16]

Electrophoresis and staining of the electropherogram were done essentially accor-ding to Beauchamp and Fridovich[17] on polyacrylamide gels (10%). Thin layer iso-electrofocusing was performed on gels prepared by mixing acrylamide (29% w/v), bisa-crylamide (0.9% w/v) and riboflavin (0.004%). For technical aspects see Manual of LKB Produkter AB. The applied pH ranges are indicated in the Figures. Staining was performed by the same method as for gel electrophoresis.

RESULTS

Electrophoresis and electrofocusing. Figure 1 shows the electrophoresis of SOD on polyacrylamide gel. As could be expected, all tested tissues show the mitochon-drial and the faster moving cytoplasmic band. Concerning the latter it looks as if this band is not homogeneous, but consists of two bands. In order to obtain more information about the isoenzyme pattern, electrofocusing was applied (pH 3.5-9.5). Figure 2 shows that cytosolic and mitochondrial SOD have a great difference in iso pH. The cytosolic enzyme is electrofocused in the acid region. Furthermore, for both (iso)enzymes several bands can be detected, especially for the mitrochondrial enzyme. In general, the pattern of bands is about the same for all tissues, although minor differences concerning the relative intensity between the bands exist. For the liver it can be seen that the cytosolic enzyme exhibits more bands. To investi-gate this further a smaller pH range (2.5-6.8) was applied. Figure 3 shows that

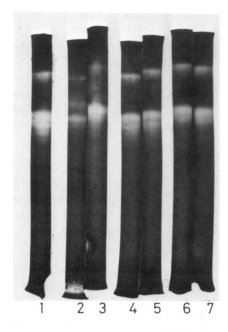

Fig. 1. Electrophoresis on polyacrylamide of SOD of various human tissues. Liver (1); brain (2,3); skeletal muscle (4,5); heart (6,7).

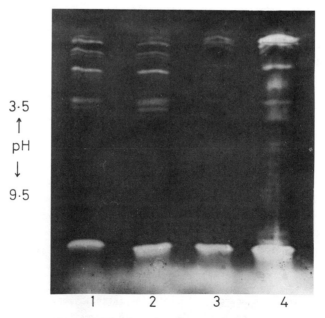

Fig. 2. Isoelectrofocusing of SOD of various human tissues (pH range 3.5-9.5). Liver (1); brain (2); skeletal muscle (3); heart (4).

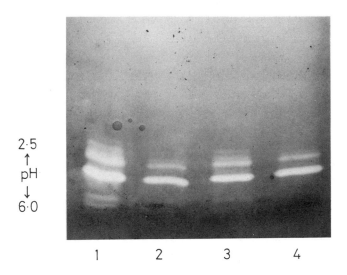

2·5
↑
pH
↓
6·0

1   2   3   4

Fig. 3. Isoelectrofocusing of SOD of various human tissues (pH range 2.5-6.0).
Liver (1); brain (2); skeletal muscle (3); heart (4).

the cytosolic enzyme from liver exhibits several bands, while brain and heart have
only two clear bands. Skeletal muscle cytosolic enzyme resembles that of liver,
although some of the bands are very faint. Also the differences in intensity between the two strong bands is different for brain and heart on the one hand and
for liver and sketal muscle on the other.

By taking advantage that the Cu/Zn-enzyme is inhibited by CN^-, it is possible
to measure the cytosolic (Cu/Zn) and the mitochondrial (Mn) SOD content in the
supernatants. The values for cytosolic and mitochondrial SOD in various human
tissues are given in Table 1, which shows that liver has the highest capacity to
dismutate O_2^-, mainly due to a high amount of mitochondrial enzyme. While the cytosolic component for heart and muscle is about the same as for liver, the mitochondrial is much less. In agreement with earlier reports[11,18] human tissues contained
much more mitochondrial enzyme than other species (liver 37%, heart 32% and muscle
23% of the total amount).

Table 2 shows the results obtained for SOD with various parts of brain together
with other tissues from one patient. In this way a direct comparison can be made
in one person, diminishing the biological variation. The amount of SOD present
in some parts of the brain are comparable to liver, while for instance the cortex
is comparable to heart and muscle. The mitchondrial enzyme for the various parts
of the brain is much less than for liver and more comparable to heart and muscle.

TABLE 1

SUPEROXIDE DISMUTASE IN HUMAN LIVER, HEART AND MUSCLE[a]

		Total SOD	SOD (CN^--insensitive)	SOD (CN^--sensitive)
Liver:	65 yr	11.4	4.0	7.4
	7 m	11.4	4.2	7.2
	84 yr	10.0	3.8	6.2
Heart:	65 yr	6.4	1.4	5.0
	84 yr	4.6	1.9	2.7
Muscle:	65 yr	6.7	1.4	5.3
	84 yr	7.2	1.8	5.4

[a]Superoxide dismutase is expressed as µg/mg protein and measured in the absence and presence of CN^- (1 mM).

The mitochondrial enzyme for the various parts of the brain is much less than for liver and more comparable to heart and muscle. In order to estimate the mitochondrial content of the tissues, another mitochondrial enzyme, namely, malonyl-CoA decarboxylase was measured. This enzyme was chosen because it is quite stable and it is also located in the mitochondrial matrix. The results are given in Table 3, together with the lipofucsin content. For most tissues it seems that they contain somewhat more mitochondrial SOD than should be expected on the activity of malonyl-CoA decarboxylase. The most striking result is that heart contains much less mitochondrial enzyme than should be expected on basis of the malonyl-CoA decarboxylase activity. Table 3 presents also the lipofucsin content. No direct correlation can be seen between the SOD and the amount of lipofucsin.

In the defence against oxygen radicals not only SOD is important, but also glutathione peroxidase and catalase in order to destroy the product of the dismutase reaction, H_2O_2. H_2O_2 can also be a factor in the lipid peroxidation. Glutathione peroxidase is located in the cytosol as well as in the mitochondria. With the frozen tissues it was impossible to measure the two components separately. Table 4 gives the activities for glutathione peroxidase and catalase. As was to be expected liver has the greatest activity for both enzymes, but no correlation with lipfucsin content can be made.

It is well-known that iron plays a role in lipid peroxidation, probably through the promotion of hydroxyl radicals, which can be formed from O_2^- (Refs. 19,20). Table 4 gives the iron contents in various tissues. Pons contains the smallest

TABLE 2

SUPEROXIDE DISMUTASE IN VARIOUS BRAIN PARTS, LIVER, HEART, AND MUSCLE OF
A 65 YEAR OLD WOMAN[a]

Organ	Total SOD	SOD CN⁻-insensitive	SOD CN⁻-senstive
Brain:			
Cortex frontalis	6.7	2.0	4.3
" occipitalis	7.6	2.3	5.3
" temporalis	6.5	1.5	5.0
Hippocampus	8.8	1.8	7.0
Corpus callosum	10.5	1.9	8.6
Nucleus caudatus	8.1	2.3	5.8
Thalamus	10.5	1.8	8.7
Corpus mamillare	12.4	2.1	10.3
Pons	10.9	2.0	8.9
Mesencephalon	11.0	2.1	8.9
Cerebellum cortex	10.6	2.1	8.8
" white	10.6	1.8	8.8
Medulla oblongata	9.0	1.9	7.1
Liver	11.4	4.0	7.4
Heart	6.4	1.4	5.0
Muscle	6.7	1.4	5.3

[a]Superoxide dismutase is expressed as μg/mg protein and measured in the absence
and presence of CN⁻ (1 mM).

amount of iron, while liver has the highest. This latter will be mainly bound
in the form of ferritin.

DISCUSSION

The electrophoretic pattern obtained shows that in contrast to rat liver[21] human
liver exhibits mitochondrial enzyme. Furthermore, the mitochondrial enzyme in
rat hepatocytes[21] has a more acid iso pH than the mitochondrial bands for human
tissues. The isoelectrofocusing pattern is for all tissues quite similar except
that for the brain and heart the cytosolic enzyme has two bands less than liver
and skeletal muscle. The meaning of the various bands is rather obscure. Henry
et al.[22] have shown that storage of pure Cu/Zn-containing SOD results in multiple
bands. For the reason that Cu/Zn-containing SOD is resistant to proteolysis[23]
they suggested that the multiple bands are due to a loss of metal. It should be

TABLE 3

MALONYL-CoA DECARBOXYLASE AND CYANIDE-INSENSITIVE SUPEROXIDE DISMUTASE ACTIVITIES
AND LIPOFUCSIN CONTENT IN VARIOUS BRAIN PARTS, LIVER, HEART AND MUSCLE OF A
65 YEAR OLD WOMAN[a]

Organ	Malonyl-CoA decarboxylase	SOD CN^--insensitive	Ratio SOD/ Mal.CoA dec.	Lipofucsin
Brain:				
Cortex frontalis	48.2	68.0	1.4	0.96
" occipitalis	49.8	69.0	1.4	1.23
" temporalis	35.2	62.2	1.8	0.74
Hippocampus	43.0	58.5	1.4	0.64
Corpus callosum	12.9	32.2	2.5	0.73
Nucleus caudatus	–	85.1	–	0.80
Thalamus	64.5	70.2	1.1	1.07
Corpus mamillare	38.7	81.9	2.1	0.86
Pons	47.7	49.0	1.0	0.71
Mesencephalon	40.0	55.6	1.4	0.94
Cerebellum cortex	65.9	77.7	1.2	0.77
" white	51.9	43.2	0.8	0.48
Medulla oblongata	48.0	80.7	1.7	0.44
Liver	438.8	368.0	0.8	7.35
Heart	635.6	62.3	0.1	2.65
Muscle	36.2	35.7	1.0	1.60

[a]Malonyl-CoA decarboxylase is expressed as mU/g wet weight, SOD as µg/g wet
weight and lipofucsin as µg quinine sulphate/mg dry weight.

mentioned that this proteolysis was performed with proteases of neutral pH, so
that the possibility of lysosomal proteolysis is still valid. For the Mn-containing
enzyme no multiple band appeared upon storage,[22] but this enzyme is much more labile
than the Cu/Zn-containing enzyme. Although care was taken by preserving the tissues
there was still 3 h between death and autopsy. Therefore, it cannot be ruled out
that the isoenzymes for Mn-containing enzyme are not due to loss of metal and/or
partial proteolysis.

    Our data confirm that human tissues have much Mn-containing enzyme. Liver contains
the greatest amount, correlating with mitochondrial content. Therefore it is surprising that human heart has much less Mn-enzyme than should be expected on the
basis of the amount of mitochondria. However, on basis of wet weight heart contains
the same amount of Mn-enzyme as the other tissues (except liver). Reiss and

TABLE 4

GLUTATHIONE PEROXIDASE AND CATALASE ACTIVITIES AND IRON CONTENT IN VARIOUS BRAIN
PARTS, LIVER, HEART AND MUSCLE OF A 65 YEAR OLD WOMAN[a]

Organ	GSH peroxidase	Catalase	Fe content
Brain:			
Cortex frontalis	78.8	9.56	0.35
" occipitalis	94.6	12.55	0.54
" temporalis	86.0	9.82	0.32
Hippocampus	63.9	11.35	0.51
Corpus callosum	87.8	28.30	0.35
Nucleus caudatus	77.3	10.08	0.49
Thalamus	70.9	14.00	0.60
Corpus mamillare	82.3	9.38	0.58
Pons	77.4	13.03	0.18
Mesencephalon	75.2	15.38	0.41
Cerebellum cortex	81.6	12.47	0.58
" white	62.4	12.02	0.57
Medulla oblongata	65.9	11.40	0.52
Liver	121.8	90.49	3.46
Heart	81.4	23.55	0.61
Muscle	23.6	21.69	0.51

[a]GSH peroxidase activity is expressed as µmol/min/mg protein, catalase activity
is expressed as µmol/min/mg protein and iron content as mmol Fe/100 g dry
weight. GSH peroxidase was measured as in Ref. 25 and catalase as in Ref. 26.

Gershon[12,13] reported that SOD changes in the process of aging, but were unable
to find any difference in isoelectrofocusing between young and old enzyme[24]. Our
data show that the capacity of O_2^--dismutation in liver of a 7 month old child is
about the same as for an adult liver (Table 1). No correlation could be found
between the amount of lipofucsin (aging pigment) and the amount of Cu/Zn- and Mn-
containing isoenzyme. Realizing also that the product of the SOD reaction (H_2O_2)
is toxic, the activities of catalase and glutathione peroxidase were measured.
Also with these data no relation with lipofucsin content could be seen. There is
strong evidence that the action of O_2^- is not radical enough for lipid peroxidation
and probably not involved in this process, at least not directly. It has been
shown that O_2^- in the presence of iron can give rise to the formation of OH˙ radicals.
These latter are much more agressive agents than O_2^-. For these reasons the iron

content was determined and liver contained the greatest amount. This tissue has also the largest lipofucsin content. It is rather doubtful if this correlation is meaningful in the present context because most of the iron is present as ferritin, which is unable to promote OH$^\cdot$ radical formation.[20]

Finally it should be mentioned that we have measured a final state and provide no information about the rate at which the lipofucsin is formed. It should be realized that the accumulation of lipofucsin is a process over years and the amount of unsaturated fatty acids in the membranes as substrate for the lipid peroxidation is important. Our data show that there is no direct correlation between the dismutating capacity of O_2^- and the H_2O_2-destroying capacity on the one hand and the amount of lipofucsin on the other.

ACKNOWLEDGEMENTS

Miss A.C. Hanson is thanked for the preparation of the manuscript and Mr. J. Fengler for the photographs. Dr. S. Gratema is acknowledged for providing the human tissues.

REFERENCES

1. Pederson, T.C. and Aust, S.D. (1972) Biochem. Biophys. Res. Commun. 48, 789-795.
2. Zimmerman, R., Flohé, L., Weser, U. and Hartmann, H.J. (1973) FEBS Lett. 29, 117-120.
3. Fong, K., McCay, P.B., Poyer, J.L., Keele, B.B. and Misra, H. (1973) J. Biol. Chem. 248, 7792-7797.
4. Tyler, D.D. (1975) FEBS Lett. 51, 180-183.
5. Strehler, B.L., Mark, D.D., Mildvan, A.S. and Gee, M.V. (1959) J. Gerontol. 14, 430-439.
6. Tygstrap, N., Schiødt, T. and Winkler, K. (1965) Gut, 6, 194-199.
7. Findor, J., Perex, V., Bruch-Igartua, E., Giovanetti, M. and Fioravantti, (1973) Acta Hepato-gastroenterol. 20, 200-204.
8. Loschen, G., Azzi, A. and Flohé, L. (1973) FEBS Lett. 33, 84-88.
9. Loschen, G., Azzi, A., Richter, C. and Flohé, L. (1974) FEBS Lett. 42, 68-72.
10. Strobel, H.W. and Coon, M.J. (1971) J. Biol. Chem. 246, 7826-7829.
11. McCord, J.M., Boyle, J.A., Day, E.O., Rizzolo, L.J. and Salin, M.L. (1977) in Superoxide and Superoxide Dismutases, Michelson, A.M., McCord, J.M. and Fridovich, I., eds., Academic Press, New York, pp. 129-139.
12. Reiss, U. and Gershon, D. (1976) Biochem. Biophys. Res. Commun. 73, 255-262.
13. Reiss, U. and Gershon, D. (1976) Eur. J. Biochem. 63, 617-623.
14. McCord, J.M. and Fridovich, I. (1969) J. Biol. Chem. 244, 6049-6055.
15. Winterbourn, C.C., Hawkins, R.E., Brian, M. and Carrell, R.W. (1975) J. Lab. Clin. Med. 85, 337-341.
16. Van Eijk, H.G., Wiltink, W.F., Bos, G. and Goossens, J.P. (1974) Clin. Chim. Acta, 50, 275-280.
17. Beauchamp, C. and Fridovich, I. (1971) Anal. Biochem. 44, 276-287.
18. Salin, M.L., Day, E.D. and Crapo, J.D. (1978) Arch. Biochem. Biophys. 187, 223-228.
19. McCord, J.M. and Day, E.D. (1978) FEBS Lett. 86, 139-142.
20. Halliwell, B. (1978) FEBS Lett. 92, 321-326.
21. Van Berkel, Th.J.C., Kruijt, J.K., Slee, R.G. and Koster, J.F. (1977) Arch. Biochem. Biophys. 179, 1-7.

22. Henry, L.E.A., Palmer, J.M. and Hall, D.O. (1978) FEBS Lett. 93, 327-330.
23. Forman, H.J. and Fridovich, I. (1973) J. Biol. Chem. 248, 2645-2649.
24. Goren, P., Reznick, A.Z., Reiss, U. and Gershon, D. (1977) FEBS Lett. 84, 83-86.
25. Paglia, D.L. and Valentine, W.N. (1976) J. Lab. Clin. Med. 76, 158-169.
26. Bergmeyer, H.U. (1955) Biochem. Z. 327, 255-260.

THE DISTINCTION BETWEEN THE ROLES OF OXYGEN AND SUPEROXIDE IN BIOLOGICAL
RADIO-DAMAGE

A. SAMUNI[+] AND G. CZAPSKI[++]
+ Department of Molecular Biology, and ++ Department of Physical Chemistry,
The Hebrew University, Jerusalem, Israel

The oxygen enhancement of the radiation-induced damage, observed in most
biological systems, combined with the unavoidable appearance of superoxide
radicals (in the presence of oxygen), readily implicates the O_2^- is 'cytotoxic".[1-3]
This implication has been further supported by the many studies of the physiological functions of the superoxide dismutases.[4-6] Protective roles have been
ascribed to SOD against the potentially deleterious effects of O_2^- radicals,
formed during normal aerobic metabolism,[7] in cancer cells,[8] ultraviolet and
ionizing radiations,[9] and by other means of free-radical formation.[9] This substantiated the attribution of the enhancement of biological radio-damage,
observed in the presence of O_2, to the O_2^- radicals (or to their potential products: singlet oxygen and/or secondary OH' radicals)(Refs.1-3,6,10).

Thus, the evidence accumulated so far, pointing at the O_2^- radicals as responsible for the enhancement of at least part of the radio-damage, was primarily
circumstantial. Yet, because of the fact that in the presence of oxygen, O_2^-
radicals are always formed, the roles of O_2 and O_2^- are prone to be intermingled.

The presence of oxygen during the irradiation of a biological system might
affect the radio-damage both directly, through the reactions of the molecular
O_2 with damaged target sites, and indirectly by the conversion of e_{aq}^- and H'
into O_2^-. In the latter, the biological damage is expected to be modified if
the specific-damaging-efficiency of O_2^- is different from that of its precursors,
and/or if the number of inactivating species hitting the biological target is
changed. A better understanding of the roles of oxygen and O_2^- radicals in
biological radio-damage, might be achieved by distinguishing between their
effects and examining them in different systems. For this purpose, a comparison
is attempted between the effects of O_2 and O_2^- in several test-systems of varying
complexity.

Test systems compared. The involvement of oxygen and of O_2^- in the radiation-
induced inactivation of an isolated enzyme (penicillinase), bacteriophage (T4),
and bacteria (E. coli B) have been compared. The simplest system for study is,
naturally, the solution of a pure enzyme (or DNA). In this case the radio-
damage, brought about via the water radicals, is exclusively an indirect effect.

The kinetics of the reactive species, inactivating the enzyme, are homogeneous. This simplifies the study of their role, since a conversion of one radical into another (by an appropriate radical-scavenger) can be readily ascertained, if the concentrations and rate constants of all the reactants are known. Moreover, repair mechanisms, which modify the damage in more complex biological systems, do not operate in this case. Thus the analysis of the experimental results is simplified.

The radiation-induced inactivation of bacteriophage is more complex compared to the enzyme. The damage is predominantly an indirect and exogenous process, originating from the water-radicals generated outside the phage.[11] But, unlike the case of an isolated enzyme, the inactivation process of the phage by the radicals does not follow homogeneous kinetics. Particularly in the presence of radical-scavengers, the effect of radicals formed near the biological target is different from that of radicals generated far away from the phage.[12] As a consequence, a change in the radio-damage resulting from a change in the number of radicals arriving at the target, might be misinterpreted as deriving from a different specific-damaging-efficiency of the active species. In addition, unlike the inactivation of the enzyme, repair mechanisms coupled with restitution processes in the phage reduce the apparent biological damage.

The interpretation of the radio-damage in bacteria is naturally more difficult. Radicals formed outside the cell hardly contribute to its inactivation, which is an endogenous effect.[13] Contrary to the other two systems, the contribution of the direct effect to the inactivation of the bacteria is certainly significant.[14] As a result, changes in the bacterial radiosensitivity, due to modifications of the spectrum of the water-radicals, might be more difficult to interpret (if the relative contributions of the direct and indirect effect are not known). Also, the high concentrations of the cellular constituents, which compete for the primary radicals within the cell, decrease the efficiency of the radical-scavengers (added in order to convert one radical into another).

Consequently, in light of the different nature of the radiation effects in the three test-systems compared, it is not necessarily anticipated that oxygen (and O_2^-) would play similar roles in all these systems.

Each of the test-systems examined was dissolved (suspended) in phosphate buffer, pH 7, and irradiated by ^{137}Cs gamma source. The dose-rate employed was 1.7 krad min^{-1}.

Enzyme inactivation. The enzyme selected for this study was the extracellular preparation of penicillinase (E.C.3.5.2.6) derived and purified from Bacillus cereus 569/H as previously described.[15] This monomeric (single-substance, single binding-site) enzyme has a molecular weight of 30,800 and its polypeptide chain is devoid of SH groups.[16] Solutions of penicillinase buffered with 50 mM phosphate, pH 7, were γ-irradiated and the catalytic activity towards benzylpenicillin was studied. The dependence of the residual enzymic activity on dose appeared to be exponential, as is generally found for isolated enzymes.[17] In Figure 1 the dose response curve obtained for penicillinase irradiated under helium is presented. Repetitions of the experiment with various enzyme concentrations (0.2 - 2 μM) showed that the inactivation constant, evaluated from the plot, is inversely dependent on the concentration of the enzyme. Hence, comparison was made only for experiments that were carried out using the same initial enzyme concentration. The value of $G_{(enzyme-inactivation)}$ calculated from the initial slope amounted only to 0.055, indicating that only a small fraction of the water-radicals formed contributed to enzyme inactivation. This is compatible with the assumption that the reaction rates of the water-radicals with <u>active</u> and with <u>inactive</u> enzyme molecules do not differ.[17]

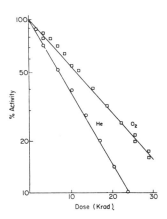

Fig. 1. Enzymatic activity of penicillinase vs. dose, after γ-irradiation at 1.7 krad min^{-1}, of 0.6 μM enzyme and in 50 mM phosphate buffer, pH 7. at room tempertaure. The enzyme solutions were saturated with: (o) helium; (□) oxygen.

<u>Effect of N$_2$O, O$_2$ and formate</u>. The method adopted was to control the predominant free radicals in the system, by an appropriate choice of the radical-scavenger, and to determine the resulting radiosensitivity. This was carried out by introducing, alternatively, into the solution: helium; oxygen; oxygen + formate; nitrous oxide; and nitrous oxide + oxygen (9:1 gas-mixture).

In phosphate buffer, under helium, the only reactive species predominant are H$^{\cdot}$ and OH$^{\cdot}$ radicals, as all the hydrate electrons react with the phosphate to yield H$^{\cdot}$ atoms[18]:

$$e^-_{aq} + H_2PO_4^- \rightarrow H^{\cdot} + HPO_4^- \quad (k= 4.9 \ 10^7 M^{-1}s^{-1}) \quad (1)$$

In the presence of oxygen, where e^-_{aq} and H$^{\cdot}$ are efficiently scavenged by oxygen, more than half of the primary water radicals are converted into O$_2^-$. The dose-

response curve for penicillinase, obtained under oxygen is also presented in
Figure 1. In the figure it is seen that oxygen does not enhance the enzyme
radiosensitivity, but rather reduces it.

In the presence of both oxygen and formate (0.2 M), practically a complete
radioprotection was obtained. Under these experimental conditions the hydroxyl
radicals, also are converted into O_2^-:

$$OH^{\cdot} + HCOO^- \rightarrow H_2O + COO^{\cdot -} \tag{2}$$

$$COO^{\cdot -} + O_2 \rightarrow CO_2 + O_2^- \tag{3}$$

The saturation of the solution with N_2O, which converts the hydrated electrons
into $OH^{\cdot}$:

$$e_{aq}^- + N_2O + H^+ \rightarrow OH^{\cdot} + N_2 \tag{4}$$

reduced the radio-damage as shown in Figure 2.

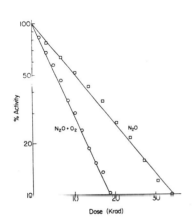

Fig. 2. Effects of N_2O with and without
oxygen on the radiation-induced enzyme
inactivation. Enzymatic activity of penicillinase vs. dose. Experimental conditions
as in Fig. 1. (□) N_2O saturated; (o) saturated with $N_2O + O_2$, 9:1 gas mixture.

These results are compared, in Table 1, with the respective findings previously reported[11,13] for T4 bacteriophage and E. coli B. It appears as if oxygen,
as well as N_2O, operate differently in the three systems. Oxygen enhances the
radiosensitivity of the phage and the bacteria, but reduces that of the enzyme
(oxygen reduces also the radiosensitivity of biologically-active DNA in solution.[19] Similarly, N_2O sensitizes the phage and the bacteria, yet protects the
enzyme. Differences appear also when the effect of formate is considered. In
the presence of both O_2 and formate the radiosensitivity of the bacteria is
greater than that observed under helium. Yet, it is lower than under O_2 in the
absence of formate. By contrast, the combined presence of oxygen and formate

considerably reduced the radiosensitivity of the phage, and completely protected the enzyme.

TABLE 1

RELATIVE RADIOSENSITIVITIES OF ENZYME, PHAGE AND BACTERIA γ-IRRADIATED IN
PHOSPHATE BUFFER[a]

Saturating gas	Penicillinase	T4 bacteriophage[b]	E. coli B[c]
He	1	1	1
O_2	0.65	1.9	3.5
O_2 + formate	0	0.18	2.5
N_2O	0.67	2.3	2.2
$N_2O + O_2$[d]	1.3	4.8	6.7

[a]Compared to the radiosensitivity measured under He, which was taken as unity.
[b]Results were taken from Ref.11.
[c]Results were taken from Ref.13.
[d]9:1 gas-mixture

Effect of $N_2O + O_2$. A significant enhancement of the radiosensitivity of
penicillinase was observed only when the enzyme was irradiated in the presence
of both N_2O and O_2. Under these experimental conditions, (9:1 gas-mixture,
where $[N_2O]$ / $[O_2]$ $\sim$ 150, most of the hydrated electrons disappear through
reaction (4) yielding OH$^\cdot$, rather than react with O_2 to give O_2^- radicals. Thus
the generation of O_2^- radicals is minimized, and the effect of O_2 can be examined
with a minimal involvement of O_2^-.

A comparison of the influence of $N_2O + O_2$ on the radiosensitivities of the
three test-systems, as well as on that of biologically-active DNA (Ref.20), shows
a similar effect. In all cases this gas-combination exerts the highest radio-
sensitization.

The primary inactivating species. The partial radioprotection afforded by O_2
to the enzyme (see Figure 1) indicates that H$^\cdot$ atoms also contribute to the
inactivation of the penicillinase. The remainder of the damage, observed under
oxygen, is certainly attributable to the OH$^\cdot$ radicals. In the presence of N_2O,
the hydrated electrons which otherwise would have been converted into H$^\cdot$ (by
reaction (1)), yielded OH$^\cdot$ radicals. Consequently, the N_2O as it were converts
H$^\cdot$ atoms into OH$^\cdot$ radicals. The partial radioprotection observed for the enzyme
under N_2O, suggests that the specific-damaging-efficiency of H$^\cdot$ exceeds that
of OH$^\cdot$ by four fold.

Role of superoxide radicals. The rate of enzyme inactivation under oxic
conditions was, as seen in Figure 1, lower than under helium. In the presence of
oxygen, H$^\cdot$ and e_{aq}^- are replaced by O_2^-. Hence, the character of the resulting

oxygen effect (protective or sensitizing) would depend on the relative damaging efficiency of O_2^- as compared with $H^\cdot$ and e_{aq}^-. In order to decide whether O_2^- radicals are also involved in the radio-damage, the effect of formate in the presence of O_2 may be considered. There, practically all the water radicals are converted into O_2^- radicals. Under these experimental conditions, actually no radiation-induced enzyme inactivation has been detected. In other words, the O_2^- radicals appear to be completely inactive towards the enzyme.

The conclusion that O_2^- radicals are actually harmless, parallels that obtained in the case of T4 bacteriophage.[11] The presence of both O_2 and formate has not rendered the bacteriophage completely radioresistant, but it eliminated practically all the <u>exogenous</u> damage. This supports the assumption that <u>exogenous</u> O_2^- radicals do not contribute to phage inactivation. In the case of the bacteria, the reduction in the radiosensitivity found in the presence of formate is attributable to the replacement of part of the highly reactive $OH^\cdot$ radicals by the harmless O_2^- species. The lack of full protection in this case is explainable assuming that 40 mM formate cannot successfully compete for all the $OH^\cdot$ radicals within the cell. In other words, the radio-damage observed in bacteria in the presence of both O_2 and formate does not necessarily suggest any damaging role for O_2^-.

<u>Role of oxygen</u>. Concluding that O_2^- radicals are apparently inactive, implies that in the presence of oxygen, which converts e_{aq}^- and $H^\cdot$ into O_2^-, a protective effect might be anticipated. Indeed, such an effect is seen in Figure 1. Yet, if the oxygen only eliminates the damage due to $H^\cdot$ atoms, the residual damage would be due only to $OH^\cdot$ radicals. Considering the respective yield and specific damaging efficiencies of the two species, the damage is not expected to exceed 25% of that observed with He. The results shown in Figure 1 prove that this is not the case, whereas in the presence of oxygen the damage was reduced to ∿65% of its value under He. These results can be explained by assuming a double role of O_2. On the one hand, oxygen reduces the radio-damage by converting active damaging species (e_{aq}^-, $H^\cdot$) into inactive ones (O_2^-), thus reducing the amount of the damaging primary hits. On the other hand, it increases the radiosensitivity by enhancing the damage already caused by $OH^\cdot$. If this is the case, the relative extent of these two opposing functions would determine the total net effect. In other words, either a sensitization or a protection or even a lack of effect might be observed. In the case of an enzyme (or DNA), the effect of protection exceeds that of sensitization. With the other two test-systems the radio-damage, in the presence of oxygen, is increased as the sensitization effect is greater than the protection.

In the presence of both N_2O and O_2 the situation that oxygen enhances the damage caused by $OH^\cdot$ is similar to that under oxygen alone, but the concentration of the $OH^\cdot$ is doubled. This explains the large enhancement of the radio-damage, observed in all the systems, under this gas-mixture.

<u>Conclusions</u>. In spite of the marked difference in the nature of the test-systems compared, it appears that the roles of oxygen, as well as of O_2^- radicals, in all these systems are the same.

Hydroxyl radicals and H' atoms are responsible for the <u>indirect</u> radio-damage under anoxic conditions. Superoxide radicals appear to be harmless in all the systems examined. Oxygen, therefore, plays a double role in all the test-systems. It protects by converting e_{aq}^- and H' into O_2^-, and enhances the damage caused by the OH' radicals. The overall result depends on the relative extent of these two opposing effects. Hence, sensitization, protection or a lack of an effect might be apparent.

Yet, concluding that the enhancement of the radio-damage, observed under oxic conditions, derives from the action of <u>molecular</u> oxygen, does not necessarily contradict the assumption of a cytotoxic role of O_2^- <u>in vivo</u>. Particularly in the presence of metal ions, where O_2^- radicals might act in some complexed form rather than as <u>free</u> O_2^- (Ref.21).

REFERENCES

1. Oberley, W., Lindgren, A.L., Baker, S.A. and Stevens, R.H. (1976) Radiat. Res. 68, 320-328.
2. Misra, H.P. and Fridovich, I. (1976) Arch. Biochem. Biophys. 176, 577-589.
3. Van Hemmen, J.J. and Meuling, W.J.A. (1975) Biochim. Biophys. Acta, 402, 133-141.
4. McCord, J.M. and Fridovich, I. (1969) J. Biol. Chem. 244, 6049-6055.
5. McCord, J.M. and Fridovich, I. (1971) Proc. Nat. Acad. Sci, USA, 68, 1024-1027.
6. Michelson, A.M. (1977) in Superoxide and Superoxide Dismutases, Michelson, A.M., McCord, J.M. and Fridovich, I., eds., Academic Press, New York, pp.245-255.
7. Fridovich, I. (1975) Annu. Rev. Biochem. 44, 147-157.
8. Oberley, L.W. and Buettner, G.R. (1979) Cancer Res. 39, 1141-1149.
9. Lavelle, F., Michelson, A.M. and Dimitrijevic, L. (1973) Biochem. Biophys. Res. Commun., 55, 350-357.
10. Petkau, A., Chelack, W.S. and Pleskach, S.D. (1978) Life Sci., 22, 867-882.
11. Samuni, A., Chevion, M., Halpern, Y.S., Ilan, Y.A. and Czapski, G. (1978) Radiat. Res. 75, 489-496.
12. Becker, D., Redpath, J.L. and Grossweiner, L.I. (1978) Radiat. Res. 73, 51-74.
13. Samuni, A. and Czapski, G. (1978) Radiat. Res. 76, 624-632.
14. Michaels, H.B. and Hunt, J.W. (1978) Radiat. Res. 74, 23-24.
15. Citri, N., Garber, N. and Sela, M. (1960) J. Biol. Chem. 235, 3454-3459.
16. Citri, N. and Pollock, M.R. (1966) Adv. Enzymol. 28, 237-324.
17. Sanner, T. and Pihl, A. (1963) Radiat. Res. 19, 12-26.
18. Anbar, M., Bambenek, M. and Ross, A.B. (1973) National Standard Reference Data System NBS: 43.
19. Held, K.D., Synek, R.W. and Powers, E.L. (1978) Int. J. Radiat. Biol. 33, 317-324.
20. Held, K.D. and Powers, E.L. (1979) 6th Int. Cong. Radiat. Res. Tokyo, Abstr. 191.
21. Halliwell, B. (1977) Superoxide and Superoxide Dismutases, Michelson, A.M., McCord, J.M. and Fridovich, I., eds., Academic Press, New York, pp.335-349.

IN VITRO RADIOPROTECTION OF ERYTHROCYTES BY SUPEROXIDE DISMUTASE

W. LEYKO, Z. JOZWIAK, Z. HELSZER AND G. BARTOSZ
Department of Biophysics, Institute of Biochemistry and Biophysics,
University of Lodz, Poland

The ubiquitous presence of superoxide dismutase (SOD) in all aerobic cells studied points to a profound importance of this protein as a protective enzyme against a highly reactive oxygen intermediate, the superoxide radical anion O_2^- and, indirectly, against other "active oxygen species" generated from it.[1] There are many intrinsic metabolic sources of O_2^- in living cells but also some environmental agents act via the superoxide radical anion. An important one among them is ionizing radiation. It is well established that irradiation of water or dilute water solutions gives rise to the following radiolysis products (in parentheses the respective radiation yields are given in molecules formed per 100 eV of radiation energy absorbed[2]):

e_{aq}^- (2.6), H· (0.6), OH· (2.6), H_2 (0.45), H_2O_2 (0.75).

In oxygenated solutions, hydrated electrons e_{aq}^- and hydrogen atoms react rapidly ($k > 10^{10}$ $M^{-1}s^{-1}$) with dissolved molecular oxygen to yield O_2^- (Ref.3). Therefore products of radiolysis of oxygenated water solutions include O_2^-.

One should expect that if O_2^- has detrimental effects on cells and their constitutuents, SOD would have a radioprotective action. Indeed, it has been demonstrated that extracellular SOD had a definite protective effect upon irradiated Acholeplasma laidlawii[4] and mammalian cells[5] and that SOD injection diminished the mortality of irradiated Mus musculus.[6-8] The latter effect seemed to be conditioned, among others, by an increased survival of red blood cells in irradiated mice.[8] In order to clarify possible mechanisms underlying this phenomenon we undertook studies on the in vitro effect of SOD on the radiation damage to mammalian red blood cells.

In our experiments, erythrocytes were isolated from porcine, bovine and human blood and suspended in physiological saline enriched with appropriate additives. Erythrocyte membranes were prepared according to Dodge et al.[9] and suspended in 10 mM phosphate, pH 7.4 with eventual additives. Erythrocyte or membrane suspensions were irradiated from ^{60}Co gamma sources at a room temperature at dose-rates of 30-86 Krad h^{-1} for lower doses and 0.5 Mrad h^{-1} for higher doses.

The levels of adenine nucleotides and "adenylate energy charge" are among the most important criteria of red cell viability. We studied the effect of radiation on the adenine nucleotides content of porcine erythrocytes in the absence and in the presence of protecting substances. In the range of low doses, radiation brings about characteristic increases in the levels of AMP and ADP (Table 1). The presence of SOD in the extracellular medium prevented these events in the majority of cases (Table 1). Catalase dismutating H_2O_2 and ethanol (an OH· scavenger) in concentrations of 50-200 µg ml^{-1} and 50 mM, respectively,

TABLE 1

EFFECT OF GAMMA RADIATION ON THE AMP AND ADP LEVELS IN PORCINE ERYTHROCYTES.[a]

Dose	AMP		ADP	
	% of control value	P	% of control value P	P
2 Krad	116.8	> 0.05	106.5	> 0.5
2 Krad + SOD	103.5		106.5	
5 Krad	119.8		108.1	
		< 0.001		< 0.02
5 Krad + SOD	102.8		101.7	

[a]The cells were irradiated in the absence and in the presence of SOD (120 µg/ml). Nucleotides were estimated chromatographically after Mills et al. (Ref.16). SOD was isolated from hog liver by a modified procedure of Fried et al (Ref.17).

not only did not protect but even potentiated the effects of radiation. This radioprotective effect of extracellular scavenging of O_2^- on the intracellular parameters is understandable taking into account the ability of O_2^- to penetrate the erythrocyte membrane.[10] A similar effect has been noted for chromosome aberrations in irradiated lymphocytes.[11] However, it seemed probable that part of the radiation-induced intracellular disturbances was caused by damage of the cell membrane. Therefore, we studied the effect of SOD and other protectors on the radiation damage to the erythrocyte membrane as evaluated by ATPase activity changes, lipid peroxidation, alterations is osmotic fragility and autohemolysis.

In the case of radiation inactivation of ATPase, the protective effect of SOD was clear-cut (Figure 1). The other proteins tested, catalase and albumin, were ineffective. In the case of radiation augmentation of osmotic fragility, both SOD and catalase afforded protection to the cells as did the chemical radioprotectors, cysteine and glutathione (Table 2).

Radiation-induced lipid peroxidation was inhibited in the presence of catalase but SOD was ineffective (Figure 2). It is known that H_2O_2 alone can induce lipid peroxidation in erythrocytes.[12] In the case of autohemolysis induced by radiation, all the three proteins tested inhibited the effect (Figure 3).

When interpreting the above results, one should take into account that all the water radiolysis products react with all components of the system. In

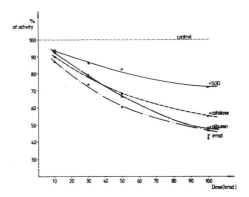

Fig. 1. Effect of gamma radiation on the $(Mg^{2+}+K^++Na^+)$-ATPase
activity of porcine erythrocyte
membranes. Protectors added at
concentrations of 200 μg ml^{-1}
(SOD was prepared, (see Table 1)
catalase was from Sigma and
albumin from International
Enzymes). Membrane protein:
3 mg ml^{-1}. ATPase activity estimated by a modified method of
Ellory and Carleton (Ref.18).

TABLE 2

EFFECT OF SUPEROXIDE DISMUTASE AND CATALASE ON THE RADIATION-INDUCED INCREASE IN
OSMOTIC FRAGILITY OF PORCINE ERYTHROCYTES[a]

Dose	Mean C_{50} S.D.	P
30 Krad	0.68 ± 0.03	
" + SOD	0.66 ± 0.02	< 0.02
" + catalase	0.66 ± 0.02	< 0.02
50 Krad	0.72 ± 0.02	
" + SOD	0.70 ± 0.01	< 0.01
" + catalase	0.69 ± 0.02	< 0.01

[a]C_{50} denotes concentration of NaCl (%) needed to cause 50% hemolysis. Erythrocytes irradiated at a concentration of 5% (v/v). Protectors as in Figure 1.

particular, SOD and other enzymes react not only with O_2^- but also with $OH^{\cdot}$, e_{aq}^-,
$H^{\cdot}$ and H_2O_2. Not to mention all of them, $OH^{\cdot}$ is known to be the most powerful
oxidant species generated in biological systems. Rate constants for reactions
with $OH^{\cdot}$ are of the order of 10^{10} $M^{-1}s^{-1}$ for all proteins studied. Therefore,
the protective effect of SOD might be partly due to reaction with other reactive

314

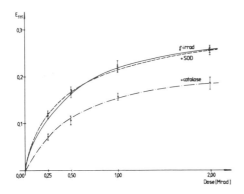

Fig. 2. Effect of SOD and catalase on radiation-induced lipid peroxidation in porcine erythrocyte membranes. Conditions as for Figure 1. Malondialdehyde determined with thiobarbituric acid.

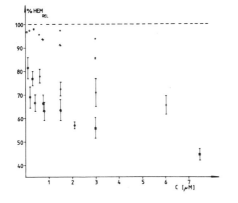

Fig. 3. Effect of proteins on the radiation-induced hemolysis of bovine erythrocytes. Hemolysis of samples devoid of protein additives assumed as 100%. Points with S.D.: proteins added before irradiation. Points without S.D. (<25% in these cases): proteins added 100 min after irradiation. Hemolysis measured after 44-h incubation at 4°C. Erythrocytes irradiated at a concentration of 5% (v/v). SOD was from Miles, catalase from Sigma and albumin from International Enzymes.

radiolytic products apart from O_2^-. We have not observed inactivation of extra-cellularly added SOD at doses up to 100 Krad in our experiments, but one can expect that the active site copper is the most reactive part of the enzyme mole-cule and that consecutive redox reactions of this copper with reducing and oxi-dizing radicals can result in deactivation of the radicals while restoring the enzymatic activity. Therefore a protective effect of SOD does not necessarily indicate the involvement of O_2^- as the main causative agent of the damage in question. To ascertain this, one should compare the effect of active and in-activated SOD or the effects of other proteins.

In the case of adenine nucleotide content and ATPase activity we observed some protective action of thermally inactivated SOD but much lower than that of active enzyme. One can conclude therefore that O_2^- is involved in the radiation damage to the red cell energy metabolism and ATPase activity. On the other hand, it is possible at present to decide unequivocally the role of O_2^- in the radiation damage causing osmotic fragility of erythrocytes and it seems that O_2^- is of minor importance in the initiation of lipid peroxidation and hemolysis in irra-diated erythrocytes.

It has been reported[13] that the SOD inhibitor, diethydithiocarbamate (DDTC) sensitized erythrocytes to the hemolysing action of ionizing radiation. However, we were unable to reproduce these results (Table 3).* Moreover, our comparison of radiation-induced hemolysis in erythrocytes from normal humans and from patients with trisomy 21, having an increased activity of SOD,[14-15] failed to reveal any differences (Figure 4) thus validating our conclusion on the small relevance of O_2^- to radiation-induced hemolysis.

It seems from the presented data that one should consider separately any event in the complex process of radiation damage to the cell as the radical mechanism. of damage may vary from one case to another. Some of these events are apparently mediated by O_2^- and they condition the radioprotective effect of SOD observed at an organismal level which may be of significant therapeutic significance.

ACKNOWLEDGEMENT

This work was supported in part by Grant R.III.13.

---

*Editors' foornote: DDTC itself is an effective radioprotector. Chinese hamster cells pretreated with DDTC are more radiosensitive than control cells. See S.L. Marklund and G, Westman, this volume.

TABLE 3

EFFECT OF DDTC ON THE RADIATION-INDUCED HEMOLYSIS OF BOVINE ERYTHROCYTES[a]

Concentration of DDTC	Relative hemolysis (%)
1 mM	42.7 ± 6.0
2.5 mM	35.8 ± 1.0
5 mM	61.4 ± 1.8

[a]Hemolysis of samples wihout DDTC assumed as 100%. Before irradiation the cells were incubated with DDTC at 37°C for 2 hr. Other details as for Figure 3.

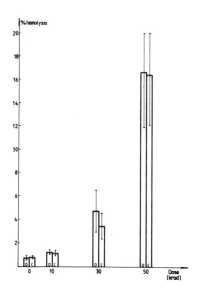

Fig. 4. Hemolysis of erythrocytes from control subjects (C) and patients with trisomy 21 (D) irradiated with different doses of gamma radiation.

REFERENCES

1. Fridovich, I. (1974) Adv. Enzymol. 41, 35-97.
2. Thomas, J.K. (1967) in Radiation Reserach, Silini, G., ed., North-Holland Publishing Co., Amsterdam, p.179.
3. Okada, S. (1970) in Radiation Biochemistry, Vol.I, Altman, K.I., Gerber, G.B. and Okada, S., eds., Academic Press, New York.
4. Petkau, A. and Chelack, W.S. (1974) Int. J. Radiat. Biol. 26, 421-426.
5. Michelson, A.M. and Buckingham, M.E. (1974) Biochem. Biophys. Res. Commun. 58, 1079-1086.
6. Petkau, A., Kelly, K., Chelack, W.S., Pleskach, S.D., Meeker, B.E. and Brady, C.M. (1975) Biochem. Biophys. Res. Commun. 65, 886-893.

7. Petkau, A., Chelack, W.S. and Pleskach, S.D. (1976) Int. J. Radiat. Biol. 29, 297-299.
8. Petkau, A. (1978) Photochem. Photobiol. 28, 765-774.
9. Dodge, J.T., Mitchell, C. and Hanahan, D.J. (1963) Arch. Biochem. Biophys. 100, 119-130.
10. Lynch, R.E. and Fridovich, I. (1978) J. Biol. Chem. 253, 1838-1845.
11. Nordenson, I. (1978) Hereditas, 89, 163-167.
12. Walls, R., Kumar, K.S. and Hochstein, P. (1976) Arch. Biochem. Biophys. 174, 463-468.
13. Stone, D., Lin, P.-S. and Kwock, L. (1978) Int. J. Radiat. Biol. 33, 393.
14. Sinet, P.M., Allard, D., Lejeune, J. and Jerome, H. (1974) C.R. Acad. Sci. Paris, 278, 3267-3270.
15. Kedziora, J., Jeske, J., Witas, H., Bartosz, G. and Leyko, W. (1977) Acta Biol. Med. Ger. (1977) 36, 779-782.
16. Milles, G.C., Burger, D.O., Schnieder, M. and Levin, C. (1961) J. Lab. Clin. Med. 58, 725-735.
17. Fried, R., Fried, L.W. and Babin, D.R. (1973) Eur. J. Biochem. 33, 439-445.
18. Ellory, J.C. and Carleton, S. (1974) Biochim. Biophys. Acta,363, 397-403.

COPPER-ZINC SUPEROXIDE DISMUTASE AND THE RADIATION SENSITIVITY OF CHINESE
HAMSTER CELLS

STEFAN L. MARKLUND AND GUNNAR WESTMAN
Department of Clinical Chemistry and Oncology, University Hospital of Umeå,
S-901 85 Umeå, Sweden

INTRODUCTION

The superoxide anion radical O_2^- and its corresponding acid $HO_2^{\cdot}$ (pK$_a$ 4.88
(Ref.1)) are formed directly and through secondary reactions by ionizing radiation. The presence of oxygen in a solution greatly increases the formation of
the radical by secondary reactions. Oxygen is also known to increase the biological effects of ionizing radiation. The mechanisms of the oxygen ehhancement
are not known. There are indications that toxic effects of O_2^- might contribute
to it. Thus superoxide dismutase, which disproportionates the radical, has
been shown to protect DNA[2], proteins[3] and cell membranes[4] against ionizing
radiation. Myoblasts[5], mycoplasma[6] and bacteria[7,8] are protected by superoxide
dismutase in the medium. Mice given superoxide dismutase parenterally show
increased radioresistance.[9-12] Variations of superoxide dismutase content in
bacteria, induced by growth in different oxygen tensions[7] or by growth in media
lacking the prosthetic metals of superoxide dismutase,[8] have indicated the
importance of endogenous enzyme in the protection against the oxygen effects.

Diethydithiocarbamate (DDTC) is an efficient inhibitor of CuZn superoxide
dismutase.[13] We have investigated its effect on mammalian cells in culture
and devised a procedure in which CuZn superoxide dismutase can be inhibited to
about 95%. This procedure was employed to test the proposition that superoxide
dismutase contributes to the radioresistance of mammalian cells.

MATERIALS AND METHODS

Sodium diethyldithiocarbamate was obtained from Sigma Chemical Company. KO_2
was a product of Alfa Europe Products. Eagle's minimal essential medium (MEM)
and fetal calf serum were obtained from Flow Laboratories. Petri dishes were
manufactured by NUNC.

Cell cutlture. Chinese hamster cells (V 79-379A) were received from the
Department of Tumor Biology, Karolinska Institutet, Stockholm, which had obtained
the cell line from Dr.M. Elkind, Bethesda, USA in 1964. The cells were grown in
Eagle's minimal essential medium (MEM) with 15% fetal calf serum, supplemented

with benzylpenicillin and streptomycin. Stock cultures were incubated in plastic Petri dishes of 9 cm diameter, at $38^{\circ}C$, in air containing 5% CO_2 and saturated with water. For the experiments, three day old non-confluent stock cultures were treated with 0.1% EDTA for two minutes and were thereafter trypsinized with 0.1% trypsin. An appropriate number of cells to obtain about one hundred colonies after irradiation was seeded into 5 cm plastic Petri dishes, with a final amount of 5 ml of medium. Three to four hours after explating the cells, the medium was exchanged for a 3 mM solution of DDTC in MEM-15% fetal calf serum. After 90 minutes incubation the dishes were rinsed four times with medium during a period of 20 minutes. In the irradiation experiments irradiation was given 30-40 minutes after the end of the DDTC-inhibition. Controls were sham inhibited and sham irradiated, and in every experiment there were also unirradiated control dishes to check plating efficiency and DDTC-toxicity. When enzymic activities and content of sulfhydryl compounds were to be analysed, 9 cm plastic Petri dishes, with three day old near-confluent cultures were inhibited for 90 minutes with the DDTC-solution, rinsed four times with medium, and then at appropriate times, washed twice with 0.15 M NaCl containing 3mM diethylenetriaminepentaacetic acid, the cells being scraped off in the second wash medium. The suspension was then centrifuged and the cells kept as pellets at $-80^{\circ}C$ until assay. Reactivation of CuZn superoxide dismutase by Cu ions was prevented by the chelator diethylenetriaminepentaacetic acid. The pellets were then sonicated in 10 mM potassium phosphate pH 7.4 containing 30 mM KCl and 0.2 mM diethylenetriamine-pentaacetic acid (0.5 ml per 10^6 cells). Determination were performed directly on the homogenates. Superoxide dismutase, catalase and protein content were analysed on the supernatants after centrifugation.

Irradiation procedure. The cells were irradiated in plastic boxes, all dishes receiving the same dose being treated at the same time. The boxes were gassed with 5% CO_2 in water saturated air, warmed to give a temperature in the box of about $30^{\circ}C$. After irradiation the cells were grown for 8-9 days, with one or two changes of medium. The cells were then fixed in methanol, and stained with Giemsa stain. All macroscopic colonies (containing at least 50 cells) were taken as viable clones. The X-ray source was a 4 MV linear accelerator (Varian Clinac 4). The dose rate at the cell surface was about 3 Gy/minute, and the source-cell distance was 80 cm. The dose was determined by means of an ionization chamber calibrated with Co^{60} at Statens Strålskyddsinstitut. Precision was better than 3%.

Calculation procedures and statistical considerations. Survival was estimated by counting the clones of four separate dishes with test cells and with control cells for each radiation dose and taking the mean values as survival for that dose. All survival figures were corrected for plating efficiency and toxicity of DDTC and were transformed to natural logarithms before regression analysis

and statistical testing. Separate graphical lin-log diagrams were drawn for
each experiment to check linearity. By means of linear regression analysis the
regression coefficient (slope), and the extrapolation number, (ordinate intercept)
were calculated for each experiment. Mean values and ratios of mean values were
pooled and tested for statistical significance using Student's t-test. Experiments were excluded if the linearity of the individual survival curves as judged
graphically was bad. The theoretical assumption of linearity was based on the
single-hit multi-target theory.

Superoxide dismutase and other analyses. Superoxide dismutase was determined
in terms of its ability to catalyze the disproportionation of O_2^- in alkaline
aqueous solution. The disproportionation was directly studied in a spectrophotometer, essentially as described before,[14] the difference being that both CuZn
superoxide dismutase and Mn superoxide dismutase were assayed at pH 9.50. One
unit in the assay is defined as the activity that brings about a decay in O_2^- at
a rate of 0.1 s^{-1} in 3 ml buffer. It corresponds to 8.3 ng human and to 4.1 ng
bovine CuZn superoxide dismutase and 65 ng bovine Mn superoxide dismutase.
When bovine and human enzymes are analysed, one unit in the present assay
corresponds to 0.024 units CuZn superoxide dismutase and 0.24 units Mn superoxide
dismutase in the "standard assay" involving xanthine oxidase and cytochrome c at
pH 7.8.[15] The chinese hamster enzymes have not been investigated, but the
activity ratios between the two assays are probably similar for these enzymes.

Cytochrome oxidase was determined as described by Wharton and Tzagoloff.[16]
Catalase was analysed with an oxygen electrode.[17] Non-protein sulfhydryl compounds were determined after precipitation of protein with 4% sulfosalicyclic
acid.[18] For protein analysis Coomassie Brilliant Blue G-250 was employed.[19]

Determination of free DDTC in cell homogenates. To a cell suspension, about
20×10^6 cells in 1 ml of 0.15 M NaCl, was added (final concnetrations) 10 mM
EDTA, 0.3 M NaCl and 20 mM $CuSO_4$. The suspension was then sonicated. Thereafter 1 ml of isoamyl alcohol was added and the tubes were extensively agitated.
After centrifugation the DDTC-Cu^{1+} complex on the isoamyl alcohol phase was
determined by reading the absorbance at 434 nm. The assay was standardized
with DDTC added to control cell suspensions. The lower limit of the assay
corresponded to about 50 µM DDTC in the cells. The volume of the cells was
estimated as the volume of cell pellets after centrifugation.

RESULTS

Effect of DDTC on Mn and CuZn superoxide dismutase, cytochrome oxidase,
catalase and content of nonprotein sulfhydryl compounds. Figure 1 shows
the inhibition of CuZn superoxide dismutase in cells exposed to 3 mM DDTC. The
CuZn superoxide dismutase remains inhibited to about 95% 30 minutes after the
DDTC is removed. The inhibition by 1.5 mM DDTC (not in Figure 1) is not as

extensive as that by 3 mM DDTC. 6 mM DDTC (not in Figure 1) does not produce
significantly more CuZn superoxide dismutase inhibition but is more toxic to
the cells. As seen, the enzyme inhibition is very reproducible, and therefore
checks of the inhibition in conjunction with each irradiation experiments was
omitted.

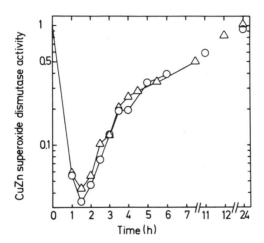

Fig. 1. Inhibition of CuZn superoxide dismutase in cells by DDTC. Cells
attached to Petri dishes were exposed to 3 mM DDTC in MEM-fetal calf serum.
The DDTC was washed away after 1.5 hours by four changes of medium during a
period of about 20 min. After various times dishes were washed with 0.15 M
NaCl with 3 mM diethylenetriamine penta-acetic acid and the cells collected
for specific CuZn superoxide dismutase analysis. The CuZn superoxide dismutase
activity in untreated cells(=1 in the figure ordinate) was 105 units per mg
protein. The Mn superoxide dismutase activity was 2.7 units per mg protein,
and it was not affected by the DDTC treatment. O and Δ show replicate
experiments carried out with an interval of four months.

The Mn superoxide dismutase in the cells was unaffected by the DDTC. Nor was
pure bovine Mn superoxide dismutase[20] or Mn superoxide dismutase in human liver
homogenates found to be affected by 3 mM DDTC.

The cytochrome oxidase activity, catalase activity and content of non-protein
sulfhydryl compounds in the cells was not affected by the DDTC treatment (Figure
2). The efficiency of the rinsing in removing the DDTC was studied in control
experiments where cells were exposed to 15 mM DDTC for 90 minutes. The remaining
DDTC after successive medium changes was estimated by analysis of cell homogenates.
After four changes less than 50 μM free DDTC could be demonstrated in the cells.

Effect of pretreatment with DDTC on cell survival after irradiation. Eighteen
identical experiments were performed, where the cells were exposed to four different radiation doses. For each dose there were four dishes in the control

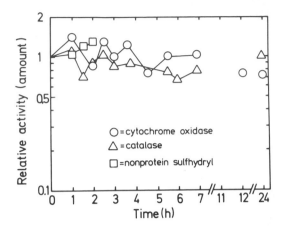

Fig. 2. Effect of DDTC treatment on cytochrome oxidase activity, catalase activity and content of nonprotein sulhydryl compounds. Cells attached to Petri dishes were exposed to 3 mM DDTC in MEM-fetal calf serum. The DDTC was washed away after 1.5 hours by four changes of medium duirng a period of about 20 minutes. After various times dishes were washed with 0.15 M NaCl with 3 mM diethylenetriamine penta-acetic acid and the cells collected for analysis as described in MATERIALS AND METHODS.

group and four in the test group. Survival curves were mapped and the parameters D_o, n and k were calculated (see Table 1). The mean values of these parameters are given in Table 1, together with the relevant ratios. As seen, the DDTC-treatment increased the radiosensitivity as indicated by an increase in slope (k). There was no effect on the extrapolation number (n).

DISCUSSION

Copper-zinc superoxide dismutase in cells is efficiently inhibited by DDTC and the inhibition is retained after the DDTC is washed away. Extensive removal of DDTC is essential, as the thiol in itself is an effective radioprotector with an efficiency about as large as that of cysteamine.[21,22] The small amount of DDTC (< 50 μM) remaining would not give a significant protection.

Our goal is to evaluate the importance of endogenous superoxide dismutase in the radioresistance of mammalian cells, tissues and tumours. The present DDTC pretreatment model should be an excellent tool for that purpose. The inhibition of CuZn superoxide dismutase decreases with time but is still significant for a few hours after the removal of DDTC. The enzyme might conceivably protect both during the irradiation by intercepting O_2^- radicals formed by radiolytic reactions as well as later by removing radicals participating in free

TABLE 1

SURVIVAL OF DDTC-TREATED AND CONTROL CELLS AFTER EXPOSURE TO X-RAYS[a]

	lnk	k^b	D_o (Gy)^c	ln n	extrapola-tion number (n)
Control group	-0.4358 ± 0.0609	0.6467 (0.6085-0.6740)	2.29 (2.01-2.67)	1.195 ± 0.4515	3.30 (2.1-5.2)
DDTC-group	-0.5352 ± 0.0809	0.5855 (0.5450-0.6348)	1.87 1.62-2.20	1.198 ± 0.5305	3.30 (2.0-5.6)

[a]Test and control cells were exposed to 0, 5.13, 6.83, 8.54 and 10.0 Gy. 18 separate experiments were performed. Survival was estimated by counting of clones after 8-9 days. Survival after DDTC treatment alone was 0.68. Mean plating efficiency was 51%. k = slope of survival curve. $D_o=(lnk)$; n=extrapolation number. Mean and S.E.M. values of experimental paramaters and ratios are given.

[b]k control/k DDTC= 1.10(1.08-1.13) [$\alpha = 0.0000015$]; α refers to one-tailed Student's t-test of significance for ratios differing from 1 (= no effect of DDTC treatment).

[c]D_o control/D_o DDTC= 1.23(1.21-1.24) [$\alpha = 0.0000024$].

radical reactions such as lipid peroxidation. Late effects of exogenous super-oxide dismutase have recently been described.[23] Both immediate and late pro-tective activity of the enzyme might be affected by the DDTC pretreament. The present results show that DDTC pretreated cells are indeed more radiosensitive than control cells.

The effects observed are not necessarily due to CuZn superoxide dismutase inhibition. DDTC is not a specific inhibitor of superoxide dismutase. It inhibits for example dopamine-β-hydroxylase[24], aldehyde dehydrogenase,[25] and cytochrome oxidase.[26] The DDTC is efficiently removed from our system, but like superoxide dismutase, other factors of importance for the radioresistance could also remain affected by the DDTC pretreatment. Cellular respiration is important for recovery after irradiation.[27,28] However, we found no inhibition of cytochrome oxidase after the DDTC pretreatment. Nor was there any inhibition of catalase which removes the potentially harmful hydrogen peroxide. The con-tent of non-protein sulfhydryl compounds appears to be of importance for cellu-lar radioresistance, at least under anoxic conditions.[29] This parameter was not affected by DDTC.

The superoxide radical is both reducing and oxidizing, but possibly not very reàctive. There are indications that most toxic effects of superoxide anion-producing systems are caused by reactive compounds derived from the radical. Often so-called hydroxyl radical scavengers as well as catalase are protective.[30,31] In other systems compounds having a high reactivity with singlet oxygen prevent

toxic effects.[32-34] The so-called Haber-Weiss reaction has been suggested as
a source of reactive toxic compounds; $O_2^- + H_2O_2 \rightarrow OH^- + OH^{\cdot} + O_2$ (singlet
form). This reaction does not occur as written, but reactions with this overall
stoichiometry are apparently catalyzed by some transition metal ions.[35-37] The
radioprotective effect of formate, which reacts with $OH^{\cdot}$ to form O_2^- is, in view
of these results, compatible with the idea that scavenging of the O_2^- may protect cells from radio-damage.[38]

In most previous radiation studies the superoxide dismutase was added to
the medium or injected into animals. It is possible that the protective effects
observed against radiation were due to interception of radicals formed in the
medium. In the animal models, the anti-inflammatory properties of the enzyme
may have contributed to the observed effects.[39] Further, the results of the
experiments with various superoxide dismutase activities in bacteria were
interpreted in terms of interception of compounds generated in the medium.[7,8]
Where stated, the media employed were simple salt solutions without much
scavenging ability. The present medium, a nutrient solution with 15% fetal calf
serum, has a fairly large radical scavenging ability. Therefore, the present
model should be more relevant for the evaluation of the importance of endogenous superoxide dismutase in mammalian tissues and tumours.

The Mn superoxide dismutase of the cells was unaffected by DDTC. It is
localized to the matrix space of the mitochondria[40] and its activity comprised
20% of the total superoxide dismutase activity in the cells. This figure is
obtained when the activity is expressed in units of the xanthine oxidase-
cytochrome c assay,[15] which operates at physiological conditions; low O_2^- concentration and neutral pH. The superoxide anion is comparatively stable, $t_{1/2}$
for the disproportionation of a $10^{-8}M$ solution at pH 7 is about a minute.[1]
It easily permeates cell membranes, apparently through the so-called anion
channels.[41] Superoxide dismutase outside a membrane-bound vesicle protects
components inside the vesicle against superoxide radicals produced within the
vesicle.[34,41] The permeability of the superoxide radical though the mitochondrial membrane is unknown, but it is possible that the Mn superoxide dismutase
in the mitochondrial matric[40] may protect other cell compartments against the
superoxide radical. This means that although most of the CuZn superoxide dismutase was inhibited, there was still a lot of fucntional superoxide dismutase
activity left within the cells in the present experiments. A more complete
enzyme inhibition might have caused even larger effects on the radiosensitivity.

SUMMARY

Copper-zinc superoxide dismutase in chinese hamster cells were inhibited to
95% by diethyldithiocarbamate treatment. The inhibition was retained after
the diethyldithiocarbamate, a potent radio-protector, was washed away. Cyto-

chrome oxidase, catalase amd the content of non-protein sulfhydryl compounds were unaffected by the inhibitor. Cells thus treated with diethyldithiocarbamate were more radiosensitive than control cells, the ratio in the presence of oxygen between the D_o values (see Table 1) of the groups being 1.23. The results are compatible with the proposition that superoxide dismutase contributes to the radioresistance of the cells.

ACKNOWLEDGEMENTS

We wish to thank Dr. E. Lundgren for his instructions and advice and for generously placing his equipment at our disposal. We are also indebted to Prof. B. Littbrand for his advice and stimulating discussions. The skilful technical assistance of Ms. K. Lindmark, Ms. M. Bäckström and Ms. B. Nilsson is gratefully acknowledged. The study was supported by grants from the Swedish Medical Research Council (04761) and the Lions Research Foundation, Department of Oncology, University of Umeå.

REFERENCES

1. Behar, B., Czapski, G., Rabani, J., Dorfman, L.M. and Schwarz, H. (1970) J. Phys. Chem. 74, 3209-3213.
2. Van Hemmen, J.J. and Meuling, W.J.A. (1975) Biochim. Biophys. Acta, 402, 133-141.
3. Lavelle, F., Michelson, A.M. and Dimitrijevic, L. (1973) Biochem. Biophys. Res. Commun. 55, 350-357.
4. Petkau, A. amd Chelack, W.S. (1976) Biochim. Biophys. Acta, 433, 455-456.
5. Michelson, A.M. and Buckingham, M.E. (1974) Biochem. Biophys. Res. Commun. 58, 1079-1086.
6. Petkau, A. and Chelack, W.S. (1974) Int. J. Radiat. Biol. 26, 421-426.
7. Misra, H.P. and Fridovich, I. (1976) Arch. Biochem. Biophys. 176, 577-681.
8. Oberley, L.W., Lindgren, A.L., Baker, S.A. and Stevens, R.H. (1976) Radiat. Res. 68, 320-328.
9. Petkau, A., Chelack, W.S., Pleskach, S.D., Meeker, B.E. and Brady, C.M. (1975) Biochem Biophys. Res. Commun. 65, 886-893.
10. Petkau, A., Kelley, K., Chelack, W.S. and Barefoot, A. (1976) Biochem. Biophys. Res. Commun. 70, 452-458.
11. Petkau, A., Kelley, K., Chelack, W.S., Pleskach, S.D., Barefoot, C. and Meeker, B.E. (1975) Biophys. Res. Commun. 67, 1167-1174.
12. Petkau, A., Chelack, W.S. and Pleskach, S.D. (1976) Int. J. Radiat. Biol. 29, 297-299.
13. Heikkila, R.E., Cabbat, F.S. and Cohen, G. (1976) J. Biol. Chem. 251, 2182-2185.
14. Marklund, S.L. (1976) J. Biol. Chem. 251, 7504-7507.
15. McCord, J.M. and Fridovich, I. (1969) J. Biol. Chem. 244, 6049-6055.
16. Wharton, D.C. and Tzagoloff, A. (1967) Meth. Enzymol. 10, 245-250.
17. Del Rio, L.A., Ortega, M.G., Lopez, A.L. and Gurgé, J.L. (1977) Anal. Biochem. 80, 409-415.
18. Ellman, G.L. (1959) Arch. Biochem. Biophys. 82, 70-77.
19. Bradford, M.M. (1976) Anal. Biochem. 72, 248-254.
20. Marklund, S.L. (1978) Int. J. Biochem. 9, 299-306.
21. Bacq, Z.M. (1953-54) Bull. Acad. Med. Belg., Ser. 6, 18-19, 226-235.
22. Van Bekkum, (1956) Acta Physiol. Neerl. 4, 508-523.
23. Nordensson, I. (1978) Hereditas, 89, 163-167.
24. Goldstein, M., Agnosta, B., Lanber, E. and McKereghan, M.R. (1964) Life Sci. 3, 763-767.

25. Dietrich, R.A. and Erwin, V.G. (1971) Mol. Pharmacol. 7, 301-307.
26. Frank, L., Wood, D.L. and Robert, R.J. (1978) Biochem. Pharmacol. 27, 251-254.
27. Littbrand, B. and Revesz, L. (1969) Brit. J. Radiol. 42, 914-924.
28. Hall, E.J. (1972) Radiat. Res. 49, 405-415.
29. Modig, H.G. (1968) Stud. Biophys. 12, 189.
30. Beauchamp, C. and Fridovich, I. (1970) J. Biol. Chem. 245, 4641-4646.
31. Grankvist, K., Marklund, S.L., Sehlin, J. and Täljedal, I.-B. (1979) Biochem. J. 182, 17-25.
32. Kellogg, E.W. and Fridovich, I. (1975) J. Biol. Chem. 250, 8812-8817.
33. Kellogg, E.W. and Fridovich, I. (1977) J. Biol. Chem. 252, 6721-6728.
34. Lynch, R.E. and Fridovich, I. (1978) J. Biol. Chem. 253, 1838-1845.
35. McCord, J.M. and Day, E.D. (1978) FEBS Lett. 86, 139-142.
36. Ven Hemmen, J.J. and Meuling, W.J.A. (1977) Arch. Biochem. Biophys. 182, 743-748.
37. Halliwell, B. (1978) FEBS Lett. 92, 321-326.
38. Samuni, A., Chevion, M., Halpern, Y.S., Ilan, Y.A. and Czapski, G. (1978) Radiat. Res. 75, 489-496.
39. Huber, W. and Saifer, M.G.P. (1977) in Superoxide and Superoxide Dismutases, Michelson, A.M., McCord, J.M. and Fridovich, I., eds., Academic Press, New York, pp.517-536.
40. Weisiger, R.A. and Fridovich, I. (1973) J. Biol. Chem. 248, 4793-4796.
41. Lynch, R.E. and Fridovich, I. (1978) J. Biol. Chem. 253, 4697-4699.

IMPORTANCE OF SUPEROXIDE DISMUTASE FOR THE ACUTE RADIATION SYNDROME

J. KRIZALA, H. KOVAROVA, V. KRATOCHVILOVA AND M. LEDVINA
Department of Biochemistry, Faculty of Medicine, Charles University,
500 38 Hradec Kralove, Czechoslovakia

INTRODUCTION

A variety of biologically important processes are accompanied by formation of superoxide radicals,[6] but these can also be generated by some exogennous factors, e.g. by ionizing radiation. After oxygenated aqueous solutions are exposed to high energy radiation (γ- or X-rays), the following primarily radiolytic products are generated from water:

$$H_2O \xrightarrow{} HO^{\cdot}, H^{\cdot}, e_{aq}^{-}, H_3O^{+}, H_2, H_2O_2$$

Subsequently, the energy-rich radicals are subjected to further reactions leading to generation of superoxide radicals in the presence of oxygen:[4]

$$H^{\cdot} + O_2 \longrightarrow HO_2^{\cdot}$$
$$HO_2^{\cdot} \longrightarrow H^{+} + O_2^{-}$$
$$O_2 + e_{aq}^{-} \longrightarrow O_2^{-}$$

The above mentioned findings indicate that superoxide dismutase (SOD) which protects living organisms against the effect of highly toxic superoxide radicals may play an important role in processes connected with radiation damage of irradiated tissues as well. This concept is in good accordance with experiments demonstrating a reliable radioprotective effect of the intravenously administered enzyme in the mouse.[18,19]

Hitherto, the influence of ionizing radiation on the activity of SOD has not been thoroughly studied. Only a post-irradiation decrease of the SOD activity in mouse spleen and thyroid has been reported,[20] and the activity has been followed in the rabbit liver.[13] One of these reports shows a decrease of the SOD activity after exposure to 1 000 R on the 1st day followed by return to normal values on the 7th day.[13] In a subsequent communication from the same research group an approximately 50 per cent decrease of the activity was described in the rabbit liver on the 1st and 3rd day after irradiation with 750 R (Ref.12).

We tried to investigate in our experiments, whether or not the physiological values of SOD in the rat and dog are altered after the whole-body irradiation.

We followed the SOD activity both in radioresistant tissues (liver) and in radio-sensitive ones (bone marrow). In addition, the activity was studied in red blood cells of the peripheral blood. We have to take into consideration that mature erythrocytes are rather radioresistant, whereas erythroblasts reveal higher sensitivity against ionizing radiation.

MATERIALS AND METHODS

The experiments were carried out partly on young male Wistar rats weighing 150 ± 5 g, partly on the dogs (non-standardized strains, males and females weighing 7 ± 1.5 kg). The experimental animals were irradiated in a homogeneous field of γ-rays (Chiso-box Chirana, CSSR, 48 sources of ^{60}Co, exposure rate 120 R/min). The rats were irradiated by whole body exposure to 650 or 800 R, the dogs were exposed to 300 R.

Rat blood was obtained from the carotid artery, dog blood from a vein in the fore limb. The red cells were isolated by centrifugation (3 000 x g, 5 min), and washed three times with 0.9 per cent NaCl. In order to lyse the cells, 1 ml of distilled water was added to the sediment of the red cells. After hemoglobin was removed by means of a mixture of chloroform with ethanol (3:5, v/v), the activity of SOD was determined in the supernatant by a method based on autoxi-dation of hydroxylamine.[10] Slightly lower values of SOD were obtained by means of this method as compared with other methods for the determination of SOD activity. Nevertheless, the method was preferred because of its simplicity and good reproducibility, and because the relative radiation-induced variations represented the main factor followed in our study. Besides this, the hemoglobin concentration in the blood was determined by the standard method based on the formation of cyanhemiglobin using a standard from Immuna (Sarisske Michalany, CSSR). Absorbances were measured using the Specol spectrophotometer (C. Zeiss, Jena, GDR).

The liver was excised after decapitation and laparotomy of the rats, and bone-marrow was taken from the femur and tibia. The samples (approx. 0.1 g) were homogenized in phosphate buffer pH 7.5, and the cytosol fraction was iso-lated by means of a 30 min centrifugation at 105,000 x g (Beckman Ultracentrifuge, L 5-50 model). The SOD activity was determined in this case with the xanthine/xanthine oxidase method of McCord and Fridovich.[14] The absorbances were measured on a Beckman model 25 spectrophotometer. The protein concentration was assayed by means of the biuret method using bovine serum albumin as a standard.[3]

Xanthine, xanthine oxidase, cytochrome c and Triton X-100 were purchased from Koch-Licht, (England). The other chemicals used were supplied by Lachema (Brno, CSSR); they were of analytic purity.

RESULTS

Activity of superoxide dismutase in the liver and bone marrow of the rats.
The SOD activity was investigated in the control non-irradiated rats and in those
on the 3rd, 7th and 14th day after whole-body exposure to 800 R. Only the
Cu-Zn containing SOD was assayed, becuase the determinations were carried out
in the cytosol fraction of the liver and bone-marrow. On the contrary, Mn-
containing SOD invariably occuring in the mitochondria of eukaryotic cells[22] was
not determined in our experiments. However, we repeatedly showed that it was
really absent in the cytosol fraction. Disc-electrophoresis in polyacrylamide
gel demonstrated that no mitochondrial enzyme penetrates into the cytosol fraction
after irradiation.[11]

Table 1 summarizes the results of the SOD activity in the liver. The enzyme
activity remained almost unchanged after irradiation. At all intervals only
insignificant differences were measured as compared with the control group of
non-irradiated rats. The cytosol fraction of the bone marrow reveals a very
pronounced decrease of the SOD activity at all the intervals investigated after
irradiation (Table 2). Regarding the fact that the protein concentration in the
cytosol fraction of the bone marrow cells is to a certain extent higher on the
3rd and 7th day after exposure, the decrease of the SOD activity measured by
means of enzyme units per mg of cytosol protein is more marked at those intervals than in the case where it is expressed in U/g of wet tissue.

TABLE 1

ACTIVITY OF SUPEROXIDE DISMUTASE IN THE LIVER OF CONTROLS AND IRRADIATED RATS[a]

Group	n	U/g tissue	Cytosol protein (mg/g)	U/mg protein
Controls	12	663.8 ± 106.9	105.3 ± 15.7	6.39 ± 1.13
3rd day	12	725.7 ± 130.4	112.5 ± 12.7	6.46 ± 0.97
7th day	11	622.2 ± 88.5	99.1 ± 9.9	6.33 ± 1.15
14th day	8	701.3 ± 135.6	95.8 ± 11.5	7.55 ± 1.63

[a]Values are mean ± S.D.

TABLE 2

ACTIVITY OF SUPEROXIDE DISMUTASE IN THE BONE MARROW OF CONTROLS AND IRRADIATED RATS[a]

Group	n	U/g tissue	Cytosol protein (mg/g)	U/mg protein
Controls	12	171.4 ± 63.5	70.5 ± 24.7	2.37 ± 0.71
3rd day	12	138.2 ± 57.2	101.0 ± 33.7[b]	1.47 ± 0.50[c]
7th day	11	75.8 ± 36.6[c]	103.6 ± 23.3[c]	0.77 ± 0.48[c]
14th day	8	80.4 ± 30.1[c]	67.6 ± 15.8	1.13 ± 0.38[c]

[a]Values are mean ± S.D.

[b]Significance, $P < 0.05$

[c]Significance, $P < 0.01$

As reproted in previous papers, degenerative changes take place in the bone marrow of the animals after high-dose γ-irradiation. This is reflected in a decrease of the number of active hemopoietic elements. Also necrotic cells and their fragments are present in the irradiated bone marrow. Thus, the determination of total protein cannot reflect the degree of cellularity of the bone marrow and scarcely can such an information be obtained by the protein estimation in the cytosol fraction, as carried out in our experiments. This was the reason, why we investigated also the cell number of the bone marrow. The SOD activity was then referred to 10^6 of those cells (Table 3). It was found that the enzyme activity in the cells of the bone marrow steeply increased on the 3rd day after irradiation. At following intervals a slow return to the normal was observed. The cause of the post-irradiation increase of the enzyme activity in a single cell cannot be unambiguously explained. Enzyme induction is one of possible hypotheses, as dealt with in DISCUSSION in detail. Under these circumstances, the increased SOD level would be followed also by an increase of the enzyme activity in the red cells of peripheral blood (though with a certain delay). From this reason we investigated the SOD activity in erythrocytes of the rats at late intervals after irradiation.

TABLE 3

NUMBER OF CELLS IN THE BONE MARROW OF CONTROLS AND IRRADIATED RATS AND ACTIVITY OF SUPEROXIDE DISMUTASE[a]

Group	n	Number of cells per mg of tissue (x 10^6)	SOD activity (U/10^6 cells)
Controls	12	1.650 ± 0.188	0.10 ± 0.04
3rd day	12	0.143 ± 0.062[b]	0.96 ± 0.13[b]
7th day	11	0.193 ± 0.052[b]	0.36 ± 0.14[b]
14th day	8	0.524 ± 0.251[b]	0.16 ± 0.03[b]

[a]Values are mean ± S.D.
[b]Significance, $P < 0.01$

Activity of superoxide dismutase in the red cells of the rats. The red cell SOD activity was assayed both in the control non-irradiated rats and in the irradiated group; six time intervals from 1 to 30 days after 650 R were investigated (i.e., the exposure corresponding approximately to the $LD_{50/30}$ dose of ionizing radiation under the given conditions). The control group consisted of 12 rats, the experimental groups amounted to 7-8 animals. The SOD activity was expressed in units/g hemoglobin. Use of the parameter enzyme activity/blood volume would not be sufficiently exact, as the development of the acute radiation syndrome is connected with electrolyte loss and water depletion, hematocrit values being thus altered.[5]

The results obtained are shown in Figure 1. The SOD activity remained almost unchanged on the 1st and 3rd day after irradiation, but it was higher from the 7th day. A statistically significant increase was found from the 14th day after exposure, and maximal values (138 per cent with respect to the controls) were reached on the 30th day. The blood level of hemoglobin varied only significantly during the whole period of the acute radiation syndrome, only on the 14th day was the hemoglobin concentration slightly lower.

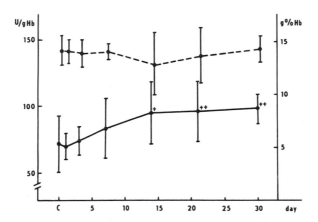

Fig. 1. Activity of superoxide dismutase (U/g hemoglobin) and hemoglobin concentration (g/100 ml blood) in erythrocytes of control non-irradiated rats (————) and rats after exposure to 650 R (– – – – –). Significance, P < 0.05(+); P < 0.01(++).

The results of the experiments performed on the rats indicate a considerable distribution range of the SOD values. This seems to be caused by the factor that the variability of physiological SOD values in individual animals is increased by a widely different individual ability to resist against the radiation attack. Moreover, the experiments had to be performed in such a manner that the rats were decapitated, therefore, each time interval represented in fact a different group of the experimental animals. In order to compensate for these disadvantages, further experiments were carried out with dogs; here multiple venepunctures were possible during the long period of an experiment.

Activity of superoxide dismutase in the red cells of the dogs. A 9-membered group of dogs was used. Before irradiation, the initial SOD activity in the red cells was determined. This group was then followed for 84 days after the whole-body exposure to 300 R. During this timespan, the SOD activity of the red cells of peripheral blood was assayed 11 times. Figure 2 demonstrates the

SOD activities in U/g hemoglobin. Regarding the fact that 7 experimental
animals died between the 14th and 21st day, no standard deviations are indicated
later than this period; the then values represent the mean value of the two
surviving dogs.

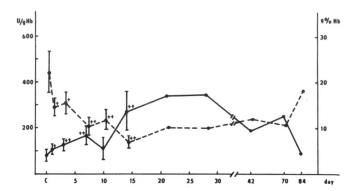

. Fig. 2. Activity of superoxide dismutase (U/g hemoglobin) and hemoglobin
concentration (g/100 ml blood) in ertyrhocytes of dogs before irradiation
(———) and after exposure to 300 R (– – – – –). Significance, $P < 0.05(+)$;
$P < 0.01(++)$.

The results obtained indicate that the 300 R exposure leads to a pronounced
and lasting increase of the SOD activity in the dog red cells. The difference
against the initial value was always significant in all the intervals, only on
the 10th day after irradiation was the activity measured unchanged. The comparison of the results on the dogs with those previously obtained on rats
indicates that the increase of the SOD activity occurs in the dogs during the
first days after irradiation. The increase of the SOD curve is much more pronounced also from the 14th day after exposure (on the 14th day the activity
in the dogs is approximately 3.5 times higher than the initial value, whereas
this increase at the same interval in the rats amounts only to 133 per cent).

The hemoglobin concentration in the dog blood is distinctly lowered at the
initial phase after irradiation. The decrease of the erythrocytes is much
faster in the dogs than in the rats. The difference may be caused not only by
the different radiation dose (it exceeds apparently the $LD_{50/30}$ in the dogs), but
also by a higher radiosensitivity of the dogs.

DISCUSSION

Our results demonstrate that in the mitotically inactive rat liver the whole-body exposure to 800 R led only to minimal alteration of the SOD activity as compared with the group of non-irradiated animals. On the contrary, the total SOD activity (U/g of tissue) decreased considerably in the bone marrow, which was in good agreement with the well-known fact that this mitotically active tissue is highly sensitive to ionizing radiation. Nevertheless, if we calculate the SOD activity in one bone-marrow cell survicing the radiation attack, it is steeply increased. The increase of the SOD activity in the circulating erythro-cytes is a consequence of this phenomenon. Till now, we have no reliable basis for the evaluation of this post-irradiation increase of the SOD activity in the bone-marrow cells, but some experimental data may help to elucidate this effect.

During the reparation phase of the radiation-damaged hemopoietic bone marrow, the number of reticulocytes in the blood increases.[7] Moreover premature erythro-cytes reveal a higher SOD activity (+ 50 per cent) than the "old" red cells.[1] However, we have to take into consideration that just the percentage of young red cells in circulating blood seems to be too low to explain fully the observed changes.

A further explanation is plausible. The increased synthesis of red cell SOD could be induced by a strong inflow of free radicals generated by radiolysis of irradiated tissue water. The resulting increase of the superoxide radical pool in the radiation-damaged bone marrow as a consequence of a marked inhibition of the "superoxide dismutase ability" may add to the SOD increase as well. Some literature data about analogous (in a certain degree) examples of the enzyme induction are available. An increased O_2 concentration in the atmosphere led to induction of SOD both in Escherichia coli[8], and in the lung of young rats and rabbits.[16,21] In addition, Bonta et al.[2] supposed an SOD induction in newly formed erythrocytes in new-born children; during the initial post-natal days the SOD activity of the red cells markedly increases. These authors think that the role of the induction factor is played by a marked increase of the partial pressure of blood oxygen, that leads also to an enhanced generation of superoxide radicals.

The protective function of SOD in the red cells is tightly connected with further enzyme systems, mainly with catalase and glutathione peroxidase.[9] In persons with inborn hypocatalasemia an increased SOD level was observed, mean-while glutathione peroxidase was within the normal level.[17] Of course, the increase of the SOD activity in the acute radiation syndrome may be connected with the previously described fall of the red cell catalase activity.[7]

Further factors play a role, but lack of experimental data prevents the formulation of other hypotheses. We have to add that experiments on dogs being performed in our laboratory support the present results. If we promote

334

the reparative ability of the radiation-damaged bone marrow by a proper therapeutic treatment, the SOD activity of the red cells increase more steeply than
described here.

The substantial decrease in the total SOD activity of the bone marrow found
in our experiments on rats after 800 R exposure is probably connected with
lethal consequences of irradiation. This statement agrees with the opinion
of Michelson et al.[15] that "levels of less that 50 per cent of the normal means
for superoxide dismutase are more or less lethal due to the increased toxicity
of uncontrolled superoxide".

REFERENCES

1. Bartosz, G., Tannert, Ch., Fried, R. and Leyko, W. (1978) Experientia, 34, 1464.
2. Bonta, B.W., Gawron, E.R. and Warshw, J.B. (1977) Pediat. Res. 11, 754-757.
3. Cleland, C.W. and Slater, E. (1953) Biochem. J. 53,547-556.
4. Czapski, G. (1971) Annu. Rev. Phys. Chem. 22, 171-208.
5. Dalrymple, G.V., Gaulden, M.E., Kollmorgen, G.M. and Vogel, H.W. (1973) Medical Radiation Biology, W.B. Saunders Co., Philadelphia.
6. Fridovich, I. (1975) Annu. Rev. Biochem. 44, 147-157.
7. Gerber, G.B. and Altman, K.I. (1970) Radiation Biochemistry, Vol.II, Academic Press, New York.
8. Gregory, E.M. and Fridovich, I. (1973) J. Bacteriol. 114, 543-548.
9. Halliwell, B. (1974) New Phytologist, 73, 1075-1086.
10. Kono, Y. (1o78) Arch. Biochem. Biophys. 186, 189-195.
11. Kovarova, H., Krizala, J. and Simsa, J. (1979) Coll. Czech. Chem. Comm., v tisku.
12. Lipecka, K., Lipinski, S. and Kanski, K. (1978) Stud. Biophys. 68, 25-30.
13. Lipinski, S., Lipecka, K., Doniec, J. and Kanski, M. (1976) Radiobiologiya, 16, 665-670.
14. McCord, J.M. and Fridovich, I. (1969) J. Biol. Chem. 244, 6049-6055.
15. Michelson, A.M., Puget, K., Durosay, P. and Bonneau, J.C. (1977) in Superoxide and Superoxide Dismutases, Michelson, A.M., McCord, J.M. and Fridovich, I., eds., Academic Press, New York, p.467-499.
16. Nerurkar, L.S., Zeligs, B.J. and Bellanti, J.A. (1978) Photochem. Photobiol. 28, 781-786.
17. Ogata, M., Mizugaki, J., Ueda, K. and Ikeda, M. (1977) Tohoku J. Exp. Med. 123, 95-98.
18. Petkau, A., Chelack, W.S., Pleskach, S.D., Meeker, B.E. and Brady, C.M. (1975) Biochem. Biophys. Res. Commun. 65, 886-893.
19. Petkau, A., Kelly, K., Chelack, W.S, and Barefoot, C. (1976) Biochem. Biophys. Res. Commun. 70, 452-458.
20. Petkau, A. (1978) Photochem. Photobiol. 28, 765-774.
21. Stevens, J.B. and Autor, A.P. (1977) Lab. Invest. 37, 470-478.
22. Weisiger, R.A. and Fridovich, I. (1973) J. Biol. Chem. 248, 3582-3591.

INTERACTION OF SUPEROXIDE DISMUTASE WITH MEMBRANES

A. PETKAU[+], K. KELLY[++] AND J.R. LEPOCK[+++]
+ Medical Biophysics Branch, Whiteshell Nuclear Research Establishment, Atomic
Energy of Canada Limited, Pinawa, Manitoba, Canada; ++ Department of Immunology,
Health Sciences Centre, University of Manitoba, Winnipeg, Manitoba, Canada;
+++ Department of Physics, University of Waterloo, Waterloo, Ontario, Canada

ABSTRACT

The uptake of bovine superoxide dismutase by egg lecithin liposomes at room
temperature is greater than that by soybean phospholipid bilayers, but for both
model membranes it increases with the enzyme concentration in accordance with
$y = a \, x^b$, where y is the uptake in µg/mg of lipid, x is the enzyme concentration,
$b \leqslant 1$ and a, the proportionality constant, is 0.36 for the liposomes and 0.15 for
the bilayers. Spin label measurements indicate that the enzyme associates with
the liposomal surface, increasing its viscosity, as well as with its hydrocarbon
regions, increasing their order. Freeze-fracture and etch electronmicroscopy of
the liposomes, before and after treatment with superoxide dismutase, shows that
the enzyme introduces changes in texture of the replicas and penetrates the
multilamellar structure of the liposomes.

INTRODUCTION

A study of the interaction of superoxide dismutase with lipids and membranes
is of interest for several reasons. Components of biological membranes, such as
lipids and sulfhydryl groups of proteins, are at risk to oxidative attack by
superoxide radicals.[1-11] These radicals may be generated within the membranes
or elsewhere in cells[12-24] and deleterious effects ensue.[1,25-27] The presence
of the active enzyme within membranes could therefore be useful.

Recently, evidence for the existence of superoxide dismutase on either side
of and within bovine erythrocyte plasma membranes was presented, on the basis
of which, it was suggested that glutaraldehyde fixation further internalized the
enzyme into the erythrocyte ghost.[28] Moreover, treatment of mouse bone marrow
stem cells with superoxide dismutase from bovine erythrocytes resulted in the
association of the enzyme with the cytosol, mitochondria, nuclei and membranes.[29]
These results suggest that the enzyme not only partitions in and out of plasma
membranes but also diffuses in and out of various cellular compartments. This
indicates that the enzyme possesses sufficient hydrophobicity to interact with
the hydrocarbon regions of lipid membranes. In this connection, it is noted

that the enzyme is often extracted from bovine erythrocytes with the use of chloroform/ethanol[30] and that about a third of its amino acids are hydrophobic.[31] Moreover its cylindrical topology[32] may confer upon it the structural features and conformational latitude which enable this water-soluble enzyme to enter membranes. Upon entering membranes, the enzyme may affect their structure which, in this communication on the enzyme's interaction with model lipid bulayers, is examined by spin probes that report changes in membrane viscosity and order. Additionally, uptake data and electronmicroscopy observations are presented which corroborate the spin labelling results.

MATERIALS AND METHODS

Lipid membrane preparation. Spherically-shaped phospholipid bilayers (PBL) were prepared in distilled water from soybean phospholipids (Azolectin, Associated Concentrates, Woodside, Long Island, NY) as previously described.[2] Multilayered liposomes of egg lecithin (Sigma) were prepared in phosphate buffered saline (PBS) pH 7.4 by suspending 80 mg of the lipid in 8 ml of buffer and vortexing the mixture for 20 min at room temperature.

Uptake of superoxide dismutase. ^{125}I-labelled bovine superoxide dismutase was prepared by the chloramine-T technique as previously described[33] using bovine superoxide dismutase (supplied by Dr. J.V. Bannister, Malta; present address: Inorganic Chemistry Department, University of Oxford, Oxford OX1 3QR, England) as starting material. For uptake measurements, the labelled enzyme was usually added to the PBL membranes to a final concentration of 0.15 µg/ml, along with varying amounts of the unlabelled enzyme (10-10000 µg/ml). After 30 min at room temperature, or at $30^{\circ}C$ and $35^{\circ}C$ in several runs, the enzyme-treated membranes were precipitated with minimal amounts of $CaCl_2$ ($\sim$ 5 mM) and centrifuged at 8000 x g for 2 min. The precipitates were dispersed in 1 ml of distilled water, incubated, precipitated and centrifuged as before. This procedure was repeated three times to remove free enzyme from the enzyme-treated membranes. The radioactivity associated with the washed precipitate was then measured, the precipitate was solubilized in 0.5% Nonidet P-40 (NP-40, Particle Data Laboratories, Elmhurst, IL) in PBS and chromotographed on a column of Bio-Gel P-60 (1 x 40 cm) equilibrated with 0.1% NP-40 in PBS (Ref.33) to determine the fractional ^{125}I-activity associated with the membrane-bound enzyme. This fractional activity was related to the total amount of superoxide dismutase initially present during the first 30 min incubation period and then used to estimate the total amount of enzyme taken up by the membranes whose lipid content had previously been determined by dry weight measurement. In one run, the amount of labelled enzyme was varied along with the total amount of enzyme, keeping the ratio by weight of ^{125}I-enzyme over unlabelled enzyme at 1/100.

Liposomal membranes were treated with labelled and unlabelled enzyme as above and washed three times in PBS.

Isolation of rabbit antibodies to superoxide dismutase. An immunosorbent was prepared by first coupling normal rabbit IgG(nIgG) to cyanogen bromide-activated Sepharose-4B (Ref.34). Superoxide dismutase was then coupled to the Sepharose-4B-nIgG using a two-step glutaraldehyde activation process.[35] The Sepharose-4B-nIgG was activated with glutaraldehyde, preheated at $50^{o}C$ for 30 min, at a final concentration of 1% in PBS. After 1 h at room temperature, the gel was filtered and washed exhaustively with PBS, then stirred for 3 h at room tempera-·· ture in a 1 mg/ml solution of superoxide dismutase in PBS. Excess glutaraldehyde-activated groups were neutralized by incubation of the immunosorbent in 2M NH_2OH in PBS. Sera from rabbits, immunized with superoxide dismutase,[36] were incubated at $4^{o}C$ for 16 h with the immunosorbent and the product, Sepharose-4B-nIgG-SOD, was subsequently filtered and washed exhaustively on a Buchner funnel with 0.5% Tween 20 in PBS to remove nonspecifically-bound protein. Adsorbed antibodies to the enzyme, Ra-SOD, were subsequently eluted by treatment with 4 M guanidine-HCl, pH 4.0.

Uptake of ^{125}I-labelled antibodies. Normal rabbit IgG (nIgG) and rabbit antibodies to bovine superoxide dismutase (Ra-SOD) were labelled with ^{125}I as previously described[36] and reacted for 30 min at room temperature with aliquots of liposomal membranes treated as above with non-radioactive superoxide dismutase. The membranes were then precipitated, centrifuged and washed twice in PBS as before and counted for [125]I. A lower uptake of ^{125}I from ^{125}I-nIgG than from ^{125}I-SOD was taken as an indication of enzyme association with the liposomal membranes.

Spin label studies. The egg-lecithin liposomal membranes were made as previously described and the spin labels added to a final concentration of 5×10^{-6} μmoles/g phospholipid (about 1 spin label/300 phospholipids). The spin labels 2,2-dimethyl-5-dodecyl-5-methyl-1-oxazoladinyloxyl (Label I), synthesized by the method of Williams et al.,[37] spin-labelled distearoylphosphatidylcholine (Label II), purchased from Syva, Palo Alto, California, and 3-(octadecylamido-methyl)-2,2,5,5-tetramethyl-1-pyrrolidinyloxyl (Label III), synthesized by the method of Lepock et al.,[38] were used as probes of membrane viscosity (Figure 1). The spin labels 2-(3-carboxylpropyl)-4,4-dimethyl-2-tridecyl-3-oxazolidinyloxyl (Label IV) and 2-(10-carboxydecyl)-2-hexyl-4,4-dimethyl-3-oxazolidinyloxyl (Label V), synthesized by the method of Hubbell and McConnell,[39] were used as probes of membrane order (Figure 1). After addition of the spin labels, samples of the liposomal membranes were halved with one half in each case serving as the control to the other half which was treated with superoxide dismutase at 1 mg/ml lipid. The samples were equilibrated at $37^{o}C$ for 6 h and run on a Varian E-12 ESR spectrometer. The rotational correlation time (τ_{c}) was calcu-

lated from the first derivative spectra of spin labels I, II and III using the
formula,

$$\tau_c = 6.9 \times 10^{-10} \; w_1 \left[\left(\frac{h_1}{h_{-1}} \right)^{\frac{1}{2}} - 1 \right]$$

where h_1 and h_{-1} are the low and high field line heights, respectively, and w_1
is the low field line width.[40]   The order parameter (S) was calculated for spin
labels IV and V using the formula,

$$S = \frac{A_{||} - A_{\perp}}{A_{xx} - \frac{1}{2}(A_{yy} + A_{zz})}$$

as described by Seelig.[41]   The hyperfine coupling constants used were those
determined by Libertini and Griffith.[42]   Correction for polarity and other factors were ignored, so the values of S are relative.   Contributions from probe-
probe interactions should be negligible because of the low ratio of spin label
to phospholipid.

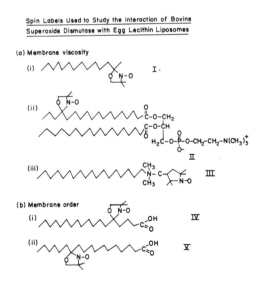

Fig. 1.   Spin labels used to study the interaction of bovine superoxide dismutase with egg lecithin liposomes: I = 2,2-dimethly-5-dodecyl-5-methyl-1-
oxazoladinyloxyl; II = distearoylphosphatidylcholine; III = 3-(octadecyl-
amidomethyl)-2-2,5,5-tetramethyl-1-pyrrolidinyloxyl; IV = 2-(3-carboxylpropyl)-
4,4-dimethyl-2-tridecyl-3-oxazolidinyloxyl; V = 2-(10-carboxydecyl)-2-hexyl-4,
4-dimethyl-3-oxazolidinyloxyl.

Electronmicroscopy. Samples of spin-labelled and enzyme-treated egg lecithin liposomes were subjected to freeze etch- or freeze fracture and etch-electron-microscopy as previously described[43] except that the 10% glycerin was omitted and droplets of the material were used. The Pt-C replicates were cleaned with full strength chromic acid and rinsed three times with distilled water.

RESULTS

The uptake at room temperature of bovine superoxide dismutase by the two model membrane systems is graphically illustrated in Figure 2 and increased as a power function of the enzyme concentration to which the membranes were exposed. With the amount of ^{125}I-labelled enzyme kept constant at the relatively low concentration of 0.15 μg/ml, the uptake of enzyme (y) by PBL membranes increased in accordance with $y = 0.15\ x^{0.96}$, where x is the total (labelled and unlabelled) enzyme concentration. Increasing the incubation temperature to 30°C and 35°C did not appreciably change the exponent of the power function correlation but progressively decreased to a small extent the proportionality constant (Table 1), indicating that a rise in tempertaure reduces uptake slightly. This decrease in associative interaction between the enzyme and membranes may be significant in terms of contributing to the increase in radiosensitivity of cells with hyper-thermia.[26]

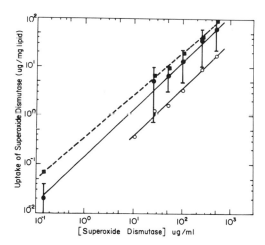

Fig. 2. Uptake of bovine superoxide dismutase by soybean phospholipid bilayers (●,○) and egg lecithin liposomes (■) at room temperature. Equations of best-fitted lines are: Egg lecithin liposomes (■----■), $y = 0.36\ x^{0.86}$; Soybean phospholipid bilayers with ^{125}I-SOD/SOD varying (●———●), $y = 0.15\ x^{0.96}$ and with ^{125}I-SOD/SOD = 1/100 (○———○), $y = 0.04\ x^{0.95}$. Error bars show standard deviation applicable to ●———●.

TABLE 1

CORRELATION BETWEEN UPTAKE OF BOVINE SUPEROXIDE DISMUTASE BY SOYBEAN PHOSPHO-
LIPID BILAYERS AND THE CONCENTRATION OF ENZYME PRESENT DURING INCUBATION AT
SEVERAL TEMPERATURES[a]

Temperature	Correlation function, $y = a\, x^b$		
$^\circ C$	a	b	Correlation coefficient
23	0.15	0.96	0.99
30	0.12	0.97	0.99
35	0.10	1.0	0.99

[a]Incubation period 30 min; enzyme uptake (y) in µg/ml lipid; enzyme concentration
in µg/ml.

The uptake of enzyme by the PBL membranes at room temperature also increased
as a power function of the enzyme concentration while maintaining the ratio of
labelled enzyme to unlabelled enzyme at 1:100 (Figure 2), However, the proportionality constant of the power function correlation, $y = 0.4\, x^{0.95}$ was about 3.7
times lower indicating reduced uptake. This reduction in uptake may be due to
experimental variation between runs (note the standard deviations associated
with the reference line discussed above; Figure 2) and the effect on membrane-
enzyme interaction of differences in the molecular structure of superoxide dismutase that may result from labelling with radioactive iodine by the chloramine-T
technique. The radioactive labelling procedure usually decreases the activity of
bovine superoxide dismutase by 10-15% (unpublished observations). It is not
known whether this drop in biochemical activity is due to conformational changes,
loss of the cupro-zinc atoms or chemical degradation.

Uptake of bovine superoxide dismutase by the egg lecithin liposomes at room
temperature also increased as a power function of the enzyme concentration
(Figure 2). In this case, the correlation $y = 0.36\, x^{0.86}$, where x and y are
defined as before, indicates a change to lower slope relative to PBL at room
temperature and a factor of about 2 increase in the proportionality constant.
Ostensibly, the bovine superoxide dismutase associates with the multilayered
egg lecithin liposomes to a greater extent than with the single bilayers of
soybean phospholipids of which lecithin is a major constitutent.[2] If the enzyme
were unable to penetrate the most superficial bilayer of the multilayered liposomes, its uptake by the liposomes on a per weight of lipid basis would be
expected to be less than for the single phospholipid bilayers. The opposite
result is observed. The greater uptake of liposomes may reflect their relatively
large size and radius of curvature, compared to the sonicated PBL whose highly
curved surface may sterically hinder entry of the enzyme.

In principle, superoxide dismutase may interact with different regions of the
liposomal membrane. If this interaction is a surface phenomemnon or the enzyme

is only partially embedded within the membrane matrix, then the membrane-associated enzyme would be accessible to specific antibodies. Table 2 gives the results of reacting the enzyme-treated liposomes with ^{125}I-labelled specific Ra-SOD and nIgG. The data show increased uptake of the specific antibody by a factor of 1.8 and 3.3 at enzyme/lipid ratios (by weight) of 1.0 and 0.3, respectively. These results suggest enzyme association with the superficial bilayer of the liposomes to an extent that leaves at least some of the enzyme's antigenic sites exposed to antigen-antibody complex formation. It is noted, however, that when the liposomes contain no superoxide dismutase, the uptake of both specific and non-specific immunoglobulin was the highest of all the measurements (Table 2). This increase in uptake of immunoglobulin under non-competitive conditions may relate to membrane damage, mediated immunologically, in which transmembrane channels are formed and removal of phospholipid from bilayers occurs.[44] Penetration of IgG antibody into living cells has recently been described[45] so that the relatively high uptake of both nIgG and Ra-SOD by non-enzyme treated liposomes comes as no surprise. It is important to note, however, that this non-specific uptake is about the same for nIgG and Ra-SOD and that it decreases with liposomes exposed to superoxide dismutase.

TABLE 2

COMPARISON OF UPTAKE BY BOVINE SUPEROXIDE DISMUTASE-TREATED EGG LECITHIN LIPOSOMES OF ^{125}I-LABELLED SPECIFIC Ra-SOD ANTIBODIES OR nIgG

Amount of Superoxide Dismutase per mg of lipid (mg)	^{125}I-Uptake		Ratio: (ii)/(i)
	(i) nIgG (x 10^4 cpm/sample)	(ii) Ra-SOD (x 10^4 cpm/sample)	
1.0	0.79	1.38	1.8
0.3	0.49	1.62	3.3
0.1	0.72	0.76	1.1
0.03	0.90	0.85	0.9
none	1.8	2.05	1.1

Both in theory and practice, spin label probes can distinguish between structural changes in the polar and apolar regions of membranes, depending on the type of membrane molecule chosen for labelling and the position of the spin label on that molecule (Figure 1). In this study, three spin labels, I, II and III, were used to report viscosity changes in liposomes following exposure to superoxide dismutase and two more, IV and V, were employed to check for changes in membrane order. The results are given in Table 3, which lists the correlation time of rotation τ_c for I, II and III. The parameter τ_c can be related to the viscosity of the spin label environment and its values are not significantly different for I and II which both probe the interior of lipid bilayers. Superoxide dismutase, therefore, does not appear to change their internal viscosity. By contrast, the difference in τ_c for III between liposomes, treated and un-

treated with the enzyme, is significant and implies that the enzyme is restrict-
ing the motion of III. This spin label probes the surface of the bilayer by
virtue of its electrical charge and indicates that superoxide dismutase increases
the viscosity at the liposomal surface by associating with it in some way.[38]
Thus, these spin label results corroborate the data on the uptake of labelled
antibodies (Table 2), in demonstrating that the enzyme interacts with the liposomal surface.

TABLE 3

SPIN LABEL STUDY OF THE INTERACTION OF BOVINE SUPEROXIDE DISMUTASE WITH EGG
LECITHIN LIPOSOMES: EFFECT ON CORRELATION TIME OF ROTATION AND ORDER PARAMETER[a]

Spin label[b]	$10^{10} \times \tau_c$ (sec)			S		
	Control	SOD	% change	Control	SOD	% change
I	6.87	6.77	− 1.5			
II	14.9	15.2	+ 2.0			
III	17.7	19.0	+ 7.3			
IV				0.56	0.63	+ 12
V				0.52	0.60	+ 15

[a] τ_c, correlation time of rotation and S, order parameter (see text)

[b] Chemical names given in text and in caption to Figure 1.

The other two labels, IV and V, measure the amount of order present at different depths in the apolar region of lipid bilayers.[46] The order parameter S
was calculated for IV from the spectrum (Figure 3) and varies from 0 for no order
(isotropic rotation of the spin label) to 1 for complete order, where the spin
label is only able to rotate about its long axis.[39] From Figure 3, it is apparent
that the enzyme increases the spacing of the outer extrema ($2A_{||}$) and decreases
the spacing of the inner extrema ($2A_{\perp}$). This is indicative of increased order.
There was a similar change in the spectrum of label V. The values of S are
given in Table 3 and show a 12–15% increase after the lipsomes have been treated
with superoxide dismutase, thus indicating that the enzyme orders the acyl hydrocarbon chains within the lipid bilayer. Collectively, the spin label data
indicate that superoxide dismutase interacts with the lipid bilayer, partially
immobilizing its surface and increasing the order of the fatty acid chains by
penetrating the hydrocarbon interior.

Replicas of the freeze-etched surfaces of the enzyme treated and untreated
liposomes show that the enzyme induces changes in texture of the liposomal
surface (Figure 4). In untreated liposomes, the surface is finely granular and
smooth (panel a, inset) whereas, after enzyme-treatment, it becomes cauliflower-
like in appearance (panel b, inset), suggesting an ordered pattern of intermediate range. This difference in pattern extends through a number of bilayers

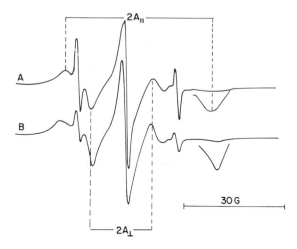

Fig. 3. 9.3 GHz ESR spectra of label IV in egg lecithin liposomes. Spectrum A is with no superoxide dismutase added, while spectrum B is for 1 mg of enzyme per mg of lipid. The sharp peaks are due to a small amount of spin label partitioned into the water. The high field extremum was scanned with a 10-fold increase in gain to better locate its position.

of the multilamellar structures (panels a and b of Figure 4), indicating that the enzyme penetrated more than the most superficial bilayer, an observation consistent with the uptake of data. Thus, the enzyme in penetrating through successive bilayers of liposomes also affects their organization.

DISCUSSION

There are many proteins which appear to immobilze or order a ring of lipids around themselves in membranes. Cytochrome oxidase has been shown to immobilize 0.16 - 0.21 mg of mitochondrial phospholipids/mg protein,[47] complement treatment appears to reduce the fluidity of sheep erythrocyte membranes,[48] and the acetylcholine receptor immobilizes a fraction of the lipid of membranes from Torpedo marmorata.[49] Superoxide dismutase appears to act similarly in that it orders the lipid and probably reduces the fluidity of a layer of boundary lipid. The order parameter cannot be directly related to fluidity but the decrease in the width of the high field extremum (Figure 3) after addition of the enzyme is indicative of a decrease in lipid fluidity.[48]

The permeability of model lipid membranes to a variety of protein molecules has previously been demonstrated by Yanopol'skaya and Deborim.[50] Using a mixture of egg phospholipids and cholesterol, these authors characterized thermodynamically the adsorption and penetration of their model membrane by water soluble

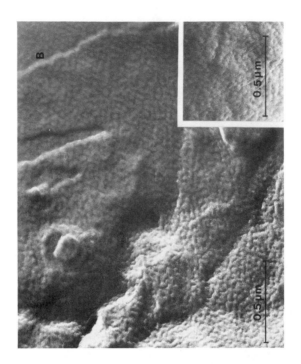

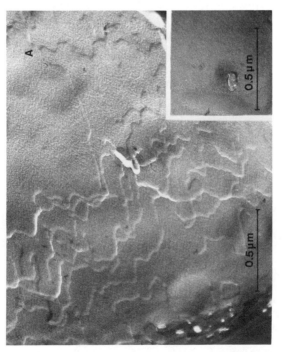

Fig. 4. Electronmicrographs of Pt-C replicas showing the multilayered structure of freeze-fractured and etched egg lecithin liposomes before (panel a) and after (panel b) treatment with bovine superoxide dismutase. Insets show texture of the enzyme-treated and untreated liposomes as revealed by Pt-C replicas of the surface after freeze etching only. Magnification 95,760X.

RNase, trypsin, amylase, aldolase, alkaline phosphatase and invertase whose molecular weights are 14,000, 24,000, 15,000, 147,000, 140,000 and 210,000, respectively. They showed that for invertase, for instance, the adsorption coefficient remained constant above $22^{O}C$, in contrast with the slight temperature-dependent decrease in uptake of bovine superoxide dismutase (mol. wt. $\sim$ 32,600) by soybean phospholipid bilayers above $\sim 23^{O}C$. This temperature dependence suggests that the initial absorptive phase is probably chemical rather than physical. The foregoing authors found that upon adsorption to a phospholipid-cholesterol membrane, invertase increased the entropy of the lipids. This also contrasts with the indication from spin label III data that superoxide dismutase increases the viscosity of the bilayer surface. Another distinction between this enzyme and invertase is that, upon interacting with the hydrocarbon regions of bilayers, the former increases order while the latter decreases it as judged by the positive change in entropy.[50] Thus, the effects of the two enzymes on lipid membranes are dissimilar for reasons that have not been clarified, although it may be assumed that the differences in their molecular size, three-dimensional structure and amino acid composition and sequence are important factors in determining the nature of their interaction, particularly with the acyl hydrocarbon chains of the constituent phospholipids.[50-52] The ability of superoxide dismutase to increase the order of the hydrocarbon region may have functional significance in that close contact of the enzyme with the fatty acids could conceivably enhance its effectiveness in protecting them from oxidative attack by superoxide radicals. Although no kinetic study appears to have been done, comparing the kinetic properties of membrane-bound and water-solubilized superoxide dismutase, it seems reasonable to expect, by analogy with other enzyme systems[53-56] that the former may possess a different specific activity or exhibit an anistrotropic characteristic of action through screening[57] and directional orientation of its catalytic site. If so, these features would augment the enzyme's ability to modulate the antioxidant status of subcellular regions by partitioning in and out of lipid membranes in response to changes in chemical potential, composition or temperature.

ACKNOWLEDGEMENTS

   We gratefully acknowledge the technical assistance of W.S. Chelack, R.A. Zepp, T.P. Copps, P. Imp and L. Arnold.

REFERENCES

1. Petkau, A. and Chelack, W.S. (1974) Int. J. Radiat. Biol. 26, 421-426.
2. Petkau, A. and Chelack, W.S. (1976) Biochim. Biophys. Acta, 433, 445-456.
3. Peters, J.W. and Foote, C.S. (1976) J. Amer. Chem. Soc. 98, 873-875.
4. Thomas, M.J., Mehl, K.S. and Pryor, W.A. (1978) Biochem. Biophys. Res. Commun. 83, 927-932.

346

5. Kellogg, E.W. and Fridovich, I. (1977) J. Biol. Chem. 252, 6721-6728.
6. Merzlyak, M.N., Yuferova, S.G. and Sobolev, A.S. (1977) Biofizika, 22, 846--849.
7. Docampo, R., Cruz, F.S., Boveris, A., Muniz, R.P.A. and Esquivel, D.M.S. (1978) Arch. Biochem. Biophys. 186, 292-297.
8. Svingen, B.A., O'Neal, F.O. and Aust, S.D. (1978) Photochem. Photobiol. 28, 803-809.
9. Borg, D.C., Schaich, K.M., Elmore, J.J. and Bell, J.A. (1978) Photochem. Photobiol. 28, 887-907.
10. Asada, K. and Kanamatsu, S. (1976) Agr. Biol. Chem., Tokyo, 40, 1891-1892.
11. Armstrong, D.A. and Buchanan, J.D. (1978) Photochem. Photobiol. 28, 743-755.
12. Misra, H.P. and Fridovich, I. (1972) J. Biol. Chem. 247, 6960-6962.
13. Debey, P. and Balny, C. (1973) Biochimie, 55, 329-332.
14. Bors, W., Saran, M., Lengfelder, E., Spöttl, R. and Michel, C. (1974) Curr. Top. Radiat. Res. 9, 247-309.
15. Bielski, B.H.J. and Chan, P.C. (1974) J. Biol. Chem. 250, 318-321.
16. Patriarca, P., Dri, P., Kakimuma, K., Tedesco, F. and Rossi, F. (1975) Biochim. Biophys. Acta, 385, 380-386.
17. Nishikimi, M. and Machlin, L.J. (1975) Arch. Biochem. Biophys. 170, 684-689.
18. Goldberg, B. and Stern, A. (1975) J. Biol. Chem. 250, 2401-2403.
19. Cassell, R.H. and Fridovich, I. (1975) Biochemistry, 14, 1866-1868.
20. Sutton, H.C., Roberts, P.B. and Winterbourn, C.C. (1976) Biochem. J. 155, 503-510.
21. Winterbourn, C.C., McGrath, B.M. and Carrell, R.W. (1976) Biochem. J. 155, 493-502.
22. Gotoh, T. and Shikama, K. (1976) J. Biochem. Tokyo, 80, 397-399.
23. Goldberg, B. and Stern, A. (1976) Biochim. Biophys. Acta, 437, 628-632.
24. Johnston, R.B., Lehmeyer, J.E. and Guthrie, L.A. (1976) J. Exp. Med. 143, 1551-1556.
25. Van Hemmen, J.J. and Meuling, W.J.A. (1977) Arch. Biochem. Biophys. 182, 743-748.
26. Lin, P.S., Kwock, L. and Butterfield, C.E. (1979) Radiat. Res. 77, 501-511.
27. Misra, H.P. and Fridovich, I. (1976) Arch. Biochem. Biophys. 176, 577-581.
28. Petkau, A. and Copps, T.P. (1979) Biophys. J. 25, 190a.
29. Petkau, A., Chelack, W.S., Kelly, K. and Friesen, H.G. (1979) in Active Oxygen and Medicine, Autor, A., ed., Raven Press, New York (in press).
30. McCord, J.M. and Fridovich, I. (1969) J. Biol. Chem. 244, 6049-6055.
31. Keele, B.B., McCord, J.M. and Fridovich, I. (1971) J. Biol. Chem. 246, 2875-2880.
32. Richardson, J.S., Richardson, D.C. and Thomas, K.A. (1976) J. Mol. Biol. 102, 221-235.
33. Petkau, A., Kelly, K., Chelack, W.S., Pleskach, S.D., Barefoot, C. and Meeker, B.E. (1975) Biochem. Biophys. Res. Commun. 67, 1167-1174.
34. Stage, D.E. and Mannik, M. (1974) Biochim. Biophys. Acta, 343, 382-391.
35. Yen, S.P.S., Rembaum, A., Molday, R.S. and Dreyer, W.I. (1976) A.C.S. Symposium Series No. 24, Emulsion Polymerization, Piirma, I. and Gardon, J.L., eds., 16, 236-257.
36. Kelly, K., Barefoot, C., Sehon, A. and Petkau, A. (1978) Arch. Biochem. Biophys. 190, 531-538.
37. Williams, J.C., Mehlhorn, R. and Keith, A.D. (1971) Chem. Phys. Lipids, 7, 207-230.
38. Lepock, J.R., Morse, P.D. II, Mehlhorn, R.J., Hammerstedt, R.H., Snipes, W. and Keith, A.D. (1975) FEBS Lett. 60, 185-189.
39. Hubbell, W.L. and McConnell, H.M. (1971) J. Amer. Chem. Soc. 93, 314-326.
40. Rule, G.S., Kruuv, J. and Lepock, J.R. (1979) Biochim. Biophys. Acta, in press.
41. Seelig, J. (1976) in Spin Labelling, Berliner, L.J., ed., Academic Press, New York, pp.373-409.
42. Libertini, L.J. and Griffith, O.H. (1970) J. Chem. Phys. 53, 1359-1367.
43. Copps, T.P., Chelack, W.S. and Petkau, A. (1976) J. Ultrastructure Res. 55, 1-3.

44. Mayer, M.M., Hammer, C.H., Michaels, D.W. and Shin, M.L. (1979)
    Immunochemistry, 15, 813-831.
45. Alarcon-Segovia, D., Ruis-Argülles, A., and Fishbein, E. (1979) Clin.
    Exp. Immunol. 35, 364-375.
46. Hubbell, W.L. and McConnell, H.M. (1969) Proc. Nat. Acad. Sci. USA, 64,
    20-27.
47. Jost, P.C., Nadakavukaren, K.K. and Griffith, O.H. (1977) Biochemistry,
    16, 3110-3114.
48. Mason, R.P., Giavedoni, E.B. and Dalmasso, A.P. (1977) Biochemistry, 16,
    1196-1201.
49. Marsh, D. and Barrantes, F.J. (1978) Proc. Nat. Acad. Sci. USA, 75, 4329-
    4333.
50. Yanopol'skaya, N.D. and Deborin, G.A. (1978) Biochemistry USSR, 43, 86-92.
51. Träuble, H. (1971) J. Membrane Biol. 4, 193-208.
52. Kimelberg, H.K. and Papahadjopoulos, D. (1971) Biochim. Biophys. Acta,
    233, 805-809.
53. Coleman, R. (1973) Biochim. Biophys. Acta, 300, 1-30.
54. Esfahani, M., Rudkin, B.B., Cutter, C.J. and Waldron, P.E. (1977) J. Biol.
    Chem. 252, 3194-3198.
55. Gazzotti, P. and Peterson, S.W. (1977) J. Bioenerg. Biomembrane. 9, 373-
    386.
56. Kurskyj, M.D., Kosterin, S.O. and Rybalchenko, V.K. (1977) Ukr. Biokhim.
    Zh. SSR, 49, 116-127.
57. Datta, C. and Gupta, M.R. (1977) Indian J. Exp. Biol. 15, 58-60.

PENETRATION OF ERYTHROCYTES BY SUPEROXIDE DISMUTASE

A.M. MICHELSON, K. PUGET, P. DUROSAY AND A. ROUSSELET
Institut de Biologies Physico-Chimique, 13, rue P. et M. Curie, 75005 Paris,
France

SUMMARY

   Fixation of native SODs, liposome encapsulated SODs and chemically modified
SOD to human erythrocytes has been studied, in particular the effect of liposomal composition. Fusion with and penetration of the cell membrane by these
various forms of SOD has been examined by use of radioactive marked SOD, by
electron spin resonance spectroscopy (using a spin label marked-lecithin in the
liposome) and by electron microscopy.

INTRODUCTION

   Superoxide dismutases are important enzymes involved in cellular defence
against uncontrolled oxidative processes which catalyze the dismutation of the
superoxide radical anion and hence diminish toxic effects due to this radical
or to other free radicals derived from secondary reactions, such as production
of OH· by a Haber-Weiss reaction.[1] The properties of superoxide and superoxide
dismutases have been extensively reviewed.[2] Since these enzymes are proteins
with molecular weights ranging from 33,000 to 80,000, cellular penetration does
not readily occur. Nevertheless, administration by various techniques (intravenous, intraperitoneal, etc.) does afford a certain protection against high
energy irradiation[3] and is effective against various acute and chronic inflammatory conditions such as rheumatoid arthritis and osteoarthritis.[4] It is likely
that in such cases, superoxide dismutase is effective in reducing extracellular
levels of O_2^- produced during phagocytosis or by other means. It may be noted
that a significant decrease in superoxide dismutase (but not of glutathione
peroxidase) has been observed in polymorphonuclear leucocytes from children
with rheumatoid arthritis compared with normal controls.[5]

   In recent years, the possible application of liposome packaged material has
begun to receive attention and an excellent review of the diverse aspects of
liposomes is available.[6] Intravenous injection of α-glucosidase containing
liposomes for the treatment of a baby suffering from type II glycogenesis (with
some success) has been described[7] as well as the oral administration of insulin
encapsulated within liposomes.[8] Liposomal methotrexate is much more effective
in reduction of solid rodent tumours than is the free drug.[9]

It was thus of interest to study the cellular penetration of diverse superoxide dismutases and the corresponding liposomal preparations. As model cell
we have used human erythrocytes. With respect to the dismutases we have chosen
those comporting a range of pI from 4.4 to 8.7 (in view of the charge characteristics of the erythrocyte), obtained from both prokaryote and eukaryote sources,
containing either copper, manganese or iron as the metal (hence covering the
three major classes of superoxide dismutases). The enzymes used were bovine
Cu-SOD (pI = 4.9), human Cu-SOD (pI = 4.6), bacterial (P. leiognathi) Cu-SOD
(pI = 8.7), human Mn-SOD (pI = 7.8 and 8.3) and bacterial (P. leiognathi) Fe-
SOD (pI = 4.4).

It was also of interest to compare a native SOD with a chemically modified
enzyme. For this purpose we treated bovine Cu-SOD with methyl 4-mercaptobutyri-
midate. This reagent[10] converts ε-lysine amine groups to terminal thiobutanol
residues, thus presenting a series of -SH groups separated from the enzyme by
four carbon spacers. These thiol groups can readily react with cysteine SH
groups in cell membranes to give an -S-S- covalent linkage, which is reversible
by the action of mild reducing agents.

With respect to cellular fixation and penetration of the liposomes (and
modified SOD), several approaches were used. As a first indication, fixation
of radioactively marked SOD which was not removed by repeated washing was examined. The time course of fixation and of penetration of human erythrocytes
(using membrane bound and soluble radioactivity to determine the proportion
of cytoplasmic exogenous enzyme after repeated washing followed by mild lysis
of the erythrocytes) was then studied.

Fusion of liposomes with the red cell membrane was followed by using bovine
Cu-SOD liposomes with a spin marker in the lecithin used. The electron spin
resonance signal is characteristic and allows a kinetic determination of fusion
(and hence entry of the encapsulated SOD into the cell). Finally, direct observation of liposome attachment to the erythrocyte membrane and fusion therewith,
was obtained by scanning electron microscopy.

It is generally accepted that intravenously injected protein-containing liposomes are rapidly concentrated in the liver and spleen and that only these
tissues have a great affinity for liposomes. Subsequent reports will describe
the organ specificity in rats and in rabbits of the liposomes studied in the
present communication.

MATERIALS AND METHODS

Carboxyl methylation of SODs. To a solution of 4 mg SOD (human Cu-SOD,
bovine Cu-SOD or bacterial Fe-SOD) in 1 ml of 10^{-2} M NaCl 5 x 10^{-3} M sodium
cacodylate pH 8.4 was added 10 μl of tri-n-butylamine and 0.3 mg of dimethyl
sulphate (^{3}H or ^{14}C). The mixture was agitated at $4°C$ for 3 hr, then passed

through a column of Sephadex G-100, to separate the methylated enzyme from low molecular weight components. The radioactive enzymes were about 85% enzymically active compared with the specific enzyme activity of the starting materials. With ^{14}C Me$_2$SO$_4$ (31 mCi/mmole) the methylated enzymes gave counts of about 1000 cpm/µg. Similar figures were obtained on methylation with ^{3}H dimethyl sulphate (150 mCi/mmole).

Lysine amino methylation of SODs. The technique used was essentially that described by Means and Feeney,[11] involving reductive methylation with radioactive formaldehyde. To a solution of the SOD (5 mg) in 0.1 M borate pH 9.0 (6 ml) was added 13.5 µmoles of ^{14}C formaldehyde (20 mCi/mmole) followed by 3.0 mg NaBH$_4$ and the mixture was kept at 0°C for 2 hr, then concentrated to about 2 ml with a Minicon. Considerable loss of enzymic activity was observed with bovine and bacterial Cu-SOD, but bacterial Fe-SOD and human Mn-SOD activities were only slightly reduced. The radioactivity of the products ranged from 1200 to 4500 cpm per µg of protein.

Treatment of SODs with methyl 4-mercaptobutyrimidate.[10] A stock solution of the mercaptobutyrimidate in 0.2 M borate pH 9.0 (255 mg in 15 ml, 0.1 M) was prepared. This was added to solutions of the various SODs in 5 x 10^{-3}M phosphate pH 7.8 at 40.5 µmoles of the reagent per mg of protein (final volume 1 ml per mg protein at pH 8.3 with borate concentration 0.08 to 0.1 M) and the solution kept at 0°C for 3 hr, then dialysed against 5 x 10^{-3}M phosphate pH 7.8 containing 10^{-4} M mercaptoethanol. In the absence of mercaptoethanol polymerisation occurs. Monomer enzyme and polymeric enzymes were readily separated by passage through Sephadex G-100.

Concentrated solutions of the modified enzyme polymerise rapidly under aerobic conditions to a solid gel. This can be reversed on addition of mercaptoethanol. Monomeric forms were enzymically active, with about 30% loss of specific activity, but polymeric forms were considerably less active, this activity being recovered on depolymerisation.

Reconstitution of human Mn-SOD. The apoenzyme was treated with a three fold excess of Mn^{2+} at pH 4.5 and the solution slowly brought to pH 7.0. At the apoenzyme stage less that 0.35% enzymic activity was retained and after reconstitution some 72% of the original enzymic activity was recovered.

Radioactive cobalt SODs. The apoenzymes were prepared by dialysis of the SOD (1 mg/ml) against 5 x 10^{-2} M acetate, 10^{-3} M mercaptoethanol, 10^{-3} M EDTA pH 3.8 for 16 hr at 4°C then overnight against 10^{-2}M acetate pH 3.8. A solution of radioactive cobalt chloride (^{60}Co^{2+} or ^{57}Co^{2+}) neutralised to pH 4.0 was then added (two to three atoms per mole of apoenzyme) and the solution brought gradually to pH 6.5 by placing it next to a solution of dilute ammonium hydroxide (0.1 N to 10^{-2}N depending on the volume of cobalt protein solution) in a closed dessicator. After neutralisation, the solution was passed through a column of Sephadex G-100 (30 cm x 1 cm), using a 5 x 10^{-3} M phosphate pH 7.8 as eluant, to separate the cobalt protein from free Co^{2+} ions. Appropriate fractions were collected and pooled. Final products were about 8.00 Ci/mmole, using ^{60}Co at a specific activity of 4.00 Ci/milli atom, since under the conditions used approximately 2 atoms of Co^{2+} were incorporated with bovine Cu-SOD, bacterial Cu-SOD, human Cu-SOD and human Mn-SOD (2.03 atoms Co). In the case of bacterial Fe-SOD, only one atom of Co^{2+} was incorporated and the specific activity was 4 Ci/mmole protein.

An alternative preparation was as follows: to a solution of 600 µCi of carrier free $^{57}Co^{2+}$ (0.08 µg, 400-500 Ci/milli atom) in 0.2 ml at pH 4.5 was added a five fold molar excess of the apoenzyme (250 µg) and the solution brought to pH 6.0 with ammonia vapour in a closed dessicator (solution of 0.1 N NH_4OH). As carrier, 2.5 mg of the native enzyme was added in 5 x 10^{-3} M phosphate pH 7.8 and the solution passed through a column of Sephadex G-100 using 5 x 10^{-3} M phosphate pH 7.8 for elution (Figure 1). Appropriate fractions were combined and used for the preparation of liposomes. In this way essentially 100% incorporation of the $^{57}Co^{2+}$ into the SOD was achieved. In general these preparations were used throughout this work.

When maximal specific radioactivity was required, 0.1% bovine serum albumin was used as a carrier instead of the native enzyme. A typical preparation is as follows. To carrier free ^{57}Co (1.2 mCi; 0.1 µg Co; specific activity 684 Ci/milli atom) in 1 ml 0.1 N HCl was added 45 µl of 2 N NaOH and 45 µl of 2 M sodium

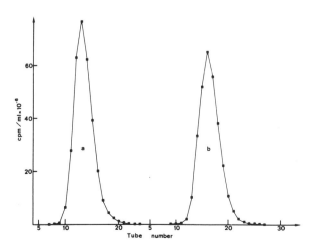

Fig. 1. Purification of ^{57}Co labelled human Cu-SOD (a) and bovine Cu-SOD (b) on Sephadex G-100.

acetate pH 8.2 to bring the solution to approximately pH 4.5. To 100 μCi (8.3 ng Co) of this solution was added 2 μg of apoenzyme from human Cu-SOD (molar ratio Co/apoenzyme = 2.44) in 26 μl of 5×10^{-3} M phosphate pH 7.8 followed by 0.1 ml of 0.1% BSA in the same buffer. After neutralisation the solution was kept at 4°C for several hours then fractionated on a column of Sephadex G-100 (10 cm x 1 cm) using the 0.1% BSA buffer as eluent. Appropriate fractions were combined and stored. Radioactivity measurements showed a specific activity of 1600 Ci/mole of protein, indicating incorporation of 2.3 atoms ^{57}Co per mole, corresponding to 94% incorporation of the input radioactive cobalt.

Preparation of liposomes. A stock solution in chloroform was prepared containing L-α-dipalmitoyl lecithin (7 μmole), cholesterol (1 μmole) and stearyl-amine (2 μmole) per ml (cationic liposomes) or 7 μmole lecithin, 2 μmole cholesterol and 1 μmole dicetyl phosphate per ml (anionic liposomes). For preparation of liposomes 2.5 ml of this solution (total 25 μmoles of the three components) was evaporated to dryness in vacuo and a solution of the SOD (1 mg) in 4×10^{-3} M phosphate pH 7.2 (1 ml) added to the residual gum. The mixture was warmed to 40°C and agitated for 5-6 min to disperse the lipids then sonicated for 30 sec. (The time of sonication at a given power input for maximum yield of liposomes is determined by preliminary experiments). The suspension was then left at room temperature for 2-4 hr then at 15°C overnight. Liposomes were separated from excess components by centrifugation in a Beckman L 65, rotor SW 65 at 40,000 rpm (139,000 g) for 2 hr at 4°C. When radioactive (^{60}Co or ^{57}Co) SODs were used, the yield was readily calculated from radioactivity present in the supernatant compared with that in the liposome pellet.

Whereas dicetyl phosphate containing liposomes could be separated from uncombined lipids and protein by passage through columns of Sepharose 6 B (apparent molecular weight in comparison with suitable markers, approx. 10^{6}) those containing stearylamine were irreversibly bound to the Sepharose.

Effects of modification of various parameters on yield of liposomes. Only slight variations in yield were observed between pH 5.5 and 8.0. With respect to ionic strength the solution should be less than 2×10^{-2} M in Na^{+} or K^{+}. Both Ca^{2+} (2×10^{-2} M) and Mg^{2+} (10^{-2} M) ions decrease the yield of dicetyl phosphate liposomes whereas 10^{-2} M EDTA is without effect. Optimum compositions of the lipid mixture was in molar ratios lecithin 7, stearylamine 2 (or dicetyl phosphate 1), cholesterol 1 (or 2 with dicetyl phosphate) with a total of 25-50 μmole lipid per mg of protein. Maximum yield was obtained at 40°C, but liposomes were also formed at 20°C as well as at higher temperatures.

Physical properties. Equilibrium centrifugation of dicetyl phosphate liposomes in sucrose gradients (0-15%) containing 4×10^{-3} M phosphate pH 7.2 at 60,000 rpm (350,000 g) for 17 h gave a band (as followed by radioactivity) at an apparent density of 1.041. A similar centrifugation of stearylamine liposomes gave a single sharp band with a density of 1.056. After storage for one month the dicetyl phosphate liposomes had liberated some 30% of free enzyme (as followed by chromatography on Speharose 6 B) and the re-isolated liposomes gave rise to two bands on sucrose gradient centrifugation with densities of 1.042 and 1.019, indicating a change in composition and structure.

Electron microscopy of the various liposomes showed that all preparations were heterogenous with respect to size, ranging from about 68 nm to 460 nm in diameter (20 different preparations were examined). In accord with this estimation, essentially all the material passes a Millipore filter of 1.2 μm, but considerable retention (in part due to adsorption processes) was observed with filters of 0.45 μm diameter pores.

Treatment of erythrocytes with the liposome preparations. Human blood (2 ml) was centrifuged at 5000 rpm (3000 g) for 10 min and the serum removed by suction. The erythrocytes (1 ml) were diluted in 3 ml of 0.9% NaCl and 400 μl of this suspension was used for each experiment. After recentrifugation of the samples at 50000 rpm for 10 min, the NaCl solution was removed and to the residual 100 μl of erythrocytes was added the different liposomes (total 500,000 cpm) in 0.15 M NaCl, 6.7×10^{-3} M phosphate pH 7.2 (0.5 ml) and the tubes agitated at room temperature for 2 hr. The same buffer (1 ml) was then added and the erythrocytes centrifuged at 2500 rpm for 5 min. This washing was repeated 3 times with one ml of buffer and after a final resuspension the erythrocytes were packed by centrifugation at 5000 rpm for 10 min. Control experiments showed that no free liposomes contaminated the final product. The cells were lysed by addition of 0.2 ml distilled water and left for 30 min at $0^{o}C$. A mixture of 3:5 (v/v) chloroform/ethanol (0.2 ml) was added and after suitable agitation the mixture was centrifuged at 5000 rpm for 10 min. The total chloroform-ethanol supernatant was used for counting in a scintillation counter. Results are expressed as number of molecules of SOD (packaged as liposomes) which penetrate (or are irreversibly bound) per erythrocyte. Percentage increase in erythrocyte SOD (erythrocuprein) content is calculated on the basis of 28,200 moelcules of erythrocuprein per nomral erythrocyte. In control experiments a solution of the same amount (500,000 cpm) of the corresponding free radioactive SOD (4.5 x 10^{6} cpm/mg) replaced the liposome suspension. This corresponds to about 110 μg SOD (free or as liposome) incubated with erythrocytes from 200 μl of total blood containing about 18.5 μg endogenous erythrocuprein (value for this subject is 92.5 μg per ml blood), and hence is approximately a six fold excess. In terms of molecules of SOD per erythrocyte, this subject had 5.88×10^{9} erythrocytes per ml of blood and hence 282,269 molecules of erythrocuprein per erythrocyte.

For the Mn-SOD and Fe-SOD preparations containing [57]Co a slightly different technique was used since these enzymes do not resist chloroform-ethanol treatment. Erythrocytes were incubated, centrifuged, washed and counted directly

in a γ-counter. This direct technique was subsequently used for all preparations
containing radioactive cobalt.

  Penentration of erythrocytes as a function of time. Radioactive preparations
(^{57}Co) of native bovine Cu-SOD, the thiobutanol modified enzyme (BuSH SOD),
stearylamine liposomes, and stearylamine α-tocopherol liposomes (molar ratios,
7 palmitoyl lecithin : 2 stearylamine : 1 cholesterol : 1 tocopherol with 27.5
μmoles total lipidic components per mg of SOD) containing about 42 μg SOD were
added to 1.2 ml human blood (total 6.9 x 10^9 erythrocytes) and diluted to 4.8 ml
with isotonic phosphate saline buffer (PBS). The mixture was incubated with
gentle agitation at 25°C and aliquots of 0.8 ml (corresponding to 0.2 ml blood)
withdrawn at different time intervals and centrifuged at 192 g for 5 mins. The
erythrocytes were resuspended in 2.5 ml PBS and recentrifuged. This washing was
repeated a total of 5 times. Hypotonic buffer (6.5% PBS, 2.5 ml) was added to
the red cells and the suspension left at 4°C for 5 min then centrifuged at
20,000 g for 40 min to separate membranes from cytoplasm. Both supernatant
(cytoplasmic SOD) and pellet (membrane bound SOD) were counted directly in a
γ counter.

  Electron spin resonance. Liposomes were prepared with a spin labelled egg
yolk lecithin esterified at the β-hydroxyl with a stearic acid residue substituted at C 16 by an oxyoxazolidine ring (1,14 PC) prepared according to Hubbell
and McConnell[12]:

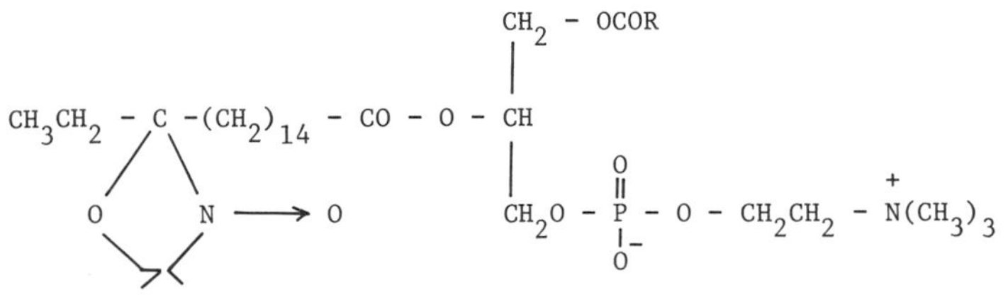

  In a typical experiment, 100 μg of 1,14 PC, 7.8 μg of cholesterol 10.5 μg of
stearylamine were dissolved in chloroform and the solution evaporated to dryness.
An aqueous solution of 6.4 μg of bovine Cu-SOD in 0.5 ml of 0.075 M phosphate
buffer pH 7.4 was added and the mixture sonicated briefly. The liposomes were
then mixed with erythrocytes from 0.2 ml human blood.

  Pure liposomes or red blood cells plus liposome suspensions were introduced
into a 50 μl quartz cell and mounted in a temperature regulated accessory of a
Varian E-9 spectrometer connected with a Tektronix 4051 computer. Spectra were
registered in the absorption mode at various intervals of time. Incubation
was at 20°C.

  Scanning electron microscopy. Preparations of stearylamine bovine Cu-SOD
liposomes at 2 mg SOD/ml (25.7 mg dipalmitoyl lecithin/ml, i.e. 35 μmole/ml,
5 μmoles cholesterol/ml and 10 μmoles stearylamine/ml) were incubated at 22°C

with human blood using two different concentrations: (a) 0.2 ml of liposome
suspension plus 2 ml blood and (b) 0.6 ml liposomes plus 2 ml blood. Samples
were withdrawn at 30, 90 and 180 min for dilution and preparation (fixation and
gold shadowing) for electron microscopy which was performed by Drs. L.G. Chevance
and M. Lesourd of the Service de Microscopie Electronique de l'Institut Pasteur,
Paris.

RESULTS AND DISCUSSION

Preliminary results using ^{14}C or ^{3}H labelled SODs (obtained by amino or carboxyl
group methylation) indicated that penentration or fixation of the free SOD by
human erythrocytes was very low but did not appear to be influenced by the pI of
the SOD. However, specific radioactivity of the proteins was very low and considerable degradation of the enzyme occured in certain cases. It was therefore
decided to use a general method valuable for all SODs which caused no degradation
and which yielded a stable marker. Iodination (^{125}I) has frequently been used
but in parallel studies involving injection of the enzyme into animals it was
found that rapid liberation of the radioactive iodine occured _in vivo_ with concentration of the radioactive label (but not the enzyme) in the thyroid. It may be
noted that this effect invalidates a certain number of studies on the metabolism
of SOD. The best approach was found to be preparation of the apo-enzyme by removal of the metal (Cu, Mn or Fe) followed by reconstitution with radioactive
cobalt (^{60}Co or ^{57}Co). This gave extremely high yields of incorporation of
radioactivity (essentially 95-100%) using a slight excess of apoenzyme. The
marker is stable. Immunochemically and physico-chemically (pI, migration on gels,
molecular weight, etc.) the marked molecule is essentially identical with the
native enzyme in each case.

Liposomes were prepared using such markers diluted with the corresponding
native enzyme. The standard procedure used dipalmitoyl lecithin, dicetyl
phosphate or stearylamine and cholesterol (in the proportions of 7:1:2 for
dicetyl phosphate liposomes or 7:2:1 for stearylamine liposomes, with a total
of either 25 or 50 μmole of these components per mg of protein. Preparations
were sonicated at 40°C (similar results were obtained by sonication at 20°C,
but technically the dispersion of lipid material is facilitated at the higher
temperature). Since the lipids were fully saturated a nitrogen or argon atmosphere was not used, and in any case the superoxide dismutase itself protects
the entire system against oxidative processes. Yields of the liposomes (in
terms of SOD encapsulated) varied between 30-85%, as measured by radioactivity
in the supernatant and in the pellet after high speed centrifugation for two
hours. Lower yields were consistently obtained with the dicetyl phosphate
liposomes (30-50%) whereas invariably the stearylamine liposomes were obtained
in yields of 55-85%. Extremely low yields were obtained in the absence of

stearylamine or dicetyl phosphate (neutral liposomes). Electron microscopy
(for which we thank Mr. J. Lepault) showed that all preparations were poly-
disperse, ranging from 68 to 460 nm in diameter.

Fixation of erythrocytes. Fixation of the liposomes on erythrocytes was
studied by incubation of red cells with the different preparations. The erythrocytes were then collected, well washed (supernatant radioactivity was controlled
at each washing) lysed and the SOD partially purified by treatment with chloroform-ethanol followed by measurement of radioactivity. In the case of ^{57}Co or
^{60}Co preparations direct counting of the washed erythrocytes was used since this
renders unnecessary the organic solvent treatment employed in the case of liquid
scintillation counting to remove the hemoglobin which absorbs light output.

Fixation by human erythrocytes of the free enzymes and of anionic and cationic
liposomes (dipalmitoyl lecithin) containing the superoxide dismutase is shown
in Table 1, both for prewashed erythrocytes and for whole blood. It can be seen
that very low levels of fixation are obtained for the free enzymes (from 100
to 3460 molecules per erythrocyte) and in general (except for bacterial Fe-SOD)
the yield is higher with washed erythrocytes than when the enzyme is diluted by
plasma proteins. Contrary to expectations, no correlation of pI of the enzyme
with fixation capacity is observed, perhaps because pI reflects total basic or
acidic amino acid content of the protein whereas for effective binding of the
enzyme to the outer membrane of the erythrocyte only external lysine amino groups
are presumably involved.

Fixation of enzymes encapsulated in negatively charged dicetyl phosphate
liposomes is not greatly increased compared with the free enzymes, as may be
expected, given the negatively charge characteristics of the outside of the
erythrocyte membrane. For both bacterial Cu-SOD and Fe-SOD containing liposomes fixation is more efficient in total blood than with washed erythrocytes.
The most striking results are seen with stearylamine liposomes where again
fixation to erythrocytes is more efficient in whole blood than with washed cells.
Extremely large increases of cellular SOD are obtained with all five enzymes
and indeed with the homologous system, human erythrocuprein liposomes and human
erythrocytes, the content can be tripled (204% increase of SOD). It could be
considered that since the protein is encapsulated by the lipid components, the
nature of the SOD should play no role in fixation. However, it can be seen
that in a general manner, liposomes containing acidic SODs (human and bovine
Cu-SOD and bacterial Fe-SOD) are more efficient that those containing basic
SODs (human Mn-SOD and bacterial Cu-SOD). This can possibly be explained in terms
of total liposome structure and composition if it is assumed that more stearylamine is incorporated per liposome unit with the acidic SODs than with those
characterized by a high pI value, thus giving a liposome with better penetration characteristics. As shown in Table 2, neutral liposomes (containing only

TABLE 1

FIXATION BY ERYTHROCYTES OF NATIVE SODs AND OF LIPOSOME ENCAPSULATED ENZYMES

Enzyme	Free SOD		Fixation: Dicetyl phosphate liposome		Fixation: Stearylamine liposome	
	Molecules per erythrocyte	As % of endogenous enzyme	Molecules per erythrocyte	As % of endogenous enzyme	Molecules per erythrocyte	As % of endogenous enzyme
Bovine Cu-SOD						
Washed erythrocytes	2236	0.79	2904	1.03	399635	141.58
Total blood	150	0.05	938	0.33	489585	173.45
Human Cu-SOD						
Washed erythrocytes	720	0.26	3033	1.07	515542	182.64
Total blood	100	0.04	-	-	576194	204.13
Bacterial Cu-SOD						
Washed erythrocytes	460	0.16	3315	1.17	158826	56.27
Total blood	130	0.05	23130	8.19	247491	87.68
Human Mn-SOD						
Washed erythrocytes	2520	0.89	-	-	-	-
Total blood	1460	0.52	10895	3.86	263445	93.3
Bacterial Fe-SOD						
Washed erythrocytes	460	0.16	7170	2.54	-	-
Total blood	3460	1.22	20565	7.29	436310	154.57

SOD, dipalmitoyl lecithin and cholesterol) are even less effective than free bovine Cu-SOD. As shown in Table 3, when excess of bovine serum albumin is present, during preparation and use of the liposomes, fixation is greatly reduced.

TABLE 2

PENETRATION OF ERYTHROCYTES BY BOVINE Cu-SOD ^{57}Co

Conditions	Molecules/cell	% exo/endo	% membrane	% soluble
1.17 x 10^9 erythrocytes[a]				
Free SOD	8527	3.02	54.4	45.6
BuSH SOD	23162	8.21	84.3	15.7
Neutral liposomes	7120	2.52	76.8	23.2
Stearylamine liposomes	794004	282	91.8	8.2
Stearylamine/tocopherol liposomes	731658	259	90.5	9.5
0.117 x 10^9 erythrocytes[b]				
Free SOD	186480	66.1	73.9	26.1
BuSH SOD	202146	71.7	83.2	16.8
Neutral liposomes	74113	26.3	85.3	14.7
Stearylamine liposomes	1.20 x 10^6	426	89.3	10.7
Stearylamine/tocopherol liposomes	1.24 x 10^6	440	89.7	10.3

[a]Ratio input exogenous/endogenous SOD = 7; 2 hr incubation.

[b]Ratio input exogenous/endogenous SOD = 70; 2 hr incubation

TABLE 3

EFFECT OF BOVINE SERUM ALBUMIN ON FIXATION OF BOVINE Cu-SOD LIPOSOMES

	Fixation as % of endogenous enzyme	
	In presence of albumin	In absence of albumin
Free SOD	0.50	0.79
Dicetyl phosphate liposome	0.40	1.03
Stearylamine liposome	11.93	141.58

It was of interest to study the effect of modification of the lecithin component in liposme preparations on their capacity of attachment to human erythrocytes. The results are shown in Table 4. With fully saturated lecithins a significant effect of chain length is seen with a marked optimum at fatty acid esters of 16 carbon atoms, both shorter or longer chain lengths giving liposomes which are less efficient. Ester linkage can be replaced by an ether linkage (dihexadecyl lecithin) with no loss whatsoever of fixation capability (indeed a slight improvement over the dipalmitoyl lecithin can be seen), However, when an unsaturated residue is introduced, for example when a palmitoyl residue is replaced by oleoyl (9,10 double bond), a large decrease in fixation capacity is observed. This is even more marked if both chains are unsaturated (di-oleoyl lecithin) since attachment is now essentially abolished. These results show the striking biological effects which result from relatively minor changes in liposome membrane structure due to the more open conformation imposed by a single double bond in the fatty acid residues.

TABLE 4

EFFECT OF COMPOSITION OF LIPOSOMAL LECITHIN ON FIXATION (BOVINE Cu-SOD, WASHED ERYTHROCYTES, STEARYLAMINE LIPOSOME)

Lecithin	Molecules per erythrocyte	As % of endogenous SOD
Dihexadecyl ether	509747	180.6
Dilauroyl (12)	279997	99.2
Dimyristoyl (14)	270469	95.8
Dipalmitoyl (16)	420135	148.8
Distearoyl (18)	155174	55.0
Dioleoyl (18)	3266	1.16
Oleoyl, palmitoyl	129796	46.0

The results presented were obtained when cells were incubated with about a six fold excess of SOD compared with the endogenous enzyme. In the case of stearylamine liposomes (bovine Cu-SOD, dipalmitoyl lecithin) increasing this excess to thirty fold did not change the amount of fixation (141.6% to 148.8%). However, in the case of the free Cu-SOD a thirty fold excess increased the fixation (as a percentage of endogenous enzyme) from 0.79% to 18.5%. Similarly, the amount of dicetyl phosphate liposome encapsulated Cu-SOD incorporated into the erythrocyte was increased from 1.03 to 46% when a thirty fold excess was used. Similar results are shown in Table 2 in which fixation is compared with a seven fold excess and a seventy fold excess of exogenous SOD over the endogenous enzyme. Again it can be seen that fixation of cationic liposomes is less than doubled, whereas for the native enzyme, the chemically modified SOD and neutral liposomal SOD, the increases range from 10 to 20 fold.

Penetration of erythrocytes. Preliminary studies having shown that SOD, liposomal SOD or chemically modified SOD, showed varying capacities of fixation to human red blood cells, it was of interest to study this association in more detail, particularly with respect to penetration of the cell membrane. Use of radioactively marked material followed by fractionation of the erythrocytes into membrane and cytosol components allows an estimation of the penetration of the cell wall. As shown in Table 2, after two hours incubation of the erythrocytes with the various forms of SOD, considerable differences in penetration are observed. Thus although native bovine Cu-SOD shows less total fixation, nearly half of the associated exogenous SOD passes through the membrane when a seven fold excess is used. Neutral liposomes show both a lower fixation and penetration. As may be expected, the butyl thiol modified SOD is attached mainly to the membrane, presumably via -S-S- linkages. Stearylamine SOD liposomes are attached to the red cells at a much higher level (with a 70 fold excess of exogenous SOD, greater than a 4 fold increase is observed) but fusion and penetration at two hours is only about 10% of the total fixed SOD.

Penetration of the membrane and passage to the inside of the red cell at $25^{\circ}C$ as a function of time is shown in Table 5. It can be seen that whereas fixation or association with the cell is rapid, passage of the membrane is relatively slow. As may be expected, the thiobutyl SOD remains essentially attached to the membrane and does not dissociate or penetrate. Indeed the 10-15% radioactivity which appears to be cytoplasmic may well simply be liberation from the membrane during lysis due to the action of liberated erythrocyte reduced glutathione. As shown in Figure 2, penetration is exponential and from the experimental results half times of passage from outside to inside with intact erythrocytes can be calculated to be 15.42 hr for free SOD, and 5.78 hr for cationic liposomal SOD. Penetration is thus considerably more rapid (2.66 fold) with the liposomal SOD compared with the free protein, in addition to the much larger quantities incorporated.

TABLE 5

PENETRATION OF ERYTHROCYTES BY SOD

Minutes	cpm (specific activity constant)			
	Membrane	Cytoplasm	Total	% cytoplasmic
Free SOD				
15	2520	693	3212	21.6
90	2197	783	2980	26.3
120	1648	702	2350	29.9
300	1760	1031	2791	36.9
BuSH SOD				
15	3071	377	3448	10.9
30	5040	835	5875	14.2
90	4141	778	4919	15.8
300	7664	1041	8705	11.6
Stearylamine liposomes				
15	35369	10374	45743	22.7
30	43241	14023	57264	24.5
90	14077	7580	21657	35.0
120	13708	7848	21556	36.4
Stearylamine/ tocopherol liposomes				
15	7777	1081	8858	12.2
30	2778	714	3492	20.4
90	1520	534	2054	26.0
300	1927	1066	2993	35.6

Electron spin resonance spectroscopy. A second approach which has previously been used to follow fusion of liposomes with cell membranes is possible by the use of electron spin markers.[13] Electron spin resonance spectroscopy allows discrimination between liposomes adsorbed on the red blood cell surface and liposomes fused with the cells. When fusion occurs the liposome lipids are integrated and diluted into the cell membrane with injection of the protein

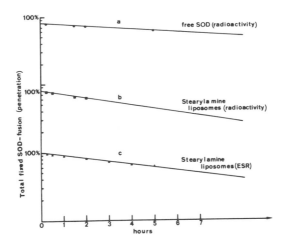

Fig. 2. Semilog plots of penetration of human erythrocytes by (a) free SOD using radioactivity measurements; (b) similarly for radioactive SOD encapsulated in cationic liposomes at 25°C; and (c) fusion of lecithin spin labelled liposomal native bovine Cu-SOD followed by electron spin resonance spectroscopy at 20°C.

content into the cell. Liposomes made with a high concentration of spin labelled lecithin generate a very broad spectrum (one wave) characteristic of strong spin-spin interaction (Figure 3,a). Cationic (stearylamine) SOD liposomes were pre-pared with a lecithin containing a suitable spin marker (oxyoxazolidine residue) at the non-ester end of a stearyl residue. When such liposomes fuse with red blood cells, the spin label is diluted into the membrane, spin spin interactions are abolished and a three line spectrum is generated (Figure 3,c). The appear-ance of a three line spectrum was measured as a function of time at 20°C. This is shown in Figure 4. Here the three line spectrum (full lines) is superimposed on the broad wave spectrum (dotted lines) provided by the non-fused liposomes present in the suspension. Determination of the fusion kinetics can be achieved in two ways. First, the sharp peak amplitude (H_2, Figure 5,a) is plotted versus time and secondly, a computer substraction of spectra allows estimation of the exact amount of spin labelled lipids introduced into the red blood cell membranes. Substracting the pure liposome spectrum (dotted lines) from the spectrum of the mixture (full lines) generates the pure spectrum of the spin labelled lecithin dissolved in the cell membrane. Integration of these differential spectra gives a value of the spin label concentration in the membrane, i.e. the percentage of fusion. Figure 5,b shows the liposome fusion kinetics as determined by this method. As shown in Figure 2, replotting these kinetics in a semi-log form gives a single exponential which passes through the origin. Calculation of the half time of fusion gives a value of 7.24 hr, remarkably similar to that determined by simple radioactivity measurements (5.78 hr). The difference in values can be readily explained by the difference in temperature for the two studies (20°C for ESR and 25°C for the radioactivity technique). In principle, a full thermodynamic study of fusion of liposomes with the erythrocyte membrane is possible using such techniques.

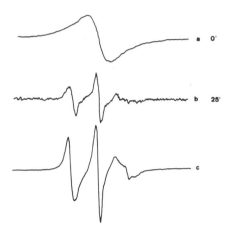

Fig. 3. Electron spin resonance spectra of (a) spin labelled liposomes; (b) after 25 min in presence of erythrocytes; and (c) spin labelled lecithin fully diluted into the membrane.

Electron microscopy. Human blood were mixed with stearylamine diplamitoyl lecithin liposomes containing native bovine Cu-SOD at 22°C and samples withdrawn at 30 min, 90 min and 180 min. Two concentrations were used (a) liposomes containing 0.2 mg SOD per ml of blood (3.5 μmoles lecithin/ml) and (b) 0.6 mg SOD (10.5 μmoles of lecithin/ml). The results are shown in Figure 6, a-h. At lower concentration, the erythrocytes appear to be perfectly normal (Figure 6,a), whereas at the higher concentration of liposomes marked erythrocyte aggregation can be seen (Figure 6,b). At 0.2 mg SOD/ml blood there appears to be an increase in fixation of liposomes with time (30-90 min) followed by a decrease due to fusion and penetration (90-180 min). Typical examples are shown in Figure 6,c and d (low liposomes concentration after 30 min incubation) and in Figure 6,c (low concentration of liposomes, after 90 min). Given the average diameter of erythrocytes as 6.68 to 7.72 μm and direct measurement coupled with the magnification used, the liposomes visibly attached to the erythrocytes range in diameter

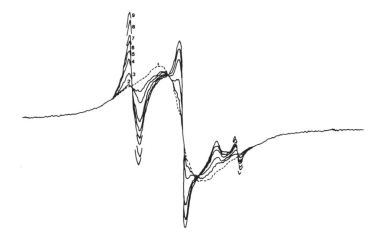

Fig. 4. Electron spin resonance spectra of human erythrocytes plus lecithin spin labelled cationic liposomal SOD at 1 - zero time (dotted line); 2 - at 6 min; 3 - at 37 min- 4 - at 89 min; 5 - at 120 min; 6 - at 150 min- 7 - at 178 min- 8 - at 255 min; and 9 - at 307 min.

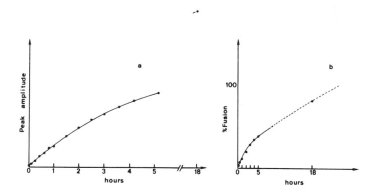

Fig. 5. (a) Peak amplitude of spin labelled liposomes containing SOD as a function of time. (b) Percentage fusion determined by integration of diff- erence electron spin resonance spectra as a function of time.

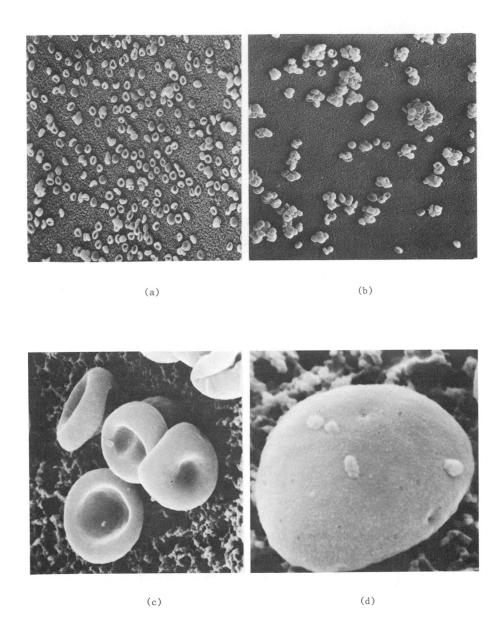

(a)

(b)

(c)

(d)

Fig. 6. Electron microscopy of human erythrocytes plus cationic liposomal encapsulated bovine Cu-SOD. (a) 3.5 µmoles lecithin (0.2 mg SOD) per ml blood
after 90 min; magnification 482. (b) 10.5 µmoles lecithin (0.6 mg SOD) per ml
blood after 180 min; magnification 482. (c) as (a) but after 30 min incubation;

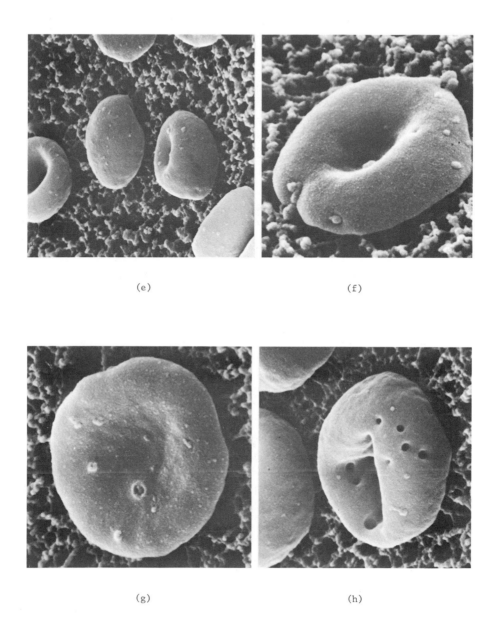

(e)

(f)

(g)

(h)

magnification 4822. (d) as (c) but magnification 10715. (e) as (a) but magnification 4822. (f) 10.5 μmoles lecithin (0.6 mg SOD) per ml blood after 30
min; magnification 9643. (g) as (a) but after 180 min; magnification 9643.
(h) as (g).

from 207-519 nm with an average diameter of 360 nm, in accord with the size
estimates described earlier (68-460 nm). In certain of the erythrocytes an
intermediate stage of fusion of liposome with the membrane can be observed for
example in Figure 6,f (high liposome concentration at 30 min) in which a liposome appears to be sinking into the surrounding membrane, and in Figure 6,g
(low liposome concentration after 180 min) in which an ulceration due to disso-
lution of the liposome can be seen. Finally, as shown in Figure 6,h (low liposome concentration, after 180 min incubation) the perturbation of the erythrocyte membrane due to local fusions are clearly visible. It may be noted that
these perturbations are not holes in the membrane since ejection of the cytoplasm does not occur, but probably indicate a process of invagination; indeed
after 180 min incubation erythrocyte ghosts were not present. Even at the high
liposome concentration after 3 hr less that 0.5% of lysis occurs as evidenced by
the appearance of ghosts. These results confirm directly fixation and penetration
of liposomes containing SOD with respect to human erythrocytes. In addition,
fusion with the membrane, while relatively slow, is clearly visible and kinetically is in accord with the results obtained by electron spin resonance spectroscopy.

ACKNOWLEDGEMENTS

This work was supported by grants from CNRS (E.R. 103), INSERM (contract
no. 77.4.0842) and the Fondation pour la Recherche Médicale Francaise.

REFERENCES

1. Haber, F. and Weiss, J. (1934) Proc. Roy. Soc. A. 147, 332-351.
2. Michelson, A.M., McCord, J.M. and Fridovich, I. (1977) Superoxide and Superoxide Dismutases, Academic Press, New York.
3. Petkau, A., Kelly, K., Chelack, W.S., Pleskach, S.D., Barefoot, C. and
   Mecker, B.E. (1975) Biochem. Biophys. Res. Commun. 67, 1167-1174.
4. Huber, W. and Saifer, M.G.P. (1977) Superoxide and Superoxide Dismutases,
   Michelson, A.M., McCord, J.M. and Fridovich, I., eds., Academic Press, New
   York, pp.517-536.
5. Rister, M., Banermeister, K., Gravert, U. and Gladtke, E. (1978) Lancet,
   1, 1094.
6. Tyrrell, D.A., Health, T.D., Colley, C.M. and Ryman, B.E. (1976) Biochim.
   Biophys. Acta, 457, 259-302.
7. Tyrrell, D.A., Keeton, B.R., Ryman, B.E. and Dubowitz, V. (1976) Brit. Med.
   J. 2, 88.
8. Patel, H.M. and Ryman, B.E. (1976) FEBS Lett. 62, 60-63.
9. Kimelberg, H.K. and Atchison, M.L. (1978) Ann. N. Y. Acad. Sci. 308, 395-409.
10. Traut, R.R., Bollen, A., Sun, T.T. and Hershey, J.W.B. (1973) Biochemistry,
    12, 3266-3273.
11. Means, G.E. and Feeney, R.E. (1968) Biochemistry, 7, 2192-2201.
12. Hubbell, W.L. and McConnell, H.M. (1978) J. Amer. Chem. Soc. 93, 314-326.
13. Rousselet, A., Guthmann, C., Matricon, J., Bienvenue, A. and Devaux, P.F.
    (1976) Biochim. Biophys. Acta, 426, 357-371.

IN VIVO EFFECTS OF PARAQUAT ON SOME OXIDATIVE ENZYMES IN MICE AND EXPERIMENTS
TO SUPPORT THE DEFENCE AGAINST THE POISONING

B. MATKOVICS[+], L. SZABO[+], SZ.I. VARGA[+], KATALIN BARABAS[++] AND GY. BERENCSI[++]
+ Biological Isotope Laboratory, "A.J." University of Szeged, Hunagary;
++ Department of Public Health of the University Medical School of Szeged,
Szeged, Hungary

INTRODUCTION

By means of a comparison with untreated controls, a study was made of the
effects of the LD_{50} and Ld_{100} of methylviologen(Paraquat;PQ) in vivo on the
enzymes superoxide dismutase (SOD), peroxidase (P-ase) and catalase (C-ase),
on tissue lipid peroxidation (LP), and on the microsomal cytochrome P-450
(c.P-450) concentrations of the main organs of mice. The results reveal that
the enzymes and the tissue properties compared are influenced by the herbicide
in different ways, depending on its concentration. Effects of such a nature
are connected primarily with the redox properties of PQ, which in mice acts
either as an electron donor or as an electron acceptor, depending on its concentration.

Paraquat products are excellent herbicides, and their use is continually
spreading. Accordingly, even with the employment of the optimum protection
methods, there is a steady increase in the number of human poisonings. We are
striving to elaborate a specific protection. This necessitates as detailed as
possibe an understanding of the mechanism of action of PQ. This latter may then
be utilized to provide a better-based protection against PQ poisoning.

In this communication we deal with the protective effects of ascorbic acid
(AA) and reduced glutathione (GSH) against PQ poisoning. Both compounds are
also employed in general methods of detoxication; their use in our case was
based on the above-mentioned redox concentrations. Both are known to be radical
scavengers. On administration of AA, it accumulates in the tissues of the lung,
but primarily in the fluid of the alveoli, in the form of dehydroascorbic acid.[1]
In the alveoli, the AA plays the role of a water-soluble antioxidant.[2] It need
not be specially emphasized here that the antioxidant effect of AA is based on
its redox properties.

In earlier morphologic and pharmacologic studies, it was demonstrated that
pretreatment with either AA or GSH, or the simultaneous administration of either
of these substances together with PQ, increases the survival time against the
LD_{50} of PQ (Ref.3). We have therefore made a study of the effects of treatment

with AA and GSH, both simultaneous with and prior to administration of the LD_{50} of PQ, on the same parameters as considered previously.[4]

MATERIALS AND METHODS

Male CFLP mice of the same age, weighing 30-35 g, and kept under identical conditions, were used in the study. They were obtained from the Laboratory Animal Breeding Center, Gödöllo, Hungary.

Paraquat is commercially available in Hungary under the name Gramoxon[R] as a herbicide (Alkaloida Pharmaceutical Co., Tiszavasvari, Hungary), with a PQ content of ca. 25%. It was administered orally (p.o.) to mice in a 2.5% sterile aqueous solution in doses of LD_{50} and LD_{100}, i.e. 120 and 300 mg/kg, referred to active ingredient. The mice were killed 24 hr after the PQ administration. Decapitation was followed by exsanguination, rapid removal and homogenization of the organs. Poisonings were carried out in parallel on at least 3 or 4 animals, and were repeated in 3 separate series, and thus the results are the means (± SD) of 9 or 12 activity values.

The SOD activity was measured by the method involving inhibition of the epinephrine-adrenochrome reaction by the enzyme. Under the conditions used, the rate of increase of A_{480} due to adrenochrome was about 0.01 units/min. One unit SOD activity was defined as the amount of the enzyme required for 50% inhibition.[5,6] Peroxidase activities were determined by the spectrophotometric guaiacol method at $25^{\circ}C$ and 470 nm.[7] The Beers and Sizer[8] method involving measurement of the rate of H_2O_2 consumption at 240 nm and $25^{\circ}C$ was used for quantitative determination of C-ase activities. The estimated C-ase activity values were converted to Bergmeyer units (BU). Lipid peroxidation was measured by the thiobarbituric acid reaction on the whole homogenates at 548 nm and the results are given in terms of malondialdehyde (MDA) nmoles/g wet tissue weight (w.t.w.)(Ref.9). Quantitative protein determinations were made by the method based on the Folin phenol reagent,[10] bovine serum albumin being used to prepare the standard curve at $25^{\circ}C$ and 675 nm. The microsomal c.P-450 values were measured by the method of Greim.[11]

Lethality studies. The LD_{50} of PQ was determined in mice by administering p.o. 2.5% Gramoxon[R] dissolved in distilled water at four dosage levels, with eight to ten mice in each dosage group.

The AA and GSH injection solutions were prepared by the Pharmacy Department of Szeged University Medical School. In the preparation of the 5% AA solution the prescriptions of the Hungarian Pharmacopoeia (VIth edition) were taken into consideration. The GSH used was a product of Sigma, USA. This was used to prepare a 1% sterile aqueous solution sealed under N_2.

Pretreatment (PRE) with AA or GSH was carried out as follows. Twelve hours before PQ administration, AA (500 mg/kg) or GSH (1 g/kg mouse weight) was in-

jected intraperitoneally (i.p.). This was followed by p.o. administration of
the LD_{50} (120 mg/kg) of PQ, by fasting then by sacrifice. The organs were then
excised and homogenized. In the simultaneous experiments (denoted by SIN) the
LD_{50} of PQ was administered p.o. at the same time as the i.p. injection of the
above dose of either AA or GSH. Subsequent treatment was as in the PRE method.
It must be noted that the PRE and SIN experiments were preceded by work aimed at
establishing the optimum doses of AA and GSH (Ref.12). Every experiment was
performed in at least 3 parallel series, on at least 3 to 5 mice.

RESULTS

The results of the measurements on organ homogenates after PQ administration are
listed in Table 1. Control values are also given. The data permit the following
general observations:

(1) The enzyme activity, lipid peroxidation and protein values for the organ
homogenates examined depend on the concentration of PQ employed.

(2) The different concentrations of PQ have different effects on the different
organs, and the intensities of these effects also differ.

(3) Paraquat is primarily toxic towards the larger organs, such as the liver,
kidney and lung, but an intense PQ effect can also be observed on the enzymes
in question in organs previously not generally studied, e.g. the brain homogenate.

(4) The parameters measured are frequently affected in opposite ways by the
LD_{50} and LD_{100} doses of PQ as compared to the controls.

The effects of AA administration are shown in Figures 1 to 6. In Figure 1
it may be clearly seen that, under the conditions employed, both the PRE and
the SIN AA treatment lead to considerable reductions in the SOD activities in
the organs , but to increases in the SOD activities on the liver, kidney and
intestines (primarily in the former two organs and in the PRE group).

Figure 2 depicts the effects of PRE and SIN AA treatment and PQ LD_{50} administration on the P-ase. Bars 2 and 3, which show the effects of AA without PQ,
are given only for the lung. The PRE AA treatment significantly increases the
P-ase activity in the lung, whereas the SIN treatment decreases it a little. As
to all exogenous materials, the organ P-ases respond to AA with an increase.
Exceptions are the spleen and the brain, where the changes are minimal.

Figure 3 shows that in those organs (liver, kidney and lung) on which the PQ
primarily acts, AA alone is also an excellent C-ase inducer. The PQ considerably
enhances the induction effect. The C-ase activities of the spleen, brain and
intestines do not vary in response to AA. It is of interest to observe the
C-ase induction manifested in the cardiac tissue.

Figure 4 illustrates the LP results; these are important as regards the
mechanism of action of PQ and have been subjected to much study. The results
are surprising and of importance from many aspects. It is striking that AA alone

TABLE 1

MEASUREMENTS ON ORGAN HOMOGENATES AFTER PARAQUAT ADMINISTRATION TO MICE[a]

Measurement and Treatment		Liver	Kidney	Lung
SOD	C	4600 ± 355.5	3250 ± 318.2	1060 ± 142.1
U/g w.t.w.	LD_{50}	3633 ± 348.4	4300 ± 425.4	703 ± 70.5
	LD_{100}	4250 ± 450.3	5633 ± 486.2	850 ± 72.0
P-ase	C	356 ± 32.4	600 ± 58.5	1579 ± 148.3
U/g w.t.w	LD_{50}	1163 ± 100.2	1650 ± 123.3	4450 ± 429.4
	LD_{100}	1650 ± 124.2	1275 ± 121.4	5138 ± 483.2
C-ase	C	0.2 ± 0.019	0.12 ± 0.011	0.13 ± 0.012
BU/g w.t.w.	LD_{50}	3.3 ± 0.279	1.4 ± 0.142	1.5 ± 0.134
	LD_{100}	0.21 ± 0.018	0.28 ± 0.019	0.22 ± 0.016
LP nmoles MDA/g w.t.w.	C	453 ± 44.4	476 ± 40.3	253 ± 24.2
	LD_{50}	669 ± 64.5	770 ± 62.4	359 ± 34.8
	LD_{100}	359 ± 34.8	377 ± 28.7	274 ± 26.3
Protein	C	224 ± 22.2	155 ± 14.3	184 ± 17.3
mg/g w.t.w.	LD_{50}	209 ± 19.8	201 ± 19.6	258 ± 22.4
	LD_{100}	214 ± 17.4	148 ± 12.4	161 ± 13.4
c.P-450	C	0.6075 ± 0.0594	0.1254 ± 0.011	
nm/mg protein	LD_{50}	0.5798 ± 0.0384	0.1544 ± 0.016	
	LD_{100}	0.2177 ± 0.0183	0.1679 ± 0.017	

TABLE 1/Contd.

Measurement and Treatment		Spleen	Brain	Intestine	Heart
SOD U/g w.t.w.	C	400 ± 31.8	392 ± 38.2	350 ± 29.3	527 ± 43.4
	LD_{50}	467 ± 41.0	387 ± 37.2	640 ± 62.5	570 ± 40.2
	LD_{100}	460 ± 42.3	1056 ± 54.8	490 ± 27.4	873 ± 72.3
P-ase U/g w.t.w.	C	2100 ± 200.2	400 ± 37.4	150 ± 14.3	1467 ± 138.5
	LD_{50}	1775 ± 114.3	1750 ± 162.8	575 ± 56.4	2525 ± 234.3
	LD_{100}	112 ± 10.2	413 ± 40.2	175 ± 16.8	2065 ± 201.4
C-ase BU/g w.t.w.	C	0.26 ± 0.014	0.26 ± 0.018	0.36 ± 0.027	0.2 ± 0.025
	LD_{50}	0.7 ± 0.06	0.47 ± 0.036	0.4 ± 0.031	1.4 ± 0.144
	LD_{100}	0.010	0.35 ± 0.028	0.2 ± 0.027	0.16 ± 0.010
LP nmoles MDA/g w.t.w.	C	137 ± 12.3	353 ± 34.9	146 ± 12.8	144 ± 13.3
	LD_{50}	195 ± 17.4	341 ± 33.3	83 ± 7.4	315 ± 30.5
	LD_{100}	169 ± 16.3	359 ± 34.8	70 ± 6.9	312 ± 29.2
Protein mg/g w.t.w.	C	128 ± 11.6	51 ± 5.2	119 ± 10.4	154 ± 14.5
	LD_{50}	274 + 26.3	98 ± 6.8	158 ⊥ 12.3	191 ⊥ 19.4
	LD_{100}	166 ± 12.4	85 ± 8.7	105 ± 9.8	250 ± 18.7

[a]Abbreviations and paraquat dosage as given in text. C = controls; values are mean ± S.D. of 10 experiments.

increases the LP in the liver, as it does in the form of PRE treatment in the kidney. In the liver this is further enhanced to a considerable extent by the simultaneous administration of the LD_{50} pf PQ. No essential difference can be observed in the other organs examined.

The AA and PQ synergism in the liver (oxidation enhancement) and the essential constancy in the kidney, lung and other organs may serve as a molecular-level explanation of the nevertheless lethal PQ LD_{50} effect in the extreme case.

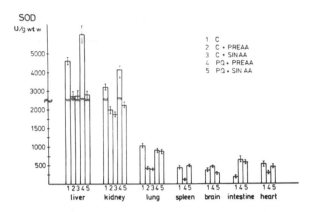

Fig. 1. Effects of AA PRE and SIN (bars 2 and 3, if shown) and of PRE and SIN AA + PQ (bars 4 and 5) treatment on tissue SOD activities. For abbreviations see text; C, control values.

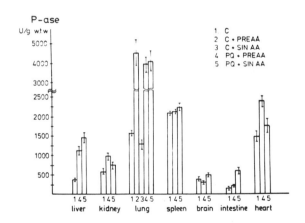

Fig. 2. Effects of AA on P-ase activities (system as in Fig.1).

Figure 5 compares protein measurements. With AA (SIN treatment), PQ may be assumed to be a fast protein synthesis inducer in most organs.

Figure 6 presents the effects of AA and AA + PQ combined on the c.P-450 measurements, primarily in the liver tissue. When administered either alone or together with PQ, AA increases the c.P-450 activity of the liver and kidney microsomes, and enhances the non-specific protection of the liver.

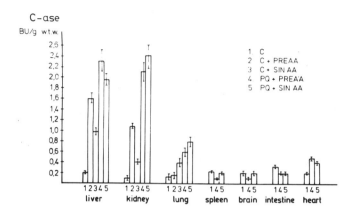

Fig. 3.   Effects of AA on C-ase activities (system as in Fig.1).

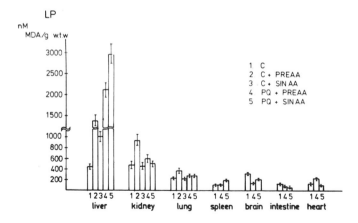

Fig. 4.   Effects of AA on tissue LP values (system as in Fig.1).

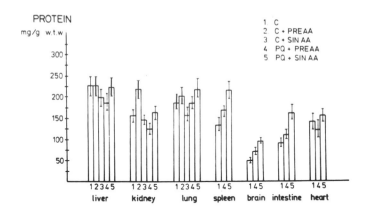

Fig. 5. Effects of AA on protein content (system as in Fig.1).

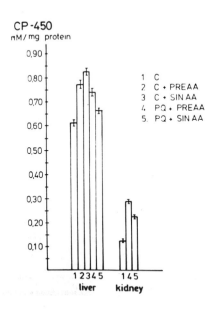

Fig. 6. Effects of AA on c.P-450 values (system as in Fig.1).

Figures 7 to 12 show the effects of GSH. Figure 7 clearly shows that in the large organs neither GSH alone nor GSH + PQ influences the SOD activity. However, the PRE GSH treatment increases the SOD activities of the lung, brain, intestines and heart.

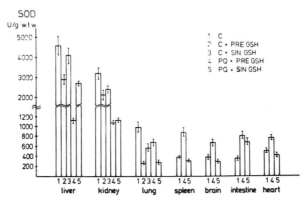

Fig. 7. Effects of GSH PRE and SIN (bars 2 and 3) and of PRE and SIN + PQ (bars 4 and 5) treatment on SOD activities. For abbreviations see text; C, control values.

The effects of GSH and PQ treatments on P-ase are depicted in Figure 8. The synergism of GSH + PQ can readily be seen, mainly in the lung, liver, kidney, intestines and heart; in a number of organs this is reminiscent of the effect of AA + PQ on P-ase

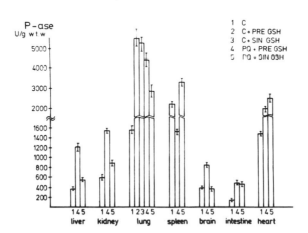

Fig. 8. Effects of GSH on P-ase activities (system as in Fig.7).

Figure 9 gives the results of GSH treatments on the tissue C-ase activities. Similarly to AA, GSH alone leads to a C-ase activity increase in the PQ "effect organs"; this increase is not reduced substantially by joint treatment with PQ. The GSH-induced C-ase changes in the other organs do not need special comment.

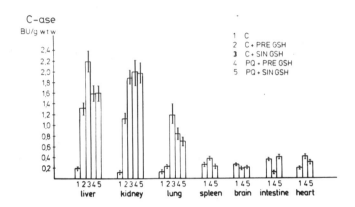

Fig. 9. Effects of GSH on C-ase activities (system as in Fig.7).

Figure 10 reveals that the GSH treatments enhance the LP in the liver and the kidney. An increase in the lung is caused only by the SIN GSH treatment. On the other hand, the GSH + PQ LD_{50} treatments give rise to a decrease of the LP in the liver and kidney, do not influence it appreciably in the lung, brain and heart, and increase it in the spleen. These results will be returned to in DISCUSSION.

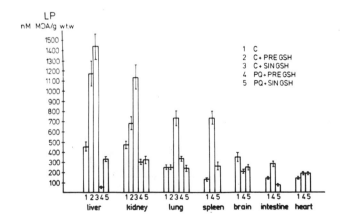

Fig. 10. Effects of GSH on tissue LP values (system as in Fig.7).

As regards protein changes (Figure 11), the protein decreases induced in the liver, kidney and heart by GSH SIN treatment, and the outstandingly high values in the kidney due to GSH PRE and SIN treatments, and more particularly that in the lung due to GSH SIN treatment, are striking. Attention may also be drawn to the GSH PRE values for the spleen and brain.

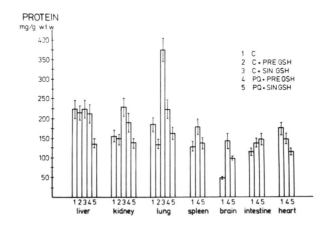

Fig. 11.   Effects of GSH on protein contents (system as in Fig.7)

In Figure 12 the GSH effects on the c.P-450 activity of the liver microsomes may be compared with the action of GSH + PQ. Here, stress may be laid in the main on the positive consequence of SIN GSH treatment.

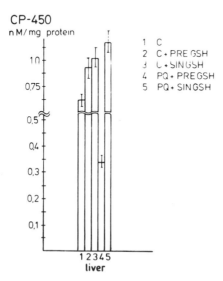

Fig. 12.   Effects of GSH liver c.P-450 values (system as in Fig.7).

A detailed survey of the PQ LD_{50} and LD_{100} data (see Table 1) leads to the following conclusions:

(a) Although it cannot be stated conclusively for the smaller PQ dose, the LD_{100} is a SOD inducer in every organ.[13] However, in the previously so-called main-effect organs (lung, liver, heart) the SOD induction cannot be detected at all; indeed, a smaller or larger SOD activity decrease can be observed.[4,14] These changes in the SOD would be even more striking of the SOD were given in units referring to mg homogenate protein.

(b) The homogenate P-ase activities appear increased in the liver, kidney and heart, but particularly in the lung. In the spleen there is a strong decrease in the P-ase activity. It increases in response to the PQ LD_{50} in the brain and intestine homogenates, but there is no variation with the higher PQ concentration. As established earlier, therefore, P-ase is one of those enzymes that responds rapidly to the actions of the external agents (microorganisms, viruses, injuries, etc.), its activity increasing.[15]

(c) The PQ LD_{50} considerably enhances the tissue C-ase values, whereas the LD_{100} causes practically no C-ase activity change; in fact, it leads to reductions in the C-ase activities of the spleen, intestine and heart.

(d) Most attention has previously been devoted to the effect of PQ on LP (Ref.16), and the LP enhancement is regarded as the decisive part of the PQ effect.[17] Naturally, we too have investigated this topic, and have come to the conclusions: (i) With the exception of the brain and intestine, the tissue LP values are enhanced appreciably by the PQ LD_{50}. The extent of the LP decrease in the intestine does not exceed 50%. The LP of the brain is not influenced by PQ in either of the concentrations used. It may be that PQ is not able to cross the blood-brain barrier. (ii) After treatment with the PQ LD_{100}, the surprising result is obtained that the LP values in the liver, kidney and intestine of the mice are lower than the control values. In contrast, a minor LP increase could be demonstrated in the lung, spleen and heart. These observations will be returned to in DISCUSSION.

With the exception of the liver, the protein contents of all the tissues are also influenced to different extents, depending on the PQ concentration. Higher tissue protein increases are produced by the smaller PQ dose than by the LD_{100}. Only the cardiac tissue forms an exception as regards such protein changes: here the protein concentration increases in parallel with the PQ concentration.

Opposite tendencies may be observed with the c.P-450 contents in the liver and kidney microsomes. Whereas the c.P-450 concentration of the liver microsomes decreases with increase of the PQ concentration, in the kidney the change is in the opposite direction.

It emerges from our results to date that PQ is a redox substance ($E_o' = -0.44$ V at pH 7.0) which is capable of interacting directly with metal-containing enzymes, i.e. enzymes that likewise have their own redox properties. This reaction depends among others on the environmental properties of the homogenates originating from the organ in question (in the case of the enzymes we examined). The above concept is well supported by the result that may be observed for the LP: at the lower concentration PQ, as an aerobic electron donor, promotes formation of the O_2^- anion radical, whereas at the higher concentration it acts as an electron acceptor, and extracts electrons from the O_2^- anion and the $HO_2^{\cdot}$ radicals playing roles in the LP.

DISCUSSION

In previous work on the effects of the LD_{50} and LD_{100} of PQ we found that primarily the various tissue redox changes caused by the different PQ concentrations are responsible for the different PQ tissue effects.[18] In our present work, besides the practical results, we have tried to confirm our earlier working hypothesis. AA, which was used as a substance to counteract the toxic effect of the LD_{50} of PQ, thereby improving the survival rate, is a redox substance with a redox potential of -0.058 V (ascorbic acid/dehydroascorbic acid system). At the same time, the redox potential between the reduced and oxidized forms of glutathione (2 GSH/GSSG), the reduced form of which also improved survival against PQ, is -0.34 V.

Our results demonstrate that the following sequence of effectiveness can be established for these two substances, when administered under various conditions to counteract the PQ LD_{50}:

PRE AA > PRE GSH > SIN AA > SIN GSH

It appears that the beneficial effects of either AA dosage on the SOD activities are correlated with the SOD inducing character of the coadministered two substances in the AA + PQ PRE treatment. This fact was also experienced in our chronic H_2O_2 drinking experiments[19] and in relation to the C-ase activities of the large organs (see Figures 1, 3, 7 and 9). The GSH + PQ treatment increases the C-ase activities only of the kidney and the lung. It has no enhancing action at all on the SOD activities; indeed, it frequently decreases them.

The combined administration of AA + PQ or of GSH + PQ increases the P-ase activities to roughly the same extent. The P-ase values generally respond with an increase to all exogenous effects on the organs, and are therefore rapid and sensitive indicators of, for instance, inflammatory processes.

Even when administered without PQ, both AA and GSH enhance the LP and the formation of MDA. When administered together with PQ, AA and GSH exhibit opposite effects on the LP in the liver. AA + PQ yields a considerable further increase in the LP, while GSH + PQ strongly decreases it. In the action of AA + PQ on the LP, we are faced with a cooperative phenomenon which, in our view, facilitates electron exchange. Through the action of AA, the O_2^- anion is converted by a further electron transfer to the O_2^{2-} (peroxy) ion, which itself also causes an intensive LP. However, if the enhancement of the LP played the decisive role with regard to the development of the symptoms, then the AA-treated mice could not belong in the first group from the aspect of survival.

The protein changes observed do not deserve any special discussion.

Our data show that the combined activity enhancement of the peroxide metabolism enzymes on the joint action of AA + PQ may be one of the reasons for the improvement in the survival, although here too the extensive increase of the LP is an advance indication of the irreparable damage that makes the poisoning fatal.

We are continuing further experiments to try to obtain a deeper understanding of the mechanism of the toxic action of PQ, and also to attempt to eliminate the toxic symptoms. AA is a known electron-donor (it is capable of transferring either one or two electrons), whereas our current knowledge indicates that the herbicidal effect of PQ can be correlated with its electron-acceptor nature.[18] It is obvious that GSH too acts as a proton- and electron-donor in the protection against PQ. Naturally, further studies are required to support these concepts.

REFERENCES

1. Willis, R.J. and Kratzing, C.C. (1976) Biochim. Biophys. Acta, 444, 108-117.
2. Willis, R.J. and Kratzing, C.C. (1974) Biochem. Biophys. Res. Commun. 59, 1250-1253.
3. Barabás, K. and Sűveges, G. (1978) Proc. 18th Hung. Annual Meeting Biochem., Salgótarján, pp.95-96
4. Matkovics, B., Szabó, L., Varga, Sz.I., Novák, R., Barabás, K. and Berencsi, G., Gen. Pharmacol., in press.
5. Misra, H.P. and Fridovich, I. (1972) J. Biol. Chem. 247, 3170-3175.
6. Matkovics, B., Novák, R., Hoang, D.H., Szabó, L., Varga, Sz.I. and Zalesna, G. (1977) Comp. Biochem. Physiol. 56B, 31-34.
7. Colowick, S.P. and Kaplan, N.O. (1975) Methods in Enzymology, Vol.2, Academic Press, New York, pp.764-775.
8. Beers, R.F.,Jr. and Sizer, I.W. (1952) J. Biol. Chem. 195, 133-140.
9. Placer, Z.A., Cushman, L. and Johnson, B.C. (1966) Anal. Biochem. 16, 359-363.
10. Lowry, O.H., Rosebrough, N.J., Farr, A.L. and Randall, R.J. (1951) J. Biol. Chem. 193, 265-269.
11. Greim, H. (1970) Naunyn-Schmied. Arc. Exp. Pathol. Pharmacol., 266, 261.
12. Berencsi, Gy. és Nagymajtényi, L. (1974) Tuberkulózis es Tüdobetegségek, 27, 225 (in Hungarian).
13. Barabás, K., Nagymajtényi, L. és Berencsi, Gy. (1978) Pneumologica Hung. 1, 24 (in Hungarian).
14. Matkovics, B., Szabó, L., Varga, Sz.I., Barabás, K. and Berencsi, G., Gen. Pharmacol., in press.
15. Kovács, K., Szonyi, D.G. and Matkovics, B., Acta Biol. Szeged, 21, 99.
16. Bus, J.S., Aust, S.D. and Gibson, J.E. (1974) Biochem. Biophys. Res. Commun. 58, 749-755.
17. Bus, J.S., Aust, S.D. and Gibson, J.E. (1975) Res. Commun. Chem. Pathol. Pharmacol. 11, 31.
18. Autor, A.P. (1977) Biochemical Mechanisms of Paraquat Toxicity, Academic Press, New York.
19. Matkovics, B. and Novák, R. (1977) Experimentia, 33, 1574.

TREATMENT OF AUTOIMMUNE DISEASES WITH SUPEROXIDE DISMUTASE AND D-PENICILLAMINE

J. EMERIT[+], J.P. CAMUS[++] AND A.M. MICHELSON[+++]
[+]Hôpital de la Salpêtrière, Université Paris VI, Paris, France; [++]Hôpital Tenon, Université Paris VI, Paris, France; [+++]Institut de Biologie Physico-Chimique, Paris, France

Genetic, immunologic and viral factors are involved in the pathogenesis of autoimmunity, interacting through complicated mechanisms still poorly understood. Recent work (I. Emerit and A.M. Michelson, this volume) suggests the involvement of O_2^- and $\cdot OH$ radicals at the origin of autoimmune reactions, in particular in the production of autoantibodies to DNA. For this reason it appeared logical to examine the use of superoxide dismutase (SOD) for treatment of these diseases. D-Penicillamine (D-PcA) mimics the action of SOD in dismutating O_2^- radicals after formation of a complex with copper.[1] This drug, introduced on the pharmaceutical market ten years ago and well-known for its efficacy in the treatment of rheumatoid arthritis,[2] was employed in cases in which the parenteral application of SOD appeared problematic. However SOD was not only used for local treatment, but also injected subcutaneously to consenting patients, since numerous experiments in animals had confirmed the non-toxicity of SOD (Ref.3). Preliminary results obtained in 5 patients with Crohn's disease, 1 patient with dermatomyositis, 1 patient with systemic lupus erythematosus and 5 patients with rheumatoid arthritis are described in this brief report.

Crohn's disease. Crohn's disease or ileocolitis is an inflammatory bowel disease frequently accompanied by systemic cutaneous and articular manifestations. We describe below the treatment of five patients with systemic D-PcA supplemented with topical application of SOD in one patient.

Case 1: This 30 year old man had Crohn's disease of the terminal region of the colon. The principal lesions were situated at 1 cm distance from the anus, such that histologic evolution could be easily followed by biopsy. After a two month treatment with D-PcA, at a daily dose of 500 mg, a considerable regression of the inflammatory infiltration was noted, with reappearance of glands.

Case 2: A 30 year old woman had very severe pancolitis since 1975. After a two year treatment with D-PcA (500 mg daily) the colon X-ray picture was normal.

Case 3: A female patient, aged 38, had Crohn's disease since 20 years. The entire colon was removed surgically in 1965 with establishment of an anastomosis between the ileum and the terminal part of the colon. The disease continued in the ileum which was considerably reduced in its diameter, and in the rectum and anus. These stenoses were favorably influenced by a two month treatment with D-PcA (500 mg daily).

Case 4: In this 29 year old man, Crohn's disease had created an enormous fistula in the anal region with deterioration of the sphincter function. Another manifestation of the disease consisted in a pyoderma gangraenosum on one ankle of the left foot. There was improvement when the patient was put on D-PcA, 500 mg daily, for two months.

Case 5: This was a female patient aged 40. D-PcA was administered alone during two years and was then supplemented by local treatment with liposome encapsulated bovine SOD. The exacerberations of the disease were generally accompanied in this patient by an edematous infiltration of the face, that responded to treatment with D-PcA. However, in spite of the continuous treatment with D-PcA, slight relapse of the disease occurred. In this case, local applica-tion of SOD gave good results. The wound after a protectomy had never healed in ten years. Under D-PcA treatment, a considerable improvement was observed with regression of the infiltration, but the sclerosis persisted and inhibited the healing of the fistula. A complete restitution was obtained by local application of SOD in liposomes.

Collagen disease. We have also employed SOD in other autoimmune diseases with connective tissue disorders, as described below.

Case 6: This was a very severe dermatomyositis in a 12 year old boy treated with high doses of cortisone and immunosuppresive drugs. SOD in liposomes was applied on the right hemiface with a favorable response.

Case 7: Severe systemic lupus erythematosus had developed since three years in a twenty year old man, whose sister had died of the disease. He was treated with high doses of prednisone (30 mg daily) without improvement. After cessation of the corticosteroid therapy, bovine SOD was given by intramuscular injection (3.9 mg) every second day over a period of 20 days. The treatment is being continued by injection of the same dose every third day. The clinical improve-ment was impressive. The general condition of the patient, very poor before treatment, improved such that he can now be followed as an outpatient. Poly-arthritis, erythema and butterfly lesion in the face, as well as alopecia have disappeared. The laboratory data are presented in Table 1. They show a con-siderable regression of the sedimentation rate, the anemia and leukopenis. There was also a decrease in the levels of antinuclear antibodies and C3d. The chromosome aberration rate, highly increased before treatment (50 aberrations per 100 mitoses) was reduced to 28 per cent after three weeks and to 8 per cent after two months of treatment with SOD.

Cases 8-12: Five cases of rheumatoid arthritis were treated with the same doses of SOD without clinical or biological improvement. However, there was a diminished complement catabolism as judged by the level of C3d. In three

cases, chromosome studies were done before and after treatment and showed reduction of the increased breakage rates to normal values.

TABLE 1

LABORATORY DATA OF A CASE OF SYSTEMIC LUPUS ERYTHEMATOSUS BEFORE AND DURING
PARENTERAL TREATMENT WITH SUEPROXIDE DISMUTASE

	Before Treatment	20 days Treatment	80 days Treatment
Sedimentation rate	83	87	23
Erythrocyte count	3.4 million	3.8 million	5.07 million
White cell count	2 900	5 200	5 000
Antinuclear factor	1/10	1/10	(−)
C3	70 mg	120 mg	92 mg
C4	15 mg	47 mg	19 mg
C3d	42%	32%	

Conclusion. The results described here in twelve patients are clearly preliminary and need to be confirmed by controlled trials. Nevertheless they confirm the good tolerance of SOD, even by intramuscular injection.

REFERENCES

1. Younes, M. and Weser, U, (1977) Biochem. Biophys. Res. Commun. 78, 1247-1253.
2. Multicentre Trial Group, NHS (1973) Lancet, i, 275.
3. Huber, W. and Saifer, M.G.P. (1977) in Superoxide and Superoxide dismutases,
   Michelson, A.M., McCord, J.M. and Fridovich, I., eds., Academic Press,
   New York, pp.517-536.

CHROMOSOME BREAKING FACTOR IN THE SERUM OF PATIENTS WITH COLLAGEN DISEASE.
PROTECTIVE EFFECT OF SUPEROXIDE DISMUTASE

I.EMERIT[+] AND A.M. MICHELSON[++]
[+]Laboratory of Cytogenetics, Institut Biomedical des Cordeliers, Pares 6ème,
France; [++]Institut de Biologie Physico-Chimique, Paris 5ème, France

INTRODUCTION

Increased chromosomal breakage is observed in blood cultures from patients with
progressive systemic sclerosis, systemic lupus erythematosus, rheumatoid arthritis,
dermatomyositis and periarteritis nodosa.[1-4] In patients with progressive systemic
sclerosis, we have also studied fibroblast cultures and direct bone marrow preparations, in which breaks and rearrangements were increased.[5] In 1973, we reported
the presence of a "breakage factor" in the serum of patients with progressive systemic sclerosis.[6] The clastogenic activity of this factor was reduced by the radioprotector L-cysteine.[7] We now present the results of a series of 40 patients with
progressive systemic sclerosis, as well as preliminary results of biochemical investigations of their sera. We also show that this breakage factor is not confined
to progressive systemic sclerosis; sera from patients with systemic lupus erythematosus and rheumatoid arthritis also contain a clastogenic factor (CF) which resembles but may not be the same as that in progressive systemic sclerosis.

The CF was found consistently, and its activity was considerably reduced or abolished in presence of the free radical scavenging enzyme superoxide dismutase (SOD).
A protective effect of SOD on spontaneous chromosome breaks in blood cultures from
patients with Fanconi's anemia and Werner's syndrome, as well as on radiation induced
breaks, has been reported.[8-10] Another reason why we tested the anticlastogenic
properties of SOD was the observation that D-penicillamine, a drug known for its
efficacy in the treatment of rheumatoid arthritis, also reduces chromosomal breakage.[11]
The copper complex of D-penicillamine (dimethylcysteine) mimics SOD in the destruction of O_2^- and the effects of D-penicillamine treatment may be due to the formation
of a complex with copper in vivo.[12,13]

MATERIAL AND METHODS

The patients studied had not been exposed to significant X-ray procedure prior
to investigation. Therapeutic agents that could produce chromosome damage had
not been administered. The criteria of the American Rheumatology Association were
adopted for diagnosis.

Healthy blood donors were recruited from the Blood Center or from among the
hospital or laboratory personnel. Mitoses were obtained by usual blood culture
procedures. The culture medium was TCM 199 from the Pasteur Institute or from
Flow Laboratories, supplemented by 20% human AB serum. Cell division was obtained
by phytohemagglutinin M and P. Duration of culture was 68-72 hours. After a hypotonic pre-treatment (75 mM KCl) for 10 minutes, the cells were fixed in 3:1 ethyl
alcoholglacial acetic acid. Iced wet slides were used for spreading of the mitoses,
which were stained with Giemsa stain. Chromosome aberrations were scored on coded
photographic prints. A minimum of 50 mitoses was examined for each experiment.
Gaps and breaks of one of both chromatids, acentric fragments and "minutes", rings,
dicentrics, translocations, tri- and quadriradial configurations were scored.

In several experiments not only chromosome aberrations were studied but also
sister chromtid exchanges. BrdU was added to the culture medium at a final concentration of 5 µg/ml during the entire cultivation period which was prolonged to 96
hours. Differential staining of chromatids was accomplished by the procedure of
Perry and Wolff[14] with Hoechst 33258 and Giemsa stain.

Cultures of patients and controls were set up simultaneously and in duplicate:
(i)   Blood from the patient was incubated for the establishment of the breakage
       rate. For these cultures the AB serum pool of the laboratory was used.
(ii)   Blood of a healthy donor was incubated with the serum of this patient and
       simultaneously with the normal AB serum for comparison of the breakage rates.
       In most experiments not only the clastogenic effect of the serum but also
       of ultrafiltrates of serum was tested. In the latter case, the quantity of
       ultrafiltrate represented one tenth of the total culture medium that was
       otherwise supplemented with AB serum.
(iii) The anticlastogenic effect of SOD was tested on cultures from patients as
       well as on cultures from healthy subjects exposed to the breakage factor
       in patient's serum or ultrafiltrates. Bovine Cu-SOD was added at a final
       concentration of 10 µg/ml medium.

In order to establish which serum fraction contains the chromosome breaking
agent, ultrafilters of different pore sizes were used. Since the activity passed
through 0.2 micron Millipore filters, the filtrate was passed through a Diaflo ultrafiltration system using XM 100, UM 10 and UM 2 membranes. Ultrafiltrates from
XM 100 and UM 10 possessed clastogenic activity, but not ultrafiltrates through
UM 2. As a routine, the serum was therefore passed through UM 10 and then concentrated twofold on UM 2. The ultrafiltrate from UM 2 was used as a control. The
breakage rate of the cultures treated with UM 2 ultrafiltrate was compared to that
in the cultures exposed to the concentrate.

In several experiments, lymphocyte cultures were set up after separation of
lymphocytes from the whole blood by standard Hypague-Ficoll gradient centrifugation.
The results were compared with simultaneous blood cultures. The cell density in

these lymphocyte cultures was 10^6/ml medium. In co-cultivations of lymphocytes
from patients and controls, the respective cell populations represented 50% each.
The breakage rate in the lymphocytes from the normal donor in these co-cultivations
was compared to that of his simultaneous lymphocyte culture. For the co-cultivations,
lymphocytes of a female patient and a male control subject were used. Only mitoses
in which the Y chromosome or 5 G-chromosomes were clearly identifiable, were scored
for aberrations.

Lymphocyte extracts were prepared by repeated freezing in liquid nitrogen and
thawing. Cell lysis was controlled under the microscope. A total of 5 x 10^7 lymphocytes separated by the Hypague-Ficoll method, was lysed in 10 ml culture medium.
After lysis of the cells the supernatant was passed through a UM 10 membrane and
concentrated by a second passage through UM 2. The clastogenic activity of these
extracts was tested in the same manner as the ultrafiltrates of serum.

RESULTS

The sera of 40 patients with progressive systemic sclerosis was tested for clastogenic activity on lymphocytes from normal subjects. The results are shown in Figure
1. The aberration rates observed in cultures set up with patient's serum was consistently higher than that in the cultures supplemented with AB serum from healthy
blood donors (P <0.001). The mean value of aberrations (7.5 per 100 cells, gaps
included) in these controls was higher than that of our present controls, since
these cultures were made with TCM 199 modified by the Pasteur Institute. This
medium gave the highest aberration rate among several tissue culture fluids tested,
probably due to its low content in L-cysteine and the presence of Tween 80 (Ref. 15).

Table 1 shows that the factor is retained by Diaflo UM 2 membrane but passes
through XM 100 and UM 10, the ultrafiltrates of which produced chromosome breaks.
Only one of 14 sera from patients with progressive systemic sclerosis was not clastogenic. On the contrary, only one of ten UM 2 ultrafiltrates from these patients
produced breakage. Similar results were obtained with the ultrafiltrates with five
patients with systemic lupus erythematosus and from five patients with rheumatoid
arthritis. This indicates that the molecular weight of the CF is between 10,000
and 1,000 daltons in all of these cases.

Figure 2 indicates the number of aberrations per 100 cells in blood cultures
from five patients with systemic lupus erythematosus and five patients with progressive systemic sclerosis (black columns). In general the breakage figures were
higher in lupus erythematosus and progressive systemic sclerosis than in rheumatoid
arthritis. The striped columns indicate the reduction of the aberration incidence
in simultaneous cultures to which 10 μg/ml of SOD were added. The difference between
treated and untreated cultures are highly significant (P <0.001). Individual results show regularly this protective effect of SOD.

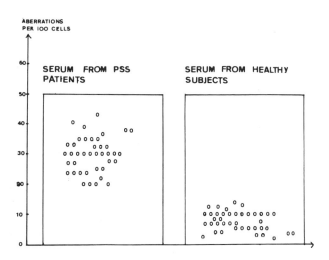

Fig. 1. Chromosome aberration rates in blood cultures supplemented with serum from patients with progressive systemic sclerosis (PSS) and serum from healthy blood donors.

TABLE 1

CLASTOGENIC EFFECT OF SERUM ULTRAFILTRATES

Ultrafiltrate[a]	Disease[b]	Breakage[c]	No Breakage[c]
	PSS	5	0
XM 100	SLE	5	0
	RA	5	0
	PSS	13	1
UM 100	SLE	5	0
	RA	5	0
	PSS	1	9
UM 2	SLE	0	5
	RA	0	5

[a] Diaflow ultrafiltration membrane as indicated.
[b] PSS = progressive systemic sclerosis; SLE = systemic lupus erythematosus; RA = rheumatoid arthritis.
[c] Number of observed results

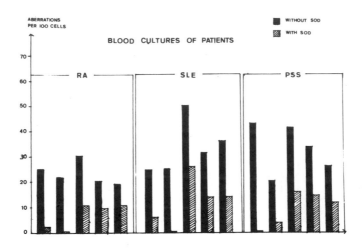

Fig. 2. Chromosome aberration rates in blood cultures from patients with rheumatoid arthritis (RA), systemic lupus erythematosus (SLE) and progressive systemic sclerosis (PSS) and the effect of adding SOD (10 μg/ml) to the culture medium.

In the five patients with rheumatoid arthritis, we also studied the effect of D-penicillamine on the breakage phenomenon (Table 2). In all five patients (but less clearly in patient 5) the number of aberrations were reduced. However, higher doses (100 μg/ml) were necessary than for SOD.

The aberration rate produced in blood cultures of healthy subjects by the CF from patients was also reduced by addition of SOD nearly to control values. The breakage incidence given in Figure 3 represents pooled data from five experiments for each disease. The individual results show consistently the protective effect of SOD.

In one patient with rheumatoid arthritis, we also studied the clastogenic effect of synovial fluid. It was found to be more significant than that produced by equal quantities of serum from this patient.

In six of seven experiments, in which the clastogenic effect of serum and the twofold concentrated ultrafiltrates (from UM 10) were studied simultaneously, the ultrafiltrate produced higher aberration rates.

The aberration rates were also found to be higher in blood cultures than in lymphocyte cultures (mean values from three patients, 40.0 and 27.0 per hundred mitoses respectively).

TABLE 2

CHROMOSOME ABERRATIONS IN BLOOD CULTURES FROM PATIENTS WITH RHEUMATOID
ARTHRITIS

Patient	Without D-PcA[a]	With D-PcA[a]
1	30.0	0.0
2	23.5	4.0
3	19.5	0.0
4	18.0	4.0
5	24.0	16.0

[a]Values are aberrations per 100 cells; D-PcA = D-penicillamine, 100 µg/ml when
added.

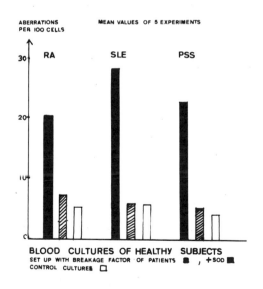

Fig. 3. Chromosome aberration rates in blood cultures of healthy subjects
incubated in the presence of the clastogenic factor (CF) from patients with
rheumatoid arthritis. (RA) systemic lupus arythematosus (SLE) and progressive
systemic sclerosis (PSS) (●); the aberration rates in simultaneous cultures
treated with CF but to which SOD was added (⊘); and Control cultures without
CF and without SOD (□).

If lymphocytes from a healthy subject were co-cultured with lymphocytes from patients with systemic lupus erythematosus (2 cases), the chromosome aberration rate increased in the control lymphocytes from 5 to 26 and 31 per cent respectively. No aberrations were observed in two co-cultivations set up with lymphocytes from two healthy subjects.

Lymphocyte extracts from two patients with systemic lupus erythematosus and one patient with rheumatoid arthritis ultrafiltered through a UM 10 membrane produced chromosome breaks, while two lymphocyte extracts from normal subjects had no clastogenic effect.

The types of aberrations observed in blood cultures from patients and in blood cultures from healthy subjects exposed to the CF are summarized in Table 3.

TABLE 3

SUMMARY OF OBSERVED CHROMOSOME ABERRATIONS[a]

	RA		SLE		PSS		Controls
	I	II	I[e]	II	I	II	III
Gaps and telomeric extrusions	10.0	8.7	9.1	9.6	11.1	8.0	2.3
Breaks	10.0	11.0	21.5	15.6	17.1	12.5	1.8
Fragments	1.8	0.8	1.2	0.4	2.3	1.5	
Dicentrics	0.5		0.4		0.5		
Rings	0.9						
Translocations			0.4	0.4	0.5		
Tri- and quadriradials				0.8			
Premature chromosome condensation			0.4	1.2	0.9		
Total	23.2	20.5	33.0	28.0	32.4	22.0	4.1

[a]Values are per 100 mitoses. RA = rheumatoid arthritis; SLE = systemic lupus erythematosus; PSS = progressive systemic sclerosis; I, blood cultures from patients; II, blood cultures from controls and clastogenic factor (CF) from patients; III, blood cultures from controls without CF.

The rate of sister chromatid exchanges (SCE) was increased in the three patients with progressive systemic sclerosis studied (0.16, 0.17 and 0.18 per chromosome) compared to 0.07, 0.08 and 0.1 in the simultaneous controls of normal subjects. Also the CF from these patients produced an increase in SCEs in blood cultures from normal subjects (0.13, 0.15, 0.16). This was the case also for the CF of two lupus erythematosus patients (0.17 and 0.19). In the second patient the UM 10 ultrafiltrate produced more SCEs (0.26). A simultaneous culture to which SOD was added together with the CF and BrdU, had a SCE frequency of only 0.13 per chromosome.

DISCUSSION

Increased chromosomal breakage is a constant finding in patients with progressive systemic sclerosis and systemic lupus erythematosus and is related to a transferable chromosome breaking agent that has also a clastogenic effect on chromosomes of normal lymphocytes. In about one third of patients sent to the laboratory with the diagnosis of rheumatoid arthritis, we did not find increased aberration rates. In the patients exhibiting chromosome breakage, the chromosome breaking factor was consistently present. However, in two patients only the concentrated UM 10 ultrafiltrate, but not the serum, produced significantly increased breakage rates. We are probably dealing with quantitative rather than qualitative differences between rheumatoid arthritis patients.

It is not possible to say at present whether the chromosome breaking agent is identical in the three connective tissue diseases studied. However, in each of these diseases, the substance has a low molecular weight, between 1,000 and 10,000 daltons. The activity of the partially purified factor is destroyed by heating above $50^{\circ}C$, but is conserved over long periods at $-30^{\circ}C$. Preliminary results indicate that exposure of the factor to pancreatic and mold ribonuclease for 3 hours at $37^{\circ}C$ reduces its clastogenic effect considerably, while proteolytic enzymes do not influence its activity. The fact that not only serum and ultrafiltrates of serum produce chromosome breaks in normal lymphocytes, but also that lymphocyte extracts behave similarly, as does contact with patients' lymphocytes in co-cultivations, is an indication that the CF is released from cells into the serum. The protective effect of SOD is an indirect proof that the O_2^- radical anion is involved in the breakage phenomenon. A previous study[7] demonstrating the efficacy of the radioprotector L-cysteine already suggested the role of free radicals.

Activated oxygen species are produced during the univalent pathway of oxygen reduction in all aerobic cells. During phagocytosis a reduced pyridine nucleotide oxidase is responsible for transferring electrons to oxygen to generate superoxide, hydrogen peroxide and hydroxyl radicals in a sequential fashion according to the following reactions:

$$O_2 + e \longrightarrow O_2^-$$
$$O_2^- + O_2^- + 2H^+ \longrightarrow H_2O_2 + O_2$$
$$H_2O_2 + O_2^- \longrightarrow OH^{\cdot} + OH^- + O_2$$

Enzymatic defences have been developed by the cell against these toxic intermediates: superoxide dismutases convert the O_2^- radical to hydrogen peroxide and molecular oxygen, while catalase and gluthathione peroxidase remove H_2O_2. The most toxic radical is perhaps the OH˙ radical for which no efficient enzymic scavenger has been identified.

Activated oxygen species play a major role in inflammation. During phagocytosis of microorganisms by polymorphonuclear neutrophils (PMNs), O_2^- is produced by a membrane bound superoxide synthetase. Under conditions of inflammation the O_2^- produced by the PMNs can escape into the cellular environment and, if excessive, can destroy normal cells. Such mechanisms could account for synovial cell lesions and synovial fluid deterioration in inflammatory arthritic conditions as suggested by McCord.[16] In this context, it is noteworthy that we found synovial fluid from rheumatoid arthritis patients to be clastogenic.

Because of the production of O_2^- by granulocytes, we wondered if their presence in the cultures set up with whole blood was responsible for the production of chromosome breaks. However, aberrations were also found in culture of separated lymphocytes at a slightly lower rate, as well in culture from patients as in those of controls exposed to the CF. Resting lymphocytes are known to produce only minimal quantities of O_2^-, but our preliminary results (5 experiments) indicate that the production of O_2^- by phytohemagglutinin stimulated lymphocytes increases under the influence of the CF, as measured by nitroblue tetrazolium reduction.

Superoxide and also OH˙ radicals react with macromolecules vital to the cell. The mutational effects of O_2^- producing systems on bacteriophage T_4 and the protection against such mutations given by SOD has been previously shown.[17] The production of single strand breaks in DNA has been reported by White et al.[18] We have shown that O_2^- produces chromosome breaks and rearrangements, if it is generated photochemically (photoreduction of flavins) or enzymatically (hypoxanthine plus xanthine oxidase) in the culture medium of normal lymphocyte cultures.[9] These observed chromosome mutations are an indication of lesions occurring at the molecular level. The modified DNA becomes antigenic and the production of DNA antibodies in connective tissue disease could be explained by analogy to that observed in patients with xeroderma pigmentosum. The observation that a concomitant decrease in antibody production and chromosome breakage rate was obtained in several patients with systemic lupus erythematosus and rheumatoid arthritis after a two week treatment with SOD suggests that the hypotheses outlined above may have a certain value as a basis for the medical application of SOD in this area (see J. Emerit et al., this volume).

A reason for increased production of O_2^- in patients with progressive systemic sclerosis, lupus erythematosus and rheumatoid arthritis remains to be determined. If our preliminary results indicating increase in the production of O_2^- by stimulated lymphocytes exposed to the CF are confirmed by further studies with the lymphocytes from patients, the resolution of this problem will be concomitant with the biochemical identification of the chromosome breaking agent. Risler et al.[20] have reported a superoxide dismutase deficiency of PMNs in juvenile rheumatoid arthritis. A decreased potential for radical scavenging could also explain the breakage phenomena since it would result in a higher steady state level of O_2^- radicals. The SOD content of lymphocytes from normal subjects and patients with systemic lupus erythematosus is presently under study. The hypothesis that the CF is an inhibitor of SOD has been tested but shown to be incorrect.

Since radiation induced chromosome breaks are probably due to the action of free radicals, it is of interest that whole body gamma or x-irradiation also gives rise to a clastogenic factor in plasma.[21,22] This is confirmed in our test system in cases of industrial accidents in which the subjects received very high doses of gamma irradiation (> 6000 rads). A clastogenic factor is present in their sera. In vitro experiments showed that gamma-irradiation of human (or bovine) whole blood, but not of plasma produces or liberates such a factor at relatively low levels of irradiation (400 - 1200 rads) (Refs. 23, 24). This breakage factor has not as yet been identified but it appears to be similar to the breakage factor in patients with progressive systemic sclerosis, systemic lupus erythematosus and rheumatoid arthritis, since preliminary results indicate that the molecular weight is also between 1,000 and 10,000 daltons.

New Zealand Black (NZB) mice develop spontaneously an immune complex nephritis similar to human lupus nephritis, as well as connective tissue lesions with enhanced production of DNA antibodies. This murine form of systemic lupus erythematosus is also accompanied by increased chromosome breakage and the presence of a breakage factor in serum and in supernatants of fibroblast cultures, that produces chromosome breaks in human lymphocytes.[25] This breakage factor has also a molecular weight of less than 10,000 daltons, thus excluding that the endogenous xenotropic C-type virus characteristic of NZB mice is identical with the factor. However, the possibility that the factor is related to the chronic virus release from the NZB mouse genome has to be considered.

Recently an increase in sister chromatid exchanges was reported in normal fibroblasts co-cultivated with fibroblasts from patients with Bloom's syndrome.[26] The exact nature of this factor produced by BS cells is not known.

McCord[27] has discovered a plasma component which is itself inactive or weakly chemotactic but after activation by O_2^- becomes very strongly chemotactic with respect to PMNs. Thus, where there is activation of this factor due to a local production of O_2^-, there is migration of the PMNs to the site of the lesion. We are presently investigating whether the CF of patients with connective tissue disease influence lymphocyte migration.

Increased understanding of the phenomenon of chromosomal breakages in collagen diseases could help to elucidate their origin, and it can be suggested treatment of collagen disease patients with SOD could be beneficial due to the scavenging of O_2^- radicals responsible for initiation of inflammation, damage to connective tissue and damage to DNA.

394

REFERENCES

1. Emerit, I., Housset, E., De Grouchy, J. and Camus, J.P. (1971) Biomedecine, 16, 684-694.
2. Emerit, I., Feingold, J., Camus, J.P. and Housset, E. (1974) Ann. Génét. 17, 251-256.
3. Pan, S.F., Rodnan, G.P. and Wald, N. (1971) C. R. Congr. Int. Genet. Paris 1971, p. 136.
4. Emerit, I. (1976) Dermatologica, 153, 145-156.
5. Emerit, I. and Housset, E. (1973) Biomedecine, 19, 550-554.
6. Emerit, I., Levy, A. and Housset, E. (1973) Ann. Génét. 16, 135-138.
7. Emerit, I., Levy, A. and Housset, E. (1974) Human Genet. 25, 221-226.
8. Nordenson, I. (1977) Hereditas, 86, 147-150.
9. Nordenson, I. (1977) Hereditas, 87, 151-154.
10. Nordenson, I., Beckman, G. and Beckman, L. (1976) Hereditas, 82, 125-126.
11. Emerit, I., Emerit, J., Levy, A. and Keck, M. (1979) Hum. Genet. 408, 1-7.
12. Younes, M. and Weser, U. (1977) Biochem. Biophys. Res. Commun. 78, 1247-1253.
13. Lengfelder, E. and Elstner, E.F. (1978) A. Physiol. Chem. 757, 751-757.
14. Wolff, S. and Perry, P. (1974) Chromosoma, 48, 341-353.
15. Keck, M. and Emerit, I. (1979) Human Genet. 435, 1-7.
16. McCord, J.M. (1974) Science, 185, 529-531.
17. Michelson, A.M., unpublished observations.
18. White, J.R., Vaughan, T.O. and Shiang, Y.W. (1971) Fed. Proc. 30, 1145.
19. Emerit, I. and Michelson, A.M. (1979) C.R. 6th Int. Congr. Radiat. Res., Tokyo, May 1979½
20. Rister, M., Baumeister, K., Gravest, O. and Gladtke, E. (1978) Lancet, 1, 1094.
21. Goh, K.O. and Summer, H. (1968) Radiat. Res. 35, 171-174.
22. Hollowell, J.G. and Littlefield, L.G. (1968) Proc. Soc. Exp. Biol. Med. 129, 240-243.
23. Scott, D. (1968) Cell Tissue Kinet. 2, 295-298.
24. Emerit, I. and Michelson, A.M., unpublished observations.
25. Emerit, I., Levy, A. and De Vaux Saint Cyr, Ch. (1979) Chromosome damaging agent of low molecular weight in the serum of New Zealand Black mice. Cytogenet. Cell Genet., in press.
26. Tice, R., Windler, G. and Rary, J.M. (1978) Nature, 273, 538-540.
27. McCord, J.M. (1979) in Active oxygen and medicine, Autor, A., ed., Raven Press, New York, in press.

SUPEROXIDE DISMUTASE PHARMACOLOGY AND ORGOTEIN EFFICACY: NEW PERSPECTIVES

WOLFGANG HUBER, M.G.P. SAIFER AND L.D. WILLIAMS
Diagnostic Data, Inc., Mountain View, California 94043, U.S.A.

INTRODUCTION

Drug versions of bovine and other Cu-Zn superoxide dismutases (SOD) were classified generically as orgotein in 1971 (Ref.1) and evdience to show that orgotein is an effective non-analgesic anti-inflammatory drug of outstanding safety has been presented.[2-4] Pharmacologically, we have found orgotein to have three specific effects, i.e., it is anti-inflammatory, anti-superoxide, and, for herpes simplex and parainfluenza-3, anti-viral. Otherwise, it is remarkably inert (Table 1). Its lack of analgesic activity sets it apart from other anti-inflammatory drugs. Yet, orgotein is clinically remarkably effective in the relief of pain, an effect which must accrue as a sequel to disease amelioration rather than suppression of symptoms.

TABLE 1

SOME PHARMOCOLOGIC EFFECTS OF ORGOTEIN

Orgotein is	anti-inflammatory anti-superoxide (O_2^-) anti-viral[a]
Orgotein is not	analgesic antipyretic immunosuppressant infection promoting
orgotein does not	potentiate effects of barbiturates or alcohol interfere with wound healing

[a]Herpes simplex; parainfluenza-3.

Following the initial observation by Babior et al.[5] it has now become clear that, at least in vitro, all phagocytosing cells generate superoxide radical which can be found in the surrounding medium.[6,7] As phagocytosing cells are invariably found at sites of both acute and chronic inflammation, these findings seemed to us to provide a potential link between anti-inflammatory effects of orgotein and its SOD activity. We reasoned that by neutralizing the toxic extracellular effects of superoxide radicals, the exogenous SOD activity could help maintain the integrity and normal function both of circulating phagocytes

and of adjacent tissue cells, and with it accelerate repair, as well as retard
spillage of lysosomal inflammants into the diseased sites.

Our efforts in this area were thus directed towards exploring whether, and if
so how, the SOD function of the orgotein molecule should be invoked as a principal in vivo action mechanism to explain its clinical efficacy. This efficacy
has been demonstrated in double-blind trials by us and others in a variety of
spontaneous and induced chronic and acute diseases in man and animals.[8-19]

With the clinical use of orgotein in mind, we have explored in blinded trials
whether and how the route of administration and/or the dosing distribution in
time are related to efficacy in animal models of induced inflammation. For
these studies we have used, among others, the carrageenan-induced paw edema in
the mouse (CPEM) and an antiserum-induced skin edema, the reverse Arthus reaction (RAR).

RESULTS

In order to correlate observed anti-inflammatory efficacy to circulating or
in situ SOD activity, it is necessary to know the absorption, distribution,
and fate of orgotein after injection by various routes. To assess this, we have
used several methods, such as measurement of bovine SOD activity by NBT-riboflavin stained zymograms after electrophoresis, radiographic tracing of orgotein
labeled with [99M]Technetium, and a radioimmunoassay. Data obtained by these
methods on the bioavailability of orgotein after administration by several
routes has been presented recently.[20] A summary of these data is presented in
Table 2., showing that serum clearance kinetics of orgotein can be strikingly
different for different routes of administration. Thus, after s.c. and i.m.
administration clearance of circulating SOD activity requires 6-7 hrs to
reduce the serum concentration of bovine Cu-Zn SOD activity to <10% of that
of the peak concentration attained at about 2 hrs. Also, 90% clearance of SOD
activity from a subcutaneous injection site required more than eight hrs. In
contrast, clearance after intravenous administration has a $t_{1/2}$ close to 6
minutes so that 99% is cleared in less than 1 hr.

Our pharmacokinetic work has shown that orgotein accumulates in the cortex
of the kidney and thereafter loses its SOD activity.[2,20] Even at the highest
clinical doses used so far (0.5 mg/kg) no measurable bovine Cu-Zn SOD activity
is found in the urine, be it in man or animal. In addition, we have shown that
clearance of orgotein from the kidney and other tissues was not influenced by
prior administration even at much higher doses, i.e., there was no accumulation
of orgotein. Also, spillage of orgotein into the urine was found to occur only
at very high doses and then, to be both dose and route dependent.

Thus, all our pharmacokinetic and bioavailability data shows that by routes
of administration other than i.v., clinically effective concentrations of bovine

TABLE 2

BIOAVAILABILITY OF ORGOTEIN SOD-ACTIVITY IN ANIMALS AND MAN[a]

Serum Level	Hours after Injection	
	i.v.	s.c.; i.m.
Maximum	At time of inj.	2-3
<10% of maximum	0.3-1.0	6-10
$t_{1/2}$[b]	0.10-0.25	3-4
	(0.5)[c]	(5-6)[c]

[a]Mouse, rat, guinea pig, dog; Thin-gel agarose electrophoresis (250 V, 10 mA, 15-20 min) followed by NBT-Riboflavin zymograms

[b]Initial half-life for i.v.; "half-life" after peak concentration for s.c. and i.m.

[c]Values for Man.

Cu-Zn SOD activity can be maintained in the circulation for several hours after parenteral adminstration of the native enzyme.

Using animal models of carrageenan and antiserum-induced inflammation, we find in addition that the efficacy of a single orgotein injection lasts well beyond the point when the peak serum activity of bovine Cu-Zn SOD has dropped from about 1 mcg/ml to below 10 ng/ml, the sensitivity endpoint of presently available methods. To demonstrate this long-lasting anti-inflammatory efficacy, we have administered orgotein at constant doses at different intervals prior to and/or after administration of carrageenan or antiserum to the animals. In the CPEM assay, the time range covers 22 hrs before to 20 hrs after plantar carrageenan administration, with sacrifice and edema measurement always at 24 hrs after carrageenan. In the standard RAR, sacrifice occurs 2 hrs after antiserum administration; hence, in this model, the orgotein dosing-time range covers 22 to 0.1 hrs prior to antiserum

Our results show clearly that inhibtion of edema in the RAR was the same whether subcutaneous orgotein is administered at 22, 2 or 0.1 hrs prior to the intradermal injection of antiserum (Table 3). The pattern is the same for the CPEM (Table 4), i.e., subcutaneous administration of orgotein 22.5, 5.0 and 1.5 hrs prior to plantar injection of carrageenan produced the same degree of edema inhibition (32.1 ± 3.7%) within the limits of error of the assay.

Orgotein administered s.c. or i.m. at 2 and 20 hrs after carrageenan also shows equal edema inhibition, though at a somewhat higher level of efficacy (42.3 ± 7.2%).

We have also explored the influence of the route of orgotein administration (s.c., i.m., i.p., i.v.) at constant times on its efficacy in both assay systems. In the RAR, orgotein when given at 2 hrs prior to antiserum injection produces the same three point dose-response whether given s.c., i.m., or i.v. (Table 5).

TABLE 3

DOSING TIME VS. EFFICACY OF SUBCUTANEOUS ORGOTEIN IN ANTISERUM-INDUCED SKIN
EDEMA IN THE RAT[a]

No. of Expts.	Orgotein Dosing Hrs. prior to antiserum	Dose[b] mg/kg	Source	% Inhibition Mean ± SD[c]
5	-2	0.6	Bovine liver	46.8 ± 10.5
4	-22	0.6	"	46.1 ± 10.3
2	-0.1	0.6	"	41.8 ± 9.8

[a]Sprague-Dawley rates (4-6/group) 155-176 g., aged 42-47 days; 0.05 ml antiserum or IgG Sorb-5% intradermally into each of 4 dorsal sites; sacrifice by
$CHCl_3$ 2 hours after antiserum administration; skin edema sites punched out and
weighed.

[b]Orgotein injected in 0.2-0.4 ml USP saline.

[c]Calculated using the results for each experiment; intra-experiment results are
statistically highly significant (P < 0.01-0.001; double-tailed test).

TABLE 4

DOSING TIME VS. EFFICACY OF A CONSTANT ORGOTEIN DOSE GIVEN BY DIFFERENT ROUTES
ON THE CARRAGEENAN-INDUCED FOOT EDEMA IN THE MOUSE[a]

Injection Time vs. Carrageenan	% Inhibition of Edema Relative to Saline Control Mean ± SD[c]					
Hours	s.c/i.m	N	i.v.	N	i.p.	N
-22.5	34.3 ± 4.7	3	nd	–	nd	–
- 5.0	29.9 ± 3.0	2	nd	–	nd	–
- 1.5	32.0 ± 3.3	3	28.6 ± 4.4	2	nd	–
+ 2.0	41.8 ± 5.4	4	49.8 ± 9.7	2	25.2 ± 5.3	2
+20.0	42.9 ± 8.9	14	38.0 ± 9.8	4	37.8 ± 7.0	2

[a]Swiss-Webster mice (8-10/group), 20-23 g, aged 42-45 days; 0.03 ml 1% carrageenan
injected into plantar area of richt hind paw; sacrifice by $CHCl_3$ 24 hr. after
carrageenan administration; both hind paws amputated at joint and weighed;
orgotein (4.7 mg/kg) injected in 0.1-0.3 ml USP/saline.

[b]Calculated using the results for each experiment; intra-experiment results are
statistically highly significant (P < 0.01-0.001).

In the CPEM, influence of route of administration upon edema suppression has
been investigated also at varying orgotein doses administered s.c., i.m., i.v.,
and i.p., respectively, with administration time either constant or varying
(Tables 4 and 6). The results for s.c. and i.m. administration were nearly
identical and have been combined for greater clarity of presentation in view of
their very similar bioavailability. It is clear that at a constant orgotein
dose (4.7 mg/kg) its administration at 1.5 hrs prior or 2 or 20 hrs after
carrageenan produces the same or very similar edema inhibition, regardless
whether the route is s.c., i.v., or i.p. At constant time, i.e., 20 hrs after

TABLE 5

DOSE RESPONSE IN THE ANTISERUM-INDUCED SKIN EDEMA (REVERSE ARTHUS) IN THE RAT TO
ORGOTEIN GIVEN BY DIFFERENT ROUTES TWO HOURS BEFORE ANTISERUM ADMINISTRATION[a]

Orgotein Dose mg/kg[b]	% Inhibition of Edema Relative to Saline Control Mean ± SD[c]					
	s.c	N	i.m.	N	i.v.	N
0.20	49.0 ± 6.9	3	46.8 ± 4.1	2	45.9 ± 7.9	2
0.40	51.9 ± 5.8	2	47.4 ± 6.5	2	nd[c]	–
0.80	51.3 ±14.8	5	58.1 ± 6.2	2	52.0 ± 5.8	2

[a]Sprague-Dawley rats (4-6/group), 155-176 g., aged 42-47 days; 0.05 ml antiserum
or IgG Sorb 5% intradermally into each of 4 dorsal sites; sacrifice by $CHCl_3$ 2 hr.
after antiserum administration; skin edema sites punched out and weighed.

[b]Orgotein injected in 0.2-0.4 ml USP saline.

[c]Calculated using the results for each experiment; intra-experiment results are
statistically highly significant ($P < 0.01$-0.001).

[d]Not determined.

carrageenan, increasing doses if orgotein (1.2, 2.8, 4.7, 7.0 mg/kg) produce
a clear-cut response of a magnitude which was closely similar, irrespective
whether administration is s.c., i.v., or i.p.

TABLE 6

DOSE RESPONSE IN THE CARRAGEENAN-INDUCED FOOT EDEMA IN THE MOUSE TO ORGOTEIN
GIVEN BY DIFFERENT ROUTES TWENTY HOURS AFTER CARRAGEENAN ADMINISTRATION[a]

Orgotein Dose mg/kg	% Inhibition of Edema Relative to Saline Control Mean ± SD[b]					
	s.c./i.m.	N	i.v.	N	i.p.	N
1.2	23.7 ± 4.2	11	26.3 ± 4.4	2	22.8 ± 5.7	3
2.8	37.0 ± 6.6	7	34.5 ±15.7	4	42.7 ± 6.2	2
4.7	43.4 ± 4.1	4	nd[c]	–	nd[c]	–
7.0	47.1 ± 8.1	5	41.7 ± 8.9	2	47.5 ± 5.3	3
18.8	55.8 ± 1.2	2	nd[c]		nd[c]	

[a]Swiss-Webster mice (8-10/group), 20-23 g., aged 42-45 days; 0.03 ml 1% carrageenan
injected into plantar area of right hind paw; sacrifice by $CHCl_3$ 24 h. after carrageenan administration; both hind paws amputated at joint and weighed; orgotein
injected in 0.1-0.3 ml USP saline.

[b]Orgotein injected in 0.2-0.4 ml USP saline.

[c]Calculated using the results for each experiment; intra-experiment results are
statistically highly significant ($P < 0.01$-0.001).

[d]Not determined.

Thus, in every experimental modality used in these two representative animal
models of induced inflammation our results clearly show that the presence of
circulating, enzymatically active bovine Cu-Zn SOD is not relevant for the
efficacy of edema suppression by orgotein.

The statistically highly significamt anti-inflammatory efficacy of orgotein
in these animal models of induced inflammation is particularly noteworthy in

view of the fact that we find that systematic orgotein administration in vivo causes, in the normal animal, a pronounced mobilization of polymorphonuclear neutrophils (PMNs) into the circulation. In addition, upon local administration into areas such as the peritoneal cavity, the joint, the skin and the anterior chamber of the eye large numbers of PMNs accumulated at these sites in response to the orgotein injection. It is remarkable that such a pronounced local accumulation of activatable PMNs does not become a pro-inflammatory event. This may be by virute of the fact that orgotein protects these PMNs against lysis and/or lysosomal leakage in situ.

To further explore this PMN mobilization, we have studied the simultaneous effects of intraperitoneally-administered orgotein in the peritoneal cavity and the peripheral blood by differential counts of white cells. Placebo and sham controls were used to account for non-orgotein related changes. Using the same system, responses to another bovine protein (bovine serum albumin, BSA) were also studied. Our results show that in the rat, i.p. injection of orgotein mobilizes PMNs from sources remote from the injection site, resulting in markedly increased peripheral blood PMN counts in addition to a pronounced influx of PMNs into the peritoneal cavity. We find that the disappearance of orgotein from the peritoneal cavity follows first-order kinetics to at least 6 hrs, when more than 99% of a 3 mg dose has been cleared (Table 7). At this dose, the orgotein content of the serum was highest between 1 and 2 hrs after i.p. orgotein injection, i.e., 32 mcg/ml at 2 hrs. By 4 hrs after the injection, the orgotein in the serum had dropped to an average of 6.7 mcg/ml. There is only 0.9 mcg/ml at 6 hrs, and at 24 hrs there is no detectable orgotein in the serum. Peak PMN concentrations lag behind the peak orgotein concentrations by several hours, with the maximum cell count in the peritoneal cavity preceeding that in the blood.

After i.p. orgotein injection the PMNs are dramatically increased to reach a level of $> 10^7$ cells per ml in the peritoneal cavity. The maximum influx of PMNs into the peritneal cavity occurs 4 to 6 hrs after the orgotein content in the cavity has begun to diminish. Bovine serum albumin also causes a significanr increase in PMNs in the peritoneal cavity, but with somewhat different kinetics. In contrast, only a very small number of PMNs are present at 4 or 6 hrs after i.p. placebo injections and intraperitoneal placebo causes no change in the PMN count up to at least 6 hrs after administration. There is no difference in the number of lymphocytes in the peritoneal cavity after either orgotein, placebo or BSA injection.

In the blood, intraperitoneal orgotein causes a concurrent increase in the number of PMNs, with the maximum PMN count (>100% increase) being reached at around 2 hrs after peak serum concentration. By 4 to 6 hrs the count is declining and at 24 hrs it approaches baseline values (Table 8). The blood PMN count 4 hrs after i.p. orgotein is significantly higher than that after placebo

TABLE 7

MEAN LEUKOCYTE COUNTS IN PERITONEAL EXUDATES OF RATS BEFORE AND AFTER INTRA-
PERITONEAL INJECTION OF ORGOTEIN, PLACEBO AND BSA[a]

Leucocyte Counts per ml (x 10^{-6})	0	Hours 4			6		
		Orgotein	Placebo	BSA	Orgotein	Placebo	BSA
Total WBcs	11.0	24	11	56	23	9.9	17
PMNs	0.3	12[c]	0.1	19[c]	12[c]	0.7	2.9[d]
Lymphocytes	9.4	10	9.1	31	16	9.2	11
[Orgotein disappearance (mcg)	3,000	179 ± 70 (5.9%)			23 ± 14 (0.75%)]		
[BSA disappearance (mcg)	6,000[b]	nd[e]			nd[e]		]

[a]Fisher-344 male rats, 95-155 g., 5-6/group; orgotein and BSA injected i.p.
(needle 27 G) in 1.8 ml USP saline; after sacrifice with ether, 10 ml 3.8% Na-citrate
pH 7.0 injected (needle 27 G) into cavity; fluid withdrawn by Pasteur pipette
after gentle but thorough massaging of cavity; recovered fluid (9.0-9.8 ml) spun
8-10 min.; orgotein recovery determined in supernatant (NBT-Zymogram); pellet
gently suspended in 50-75 mcl fetal calf serum; smears stained with May-Grünwald-
Giemsa.

[b]Approx. equimolar to orgotein.

[c]$P < 0.05$

[d]$P < 0.01$

[e]Not determined.

and in sham controls (Table 9), while there is no significant difference in PMN
behaviour in blodd between placebo-treated animals and sham controls. However,
sham controls and placebo recipients do show an increase in PMNs at 24 hrs.
Such an increase is not seen in the orgotein-injected animals, with the difference being statistically significant (Table 10). Contrary to the effect in the
peritoneal cavity, BSA recipients behave like placebo or sham-treated rats,
showing no significant increase in peripheral blood PMNs from 1 to 6 hrs after
injection.

In orgotein-treated animals, as well as in placebo and sham controls, the
lymphocyte count in the blood (Table 11) drops generally below the zero hour
level at 2, 4 and 6 hrs after injection. It is not clear whether this is an
effect of the procedures to which the rats are subjected or a diurnal variation.
At 24 hrs after i.p. orgotein,, the lymphocytes are at the zero hour level. In
contrast, the lymphocyte count in the placebo and the sham-control animals is
significantly higher than in the placebo controls. This effect may well be due
to stress of the procedures employed on the rats, an event that seems to be
normalized in the orgotein recipients. In other words, the rats respond to the
injection and bleeding procedures employed by increasing PMN and lymphocyte
counts at 24 hrs after the start of the experiment. Orgotein injection appears

TABLE 8

INCREASE OVER BASELINE VALUES OF PERIPHERAL PMN COUNTS IN RAT BLOOD AFTER INTRAPERITONEAL INJECTION OF ORGOTEIN, BSA AND PLACEBO[a,b]

	Hours after i.p. Injection			
	2	4	6	24
Orgotein[c]	1.47 ± 0.40 (20)[g]	2.20 ± 0.63 (22)	1.52 ± 0.51 (13)	1.23 ± 0.13 (4)
Placebo[d]	1.01 ± 0.19 (16)	0.96 ± 0.25 (15)	1.00 ± 0.20 (10)	1.65 ± 0.21 (2)
BSA[e]	1.16 ± 0.18 (4)	1.14 ± 0.24 (4)	1.06 ± 0.16 (4)	nd[h]
Sham[g]	1.06 ± 0.29 (8)	1.09 ± 0.33 (7)	1.15 ± 0.23 (8)	2.11 ± 0.54 (5)

[a]Increase: $\dfrac{\text{Mean counts at 2,4,6 or 24 hrs}}{\text{Mean counts at 0 hrs}}$

[b]Free flowing blood obtained at 0, 2, 4, 6 and 24 hrs by razor cuts of tail, preheated in wamr water; tail protected by surgical tape between cuts; smears stained by May-Grünwald-Giemsa for differential counting.

[c]3.0 mg protein with 6.0 mg sucrose USP in 1.8 ml saline injection USP

[d]9.0 mg sucrose USP in 1.8 ml saline injection USP

[e]6.0 mg BSA (Pentex, Cryst.) which is approx. equimolar to the orgotein, in 1.8 ml saline injection USP

[f]Uninjected; bled only.

[g]Number of experiments in parentheses.

[h]Not determined.

TABLE 9

CHANGE FROM BASELINE OF MEAN LEUKOCYTE AND DIFFERENTIAL COUNTS IN PERIPHERAL BLOOD OF RATS AFTER INTRAPERITONEAL INJECTIONS OF ORGOTEIN AND PLACEBO[a]

Leukocytes	Hours	
	2	4
Total WBCs		
Orgotein	0.94 ± 0.11	1.30 ± 0.10[b]
Placebo	0.85 ± 0.19	0.80 ± 0.18
PMNs		
Orgotein	1.40 ± 0.42[b]	2.06 ± 0.31[c]
Placebo	0.90 ± 0.17	0.86 ± 0.06
Lymphocytes		
Orgotein	0.79 ± 0.08[b]	0.96 ± 0.15
Placebo	0.84 ± 0.19	0.89 ± 0.26

[a]Fisher-344 male rats, 95-115 g., 5-6/group; orgotein (3mg + 6mg sucrose stabilizer) and placebo (9 mg sucrose) injected i.p. (needle, 27 G) in 1.8 ml USP saline.

[b]$p < 0.05$ Versus baseline and/or placebo values

[c]$p < 0.01$

TABLE 10

STATISTICAL ANALYSIS OF CHANGES IN PERIPHERAL PMN COUNTS IN RATS AFTER INTRA-
PERITONEAL INJECTION OF ORGOTEIN, BSA AND PLACEBO[a]

| | Hours after i.p. Injection | | | | | | | |
| | 2 | | 4 | | 6 | | 24 | |
	Δ of Means	P	Δ of Means	P	Δ of Means	P	Δ of Means	P
Orgotein/Placebo	0.46	<0.001	1.14	<0.001	0.52	<0.01	−0.42	<0.05
Orgotein/uninjected	0.41	<0.02	1.01	<0.001	0.37	<0.01	−0.88	<0.02
Orgotein/BSA	0.31	<0.2	0.96	<0.01	0.46	<0.01	nd[b]	
BSA/Placebo	0.15	ns[c]	0.18	ns	0.06	ns	nd	
Placebo/Uninjected	0.05	ns	−0.13	ns	−0.15	ns	−0.46	ns

[a]Student's t-test

[b]Not determined

[c]Not significant

to diminish these responses. Since placebo-injected and uninjected rats have
very similar PMN and lymphocyte responses, the stimulus for the observed leukocytosis at 24 hrs must be the tail-bleeding procedure rather than the injection
procedure.

TABLE 11

INCREASE OVER BASELINE VALUES OF PERIPHERAL LYMPHOCYTE COUNTS IN RAT BLOOD AFTER
INTRAPERITONEAL INJECTION OF ORGOTEIN AND PLACEBO[a,b]

| | Hours after i.p. Injection | | | | |
	2	4	6	24	48
Orgotein[c]	0.79 ± 0.08 (20)[f]	0.96 ± 0.15 (22)	0.86 ± 0.19 (13)	0.98 ± 0.15[g] (7)	0.98 ± 0.17 (3)
Placebo[d]	0.84 ± 0.19 (16)	0.89 ± 0.26 (15)	0.86 ± 0.21 (10)	1.16 ± 0.10[g] (5)	1.11 ± 0.28 (3)
Sham[e]	0.73 ± 0.25 (8)	0.83 ± 0.22 (7)	0.79 ± 0.18 (8)	1.31 ± 0.10[g] (5)	nd[h]

[a]Increase: $\frac{\text{Mean Counts at 2,4,6 or 24 hrs}}{\text{Mean Counts at 0 hrs}}$

[b]Free flowing blood obtained at 0,2,4,6,24 and 48 hrs by razor cuts of tail,
preheated in warm water; tail protected by surgical tape between cuts; smears
stained by May-Grünwald-Giemsa for differential counting.

[c]3.0 mg protein with 6.0 mg sucrose USP in 1.8 ml saline injection USP

[d]9.0 mg sucrose USP in 1.8 ml saline injection USP

[e]Uninjected; bled only

[f]Number of experiments in parentheses

[g]Orgotein vs. placebo: P <.02; orgotein vs. sham: P <.01; placebo vs. sham:
P <.05; t-test, significance of difference between means.

[h]Not determined

DISCUSSION

The results presented above permit a number of conclusions relevant to the mechanism(s) by which orgotein exerts its drug effects.

Our bioavailability studies show that clearance rates of SOD-active orgotein from tissues and circulation depend markedly on the route of administration. After s.c. or i.m. administration clinically significant levels of local or systemic, enzymatically active Cu-Zn bovine SOD are maintained for several hours, while after i.v. administration such levels are maintained only for minutes, even though the kidney is the organ of clearance in both instances.

The anti-inflammatory efficacy of orgotein persist far longer than that indicated by the bioavailability of native bovine Cu-Zn enzyme activity when given s.c. or i.m. Clearly, the statistically highly significant anti-inflammatory effects of orgotein in two representative animal models of induced inflammation can not be ascribed to the presence many hours (up to 46 hrs) after administration of effective quantities of enzymatically active bovine Cu-Zn SOD in the circulation or at sites of injury.

The situation in vitro appears to be different than in vitro. Salin and McCord[21] have shown 80% protection in vitro of human PMNs against death by 0.3 mg/ml of native bovine Cu-Zn SOD in the incubation mixture. This protection manifested itself, however, only if the exogenous SOD was present during the 20-30 minutes of the oxygen burst following stimulation of the cells by endotoxin. Our in vivo results show that the presence of measurable bovine Cu-Zn SOD activity at the time of induction of inflammation and the attendant oxygen burst of the phagocytosing cells is not essential for the efficacy of suppression by orgotein of the resulting edema.

We find that the efficacy of edema suppression by orgotein in the CPEM and RAR models is for all practical purposes independent of route, be it s.c., i.m., i.p. or iv. Concerning efficacy after intravenous administration, our results are in contrast to those by McCord and Wong,[22] who reported a complete lack of edema suppression by native bovine Cu-Zn SOD in the rat, both in the reverse passive Arthus reaction and the carrageenan-induced foot edema. In the latter model, species difference (rat vs. mouse) could provide a rationale for the difference in results were it not for the fact that Oyanagui[23] has demonstrated efficacy for the i.v. route in the rat. We have at present no explanation for the startling divergence between theirs and our observations for the i.v. route in the RAR model.

That the anti-inflammatory effects or orgotein dosing can be long lasting has been further substantiated in double-blind trials in humans, both in inflammatory rheumatoid and osteoarthritis. The reality of this effect was shown both by the continuing clinical efficacy of intermittent dosing (once per week) and the maintenance of the clinical response of orgotein patients during a no-medication period (1 month) in contrast to placebo patients who deteriorated during this interval (Ref.20; see also L. Flohé et al., this volume).

Overall it appears that the edema suppressing effects of orgotein in the two animal models and the anti-inflammatory effects seen in the clinic are functions of the orgotein molecule, which at least in part are not related to its SOD activity.

The mobilization of PMNs into the circulation from reservoirs distant from the site(s) of injection is an interesting and somewhat unexpected discovery, which may provide valuable insight into the action mechanism of exogenous SOD.

In our experiments, i.p. injection of orgotein produced a transient increase of PMNs in the peripheral blood, which was highly statistically significant 4 and 6 hrs after injection and had returned to nearly normal at 24 hrs. The large number of PMNs which enter the peritoneal cavity do not deplete the peripheral blood of neutrophils, as the number of PMNs in the blood doubles between zero and 4 hrs after i.p. orgotein injection. The kinetics of appearance of PMNs in the peritoneal cavity and blood are not simply related to the concentration of orgotein present. The peak number of peritoneal PMN occurs several hours after orgotein injection when its content in the peritoneal cavity has dropped to a few percent of the initial level. In the blood, the peak PMN level occurs at 4 hrs, which is about 2 hrs after the serum concentration of orgotein has already begun to subside. Whether PMNs continue to be attracted to the residual, enzymatically active orgotein, present at each instant is not yet clear.

Bovine serum albumin shares with orgotein the property of attracting PMNs into the peritoneal cavity, but does not have the effect of raising the peripheral blood neutrophil count when injected at an equimolar dose. Hence, PMN mobilization by orgotein into the circulation appears not to be a nonspecific protein effect. In addition, our above results confirm prior findings which demonstrated PMN mobilization following local orgotein administration into the guinea pig skin, canine eye and equine carpal joint.[2] A transient leukocytosis during treatment of paraquat poisoning with large doses of orgotein (i.v., i.m. and aerosol) has also been observed.[24]

Orgotein does not exhibit any chemotactic properties when evaluated in the Boyden chamber using the methods described in the literature.[25] Our findings have been confirmed independently by Dr. P.A. Ward, who at concentrations as high as 50 mg/ml found orgotein not to interfere directly or indirectly with the ability of neutrophils to respond to a chemotactic stimulus.[26] The PMN mobilizing effect of orgotein therefore appears to be mediated indirectly through interaction with and/or release of products present in the fluids of body cavities or the circulation. Recent papers[27,28] have suggested a role for superoxide anion in the formation of potent attractants for neutrophils.

In vitro, addition of SOD inhibited the development of this chemotactic activity. If the same situation exists in vivo, the addition of exogenous orgotein should reduce PMN mobilization. Our results show that this is not so and this seeming discrepancy needs to be further explored. If PMN mobilization were dose-dependent both for the factor generated by O_2^- and for SOD, then SOD concentrations above certain threshold values could be responsible for the PMN mobilization observed by us, with the net result representing mobilization by SOD less inhibition by the O_2^- induced chemotactic factor.

It must be kept in mind that our studies to date were carried out in normal rats. However, clinical observations indicate that a similar PMN mobilizing pattern can occur in patients receiving orgotein therapy. In this connection, the findings of Walker et al.[29] are of interest. They reported recently that PMNs of patients with rheumatoid arthritis show in vivo a significant decrease in their chemotactic response as compared to normal controls. To further explore these aspects, studies in animals while under the influence of induced inflammation are being initiated.

Polymorphonuclear neutrophils, of course, fulfill an essential function in clearing up debris by phagocytosis at a site of injury, even though they can exert pro-inflammatory effects by O_2^- production and spillage of lysosomal inflammants upon their lysis or leakage in situ. Stabilization of cell and lysosome membranes by orgotein or SOD has been demonstrated (Table 12). It is intriguing to speculate that the anti-inflammatory efficacy of orgotein is the product of providing ample numbers of PMNs at the injured site while protecting these cells against premature lysis or leakage in situ during phagocytosis.

TABLE 12

EFFECT OF ORGOTEIN ON SOME BIOLOGICAL MEMBRANES

System	Effective Dose	Efficacy Parameter
Lysosome Dysfunction[a]		
Heat-induced-37°C (rat liver)	3.0 mg/ml	Acid phosphate release
Histamine-induced-10^{-4} M	0.01 mg/ml	Naphthylamidase release 20 hours after orgotein addition
Muscle Cell Dysfunction		
Human Duchenne Dystrophy[b]	0.30 mg/ml	^{3}H-Leucine incorporation into myosin 24 hours after orgotein addition
Leukocyte Dysfunction		
Human PMNs[c]	0.30 mg/ml	Survival 10-35 hours after SOD addition

[a]W. Huber and M.G.P. Saifer (Ref.2).

[b]Ionasescu, V. et al. (1979) Ann. Neurol. 5, 107.

[c]M.L. Salin and J.M. McCord (Ref.21)

ACKNOWLEDGEMENTS

The authors wish to thank Peter H-C Dang, Ouida Y. Dotson, Eugenia Le Mat, Juliet W. Smock, Elke L. Thiemann and Joan A. Waranoff for their capable technical assistance.

REFERENCES

1. United States Adopted Names Council (1971). New Names List No. 106. J. Amer. Med. Ass. 218, 1936.
2. Huber, W. and Saifer, M.G.P. (1977) in Superoxide and Superoxide Dismutases, Michelson, A.M., McCord, J.M. and Fridovich, I., eds., Academic Press, New York, pp.517-536.
3. Menander-Huber, K.B. and Huber, W. (1977) in Superoxide and Superoxide Dismutases, Michelson, A.M., McCord, J.M. and Fridovich, I., eds., Academic Press, New York, pp.537-549.
4. Huber, W., Menander-Huber, K.B., Saifer, M.G.P. and Dang, P.H-C. (1977) in Perpsectives in Inflammation; Future Trends and Developments, Willoughby, D.A., Giroud, J.P. and Velo, G.P., eds., MTP Press, Lancaster, pp.527-540.
5. Babior, B.M., Kipnes, R.S. and Curnutte, J.T. (1973) J. Clin. Invest. 52, 741-744.
6. Johnson, R.B.,Jr., Keele, B.B.,Jr., Misra, H.P., Lehmeyer, J.E., Webb, L.S., Baehner, R.L. and Rajagopalan, K.V. (1975) J. Clin. Invest. 55, 1357-1372.
7. Cheson, B.D., Curnutte, J.T. and Babior, B.M. (1977) in Progress in Clinical Immunology, Vol. 3, Schwarz, R.S., ed., Grune and Stratton, New York, pp.1-65.
8. Lund-Olesen, K. and Menander, K.B. (1974) Curr. Ther. Res. 16, 706-717.
9. Lund-Olesen, K. and Menander-Huber, K.B. (1977) 14th Int. Congr. Rheum., San Francisco, Abstr.892.
10. Restifo, R.A. 14th Int. Congr. Rheum., San Francisco, Abstr.896.
11. Commandré, F. (1977) 14th Int. Congr. Rheum., San Francisco, Abstr.876.
12. Menander-Huber, K.B., Huskisson, E.C. and Huber, W. (1978) Int. Congr. Inflammation, Bologna, Abtrs.P2/44.
13. Ahlengard, S., Tufvesson, G., Petterson, H. and Andersson, T (1978) Brit. Equine Vet. J. 10, 122-124.
14. Coffman, J.R., Johnson, J.H., Tritschler, L.G., Garner, H.E. and Scrutchfield, W.L. (1978) J. Amer. Med. Vet. Ass. 174, 261-264.
15. Edsmyr, F., Huber, W. and Menander, K. (1976) Curr. Ther. Res. 19, 198-211,
16. Edsmyr, F., Menander-Huber, K.B. and Huber, W. (1979) Proc. Symp. Active Oxygen and Medicine, Honolulu.
17. Marberger, H., Bartsch, G., Huber, W., Menander, K.B. and Schulte, T.L. (1975) Curr. Ther. Res. 18, 466-475.
18. Marberger, H., Huber, W., Bartsch, G., Schulte, T.L. and Swoboda, P. (1974) Int. Urol. Nephrol. 6, 61-74.
19. Marberger, H., Menander-Huber, K.B., Huber, W. and Bartsch, G. (1979) Proc. Symp. Active Oxygen and Medicine, Honolulu.
20. Huber, W., Menander-Huber, K.B., Saifer, M.G.P. and Williams, L.D. (1979) Proc. Int. Meeting Inflammation, Verona.
21. Salin, M.L. and McCord, J.M. (1975) J. Clin. Invest. 56,1319-1323.
22. McCord, J.M. and Wong, K. (1978) in Oxygen Free Radicals and Tissue Damage Ciba Foundation Symp. 65 (New Ser.) Excerpta Medica, Amsterdam, pp.343-360.
23. Oyanagui, Y. (1976) Biochem. Pharmacol. 25, 1465-1472.
24. Rosen, S.M. (1974): Personal Communication.
25. Snyderman, R., Altman, L.C., Hausman, M.S. and Mergenhagen, S.E. (1972) J. Immunol. 108, 857-882.
26. Ward, P.A. (1974): Personal Communication.
27. Simchowitz, L., Mehta, J. and Spilberg, I. (1979) Arthritis Rheum. 22, 755-763.
28. McCord, J.M., English, D.K. and Petrone, W.F. (1979) Proc. Int. Meeting, Verona.
29. Walker, J.R., James, D.W. and Smith, J.H. (1979) Amer. J. Rheum. Dis. 38, 215-218.

DOUBLE-BLIND CONTROLLED CLINICAL TRIALS IN MAN WITH BOVINE COPPER-ZINC
SUPEROXIDE DISMUTASE (ORGOTEIN)

KERSTIN B. MENANDER-HUBER
Diagnostic Data, Inc., Moutain View, California 94043,
USA

At the first International Meeting on Superoxide (O_2^-) and Superoxide Dismutases
(SODs) at Banyuls (France)[1] in 1976 we presented our first overview on the established and potential clinical uses in man and animals of orgotein, the generic
classification[2] of the drug version of bovine Cu-Zn SOD. In the intervening years,
double-blind and controlled clinical trials in man and animals have been completed
all over the world.[3-13] They confirm that orgotein is an effective, non-analgesic
anti-inflammatory drug of outstanding safety. Its lack of analgesic and anti-pyretic
activity sets it apart from other anti-inflammatory agents. Yet, orgotein is clinically effective in the relief of pain, an effect which must accrue as a sequel to
disease amelioration rather than suppresion of symptoms and shows that it is a
true anti-inflammatory drug.

In order to show the potential of orgotein as a clinically useful anti-inflammatory drug, we present here the results of a series of double-blind, controlled
trials in rheumatoid arthritis (RA), osteoarthritis (OA), and amelioration of adverse
effects due to radiation therapy.

The double-blind, controlled trial has become a major tool in clinical research
to assess the efficacy or the lack of it of a new drug in a particular disease.
Thus, it seems worthwhile to summarize briefly the salient features of such trials.
Double-blind means that neither physician nor patient know whether the test drug
or the control substance is being administered. Every effort has to be made so
that drug and control substance will be identical in appearance and presentation.
Where differences between test drug and control substance cannot be compensated,
one uses either a "blind" observer for the assessment of effects or an individual
removed from all other aspects of the trial for administration.

A further prerequisite for the evaluation of a new drug are clinical trials
which compare the observed effects against placebo or a well-known drug used for
the treatment of the specific disease. As all patients under treatment also respond to the TLC (tender loving care) they are receiving, this attention can and
does bring about favorable responses in patients on inactive (placebo) treatment,
which express themselves as improvement not only in subjective but also in objec-

tive assessment parameters. That these placebo effects are real has been amply documented.[14,15] In consequence, it becomes apparent that the double-blind trial becomes essential both for the unbiased performance of the clinical study and the analysis of the results, which then established the real efficacy of the drug. The protocols for these trials must be rigidly structured both for admittance criteria and execution, to assure that the patient population at baseline is of sufficient uniformity both in demographic and disease parameters distribution. The individual patients are then randomly assigned to various teatment groups without built-in bias. Each patient admitted to the trial will have been and continues to be examined exactly the same way and by the same clinician, both prior to and throughout the trial.

Exogenous Cu-Zn SOD, in the form of orgotein, has been shown to be a potent anti-inflammatory agent in a number of animal models of induced inflammation.[16] Human diseases with inflammatory compnents are, therefore, prime candidates for testing. Examples are diseases such as RA and OA. These diseases have been shown to be prone to placebo responses and also to spontaneous remissions. In view of this we feel it imperative that any discussion on the role of exogenous SOD in the clinic be based on results from randomized double-blind trials. "Open drug" trials, in our opinion, have no place in the assessment to establish efficacy of any drug or treatment.

Against this background, the design and execution of individual double-blind trials covering the evaluation of orgotein as a useful and safe drug in the treatment of individual diseases will be discussed.

Rheumatoid Arthritis. In the United States, a multicenter placebo-controlled, double-blind study was performed using the protocol outlined in Table 1. One hundred and one patients with active and definite RA and on a fixed-dose regimen of aspirin (ASA) or ASA plus corticosteroids were entered into this 16-week randomized study. The trial consisted of six different sub-groups, with the patients treated at six different centers using the same protocol and prescribing the following dose regimen: Each injection should consist of 8 mg and a total of 40 injections should be given deep subcutaneously over a period of 12 weeks. No experimental drug was to be administered during the following 4 weeks. During the trial, the parameters shown in Table 1 were followed, as no single parameter can be considered as indicative of disease activity. All groups entered into the study were comparable in distribution of age, sex, and duration of disease (Table 2).

Figure 1 illustrates the time course of the mean response in the orgotein and placebo group during the 16 weeks of the trial for pain, use of analgesics, morning stiffness, knee circumference, grip strength, and evaluation by the patient.

Orgotein efficacy over placebo of clinical and statistical significance was found in almost all parameters, as illustrated by the data in Table 3. No serious side effects were found during or after this trial in which the patients had received

TABLE 1

PROTOCOL PARAMETERS OF DOUBLE-BLIND, PLACEBO-CONTROLLED TRIALS WITH ORGOTEIN
IN ACTIVE ADULT RHEUMATOID ARTHRITIS

Selection Criteria	A. Disease Status:	Must have definite, active, adult RA (ARA and FDA criteria)
	B. Other:	Must belong to functional, anatomi-classes 1, 2 or 3 (ARA)
	C. Other Medications:	Constant for background (ASA, Corti-costeroid, Gold); as needed for propoxyphene
Exclusion Criteria	A. Disease:	As specified by ARA, the ARA Coopera-Clinics Commitee and the FDA guide-lines
	B. Other Therapy (Wash-Out Time):	NSAI (2 wks); antimalarials (6 mos); immunosuppressives (6 mos); Gold (6 mos) except as background therapy, Corticosteroids (2 mos) except as background therapy
Dosage (Orgotein or Placebo)	8 mg x 4 x 4 wks; 8 mg x 3 x 8 wks; Omg x 4 wks; (Total dose 320 mg)	
Efficacy Evaluation	A. Pain, tenderness, use of analgesics	
	B. Swelling, circumference	
	C. Number of active joints	
	D. Morning stiffness, walking time, grip strength	
	E. Global assessment by doctor and patient	
	F. ESR	
Safety Evaluation	A. Clinical Chemistry:	CBC, SMA-12, Urinalysis
	B. Other:	Ophthalmology, adverse reactions (direct and indirect questioning)

TABLE 2

DOUBLE-BLIND, PLACEBO-CONTROLLED TRIALS WITH ORGOTEIN IN ACTIVE ADULT
RHEUMATOID ARTHRITIS: DEMOGRAPHIC INFORMATION

	Aspirin Regimen		Aspirin + Corticosteroid Regimen	
	Orgotein	Placebo	Orgotein	Placebo
Number of Patients in Efficacy Analysis	17	22	30	32
Age (Mean + SD)	54.4 ± 11.2	52.7 ± 12.3	55.0 ± 12.4	52.0 ± 12.7
Sex Distribution (M/F)	5/12	4/18	7/23	9/23
Duration of Disease, Years (Mean + SD)	9.7 ± 6.4	8.9 ± 6.0	9.7 ± 9.3	9.0 ± 7.3

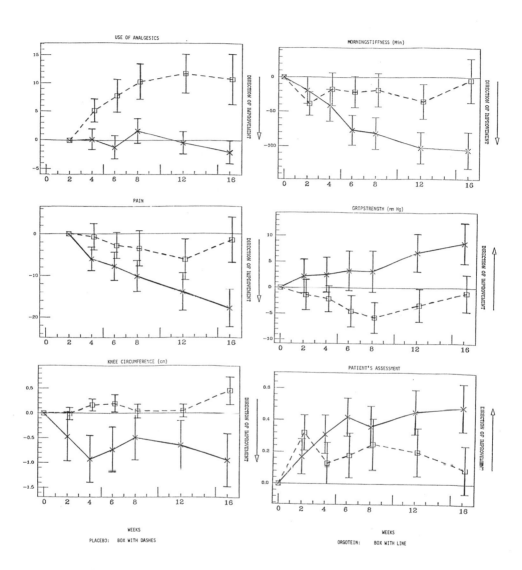

Fig. 1. Multicenter rheumatoid arthritis trials: Time course of efficacy
of orgotein over placebo in six clinical parameters.

orgotein injections at a rate of 3-4 per week.

The results from this double-blind study establish that orgotein does change all abnormal disease parameters in the direction of clinical improvement and that the correlation between parameters is very good. This should make orgotein a useful drug for the treatment of RA, as it has been said[17] that a drug for this disease would be the more useful the more it changed all abnormal parameters in the direction of improvement.

TABLE 3

DOUBLE-BLIND, PLACEBO-CONTROLLED TRIAL WITH ORGOTEIN IN ACTIVE ADULT RHEUMATOID ARTHRITIS

	Analysis of Slopes[a]	Analysis of Change[a]	Analysis of Clinical Significance[a]
Morning Stiffness	$P < 0.05$	$P < 0.05$	$P < 0.05$
Grip Strength	$P < 0.01$	$P < 0.01$	$P < 0.01$
Use of Analgesics	$P < 0.05$	$P < 0.01$	$P < 0.05$
Painful Joint Score	NS[b]	NS[b]	$P < 0.05$
Tender Joint Score	$P < 0.05$	$P < 0.01$	$P < 0.05$
Swollen Joint Score	$P < 0.05$	$P < 0.01$	$P < 0.05$
Number of Active Joints	$P < 0.05$	$P < 0.05$	ND[c]

[a]One-tailed Mann-Whitney test on change between 0-16 weeks, without removing clinic and clinic x treatment interaction variability from error variance. Significance levels at last visit.
[b]Not significant.
[c]Not determined.

Controlled trials at different centers, using blind observers comparing the effects of orgotein against gold therapy, have also been concluded and evaluated. In these trials, the efficacy of orgotein-dose regimen was evaluated by using 4 and 12 mg doses per injection, respectively, administered 3 times per week for the first 4 weeks, 2 times per week for the next 12 weeks and once a week for the last 10 weeks. Gold dosing was kept constant at 75 mg per week after an initial loading dose of 50 mg per week. This approach allowed for an efficacy evaluation of orgotein both in terms of dose and dose regimen against gold, which is recognized as representative for the group of slow-acting, disease-modifying, anti-rheumatic drugs. With the exception of dose regimen, the protocol for the RA comparison trials was the same in all essential aspects, as outlined in Tables 1 and 2. All patients entered into these trials were comparable in distribution of age, sex and duration of disease (Table 4).

TABLE 4

BLIND OBSERVER COMPARISON TRIAL OF ORGOTEIN AGAINST GOLD:

DEMOGRAPHIC INFORMATION

Test Group	Gold	12 mg Orgotein	4 mg Orgotein
Sex (M/F)	2/8	3/8	3/11
Age (Mean ± SD)	53.8 ± 10.4	57.6 ± 16.0	58.9 ± 11.0
Duration, Years (Mean ± SD)	8.0 ± 8.4	7.2 ± 4.0	7.3 ± 4.0

The results of these trials also show that orgotein is efficacious in changing all measured parameters in the direction of clinical improvement, with a good correlation between parameters. Within the dosing regimen used, the response was best for the highest dose, i.e., 12 mg per injection. At this level, clinical improvement at the end of treatment was as good or better than that for gold (Table 5). In contrast, in the low-dose orgotein group (4 mg per injection), the degree of improvement versus gold decreased as the frequency of orgotein injections was reduced. In fact, two patients dropped out because of increase in disease activity. This effect was not observed in the 12 mg group. Hence, the results indicate a dose dependency in the orgotein response. As the data obtained in the United States double-blind trials indicates a similar pattern, one should explore whether a constant dose regimen would improve the possibilities for beneficial results, especially when a lower dose per injection is given.

In the gold group, the degree and the rate of improvemnt corresponded to that described in the literature for similar dose regimens.[18] The same was true for the frequency and severity of side effects, as about one-third of the patients experienced adverse effects severe enough to force them to drop out of the trial. In contrast, none of the patients in the two orgotein groups experienced any disturbing adverse effects, overt toxicity, significant sensitization or significant changes in clinical biochemistry, hematology and urinalysis.

Osteoarthritis. It has been recognized for some time that inflammation is a regular component of osteoarthritis in addition to the degenerative characteristics of the disease.[19] Recent investigations[20] have shown that the presence of hydroxyapatite crystals in the joint may be causally related to the disease. In animal models, hydroxyapatite crystals caused a pronounced inflammatory response, accompanied by an influx of phagocytizing cells.[21] Such phagocytosis is accompanied, by O_2^- radicals generated from the plasma side of the cell membrane, often leading to granulocyte lysis in situ, with release of lysosomal inflamants into the injured

TABLE 5

RESPONSES IN BLIND OBSERVER TRIAL OF ORGOTEIN AGAINST GOLD

Parameter	Medication	Start of Treatment Mean ± SE	End of Treatment Mean ± SE
ESR (mm/1st hr)	Orgotein - 12 mg	47.2 ± 6.2	30.3 ± 6.4
	Gold	61.8 ± 8.0	35.5 ± 9.8
Morning Stiffness (min)	Orgotein - 12 mg	98.2 ± 25.0	40.8 ± 20.9
	Gold	171.0 ± 12.0	116.7 ± 54.6
Use of Analgesics (No. Tables/14 days)	Orgotein - 12 mg	39.6 ± 9.4	10.9 ± 6.6
	Gold	78.0 ± 7.4	38.3 ± 8.0
Grip Strength (mm Hg)	Orgotein - 12 mg	16.5 ± 4.9	37.4 ± 5.3
	Gold	13.1 ± 2.1	28.2 ± 6.5
Pain (Visual Analogue Scale)	Orgotein - 12 mg	69.6 ± 4.8	25.0 ± 7.8
	Gold	61.8 ± 4.1	30.5 ± 8.3
Articular Index (Ritchie Units)	Orgotein - 12 mg	29.7 ± 4.2	11.5 ± 4.1
	Gold	36.8 ± 2.7	20.8 ± 7.1

site. As in vitro SOD has been shown to inhibit this lysis, we have developed the concept of using patients with active, inflammatory osteoarthritis as a representative in vivo model disease in man for the clinical evaluation of orgotein and other non-analgesic anti-inflammatory drugs.[22] The validity of our concept to date has been confirmed by the results of six individual double-blind, placebo-controlled trials using a total of 157 patients suffering from acute inflammatory osteoarthritis in their knee joints, of which 85 received orgotein and the balance placebo. In the different trials various dose regimens were explored, using 2 and 4 mg per injection, an injection frequency of once per week or per fortnight, with total of 4, 8 and 12 injections, respectively. The protocol used in these trials was based on the guidelines for controlled trials in osteoarthritis of the FDA and the American Rheumatism Association (ARA). The data in Table 6 are representative of all our OA trials, with the exception, of course, of dosage and number of patients. To illustrate the response pattern to orgotein therapy in osteoarthritis of the knee, we present here relevant details for two of these trials, representing opposite approaches in terms of dose, dose regimen, and concomitant medication. The results of trials conducted at four different university hospitals in West Germany and Austria which all used the same protocol, with a dose of 4 mg orgotein or placebo given intra-articularly once a week at weekly intervals for 8 weeks are presented in this volume.[23]

In the first trial a group of elderly patients (age 65.5 ± 5.9 years) with definite, active osteoarthritis of the knee(s) were treated with 2 mg orgotein or placebo in 2 ml saline administered intra-articularly every 2 weeks for 20 weeks. Acceptance and exclusion criteria were in keeping with FDA/ARA Guidelines.

TABLE 6

PROTOCOL PARAMETERS OF DOUBLE-BLIND, PLACEBO-CONTROLLED TRIAL WITH
ORGOTEIN IN OSTEOARTHRITIS

Patient Selection	X-ray evidence of osteoarthritis of knee joints combined with active inflammation (2 cardinal signs or symptoms of inflammation)
No. of Patients	45
Dosage of Experimental drugs (Orgotein, Placebo)	2 mg injected into each knee joint every 2 weeks for 6 months
Parameters Followed:	
Pain	Day pain, night pain, pain on walking, use of analgesics
Function	Longest distance walked, limp, use of aids, stair climbing
Overall Evaluation	Patient's evaluation, doctor's evalutation
Safety	Patients asked about possible side effects

Three pain and three functional parameters, as well as global evaluation by patient
and doctor, were assessed at biweekly intervals (Table 6). The same physician performed all examinations for a given patient, both directly and indirectly, about
adverse reactions and also determined consumption of analgesics. No anti-inflammatory drugs other than the experimental medication were permitted.

The results (Table 7) show that orgotein when injected into the knee joint caused
a pronounced improvement in the combined pain parameters, the combined function
parameters, the use of analgesics, and the overall status of the disease as judged
by physician and patient. The time course of response to orgotein and placebo is
shown in Figure 2, which demonstrates for four clinically important parameters that
improvement after orgotein was both clinically and statistically significantly superior over placebo. Of further interest is the prolonged placebo response which
emphasizes the aforementioned need for double-blind trials in evaluating the clinical uses of SOD.

The second double-blind OA trial followed a somewhat different protocol (Table 8)
employing a group of patients who used concomitant anti-inflammatory medication
throughout the medication period (weeks 0-4) and the blind follow-up period (weeks
5-12). Concomitant anti-inflammatory medication included indocin, naprosyn, nalfon,
motrin and clinoril and was kept constant for each patient at the level present
upon entry into the trial. This approach permitted an assessment of the long-lasting
effects of the intra-articular therapy. The results (Table 9) show that at the end
of treatment, response to orgotein was more pronounced than to placebo. The values
at week 12, i.e., 8 weeks after end of therapy, show that a total dose of 16 mg
orgotein given during a 1 month period provides long-lasting benficial effects.

TABLE 7

RESPONSE TO MEDICATION IN DOUBLE-BLIND, PLACEBO-CONTROLLED OSTEOARTHRITIS
TRIAL AT END OF MEDICATION

	Mean change from wk 1 to wk 20		
	Pain Score[a]	Function Score[b]	Physician's Global Evaluation
Placebo (17)	-1.6	-0.3	+0.5
Orgotein (13)	-5.9	-2.3	+1.2
P (t-test)	< 0.01	< 0.025	< 0.0025

[a]Pain: Combined scores for day-night pain; pain on walking; score: 1 = none, 2 =
mild, 3 = moderate, 4 = severe, 5 = very severe.
[b]Function: Combined scores for use of aids; limp, score: 1 = none, 2 = mild,
3 = moderate, 4 = severe, 5 = very severe; maximal distance walked outdoors
without stopping, score: 1 = 500 m, 2 = 300-500 m, 3 = 100-300 m, 4 = 50-100 m,
5 = indoors only; and ability to climb stairs, score: 1 = normal, 2 = with help
of banister, 3 = with walking aid, 4 = with assistance, 5 = impossible. Stair
climbing: 1 = normal, 2 = normal with banister, 3 = able with cane, 4 = able
with crutches, 5 = able when aided by other person, 6 = unable.

TABLE 8

PROTOCOL PARAMETERS OF DOUBLE-BLIND TRIAL WITH ORGOTEIN IN OSTEOARTHRITIS

Patient Selection	X-ray evidence of osteoarthritis of knee joints combined with active inflammation (> 2 cardinal signs or symptoms of inflammation)
No. of Patients	40
Dosage of Experimental drugs (Orgotein, Placebo)	4 mg injected into each knee joint every week 4 times
Parameters followed	Pain (visual analogue scale)
	Function (visual analogue scale)
	Effusion (5-point scale)
	Morning stiffness (min)
	Inactivity stiffness (min)
	Success (visual analogue scale)

Interestingly, the placebo response in this trial was remarkably high at the
end of the treatment period. However, the retention of the degree of involvement
at week-12 was more often less than that seen with orgotein. It is inviting to
speculate that the combination of TLC, injections, and concomitant medication is
the reason for this extraordinary placebo response. Orgotein-related adverse reactions were not observed in either of these trials.

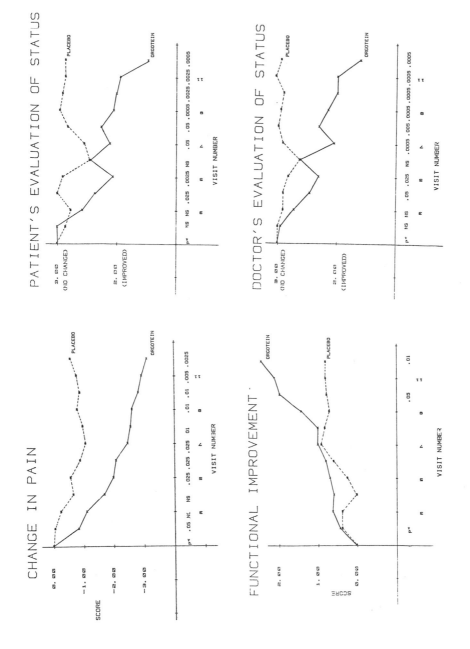

Fig. 2. Osteoarthritis trial: Time course of efficacy of intra-articular orgotein (2 mg) over Placebo in four clinical parameters.

TABLE 9

RESPONSE TO MEDICATION IN DOUBLE-BLIND OSTEOARTHRITIS TRIAL AT END OF
TREATMENT AND AT FOLLOW-UP

Parameter	Medication	At End of Treatment Week-4 Mean ± SE	At Follow-up Week-12 Mean ± SE
Pain[a]	Orgotein	-35.1 ± 6.8	-26.4 ± 4.7
	Placebo	-17.5 ± 4.7	- 8.9 ± 4.8
Function[a]	Orgotein	+17.1 ± 5.3	+15.4 ± 6.6
	Placebo	+12.9 ± 7.3	+11.7 ± 7.8
Morning Stiffness	Orgotein	-76.3 ± 13.4	-73.8 ± 13.3
(min)	Placebo	-12.3 ± 5.7	-15.9 ± 4.2
Inactivity Stiffness	Orgotein	- 4.8 ± 1.3	- 4.4 ± 2.1
(min)	Placebo	- 5.6 ± 2.2	- 5.4 ± 2.7
Effusion	Orgotein	- 0.7 ± 0.1	- 0.7 ± 0.1
(score)	Placebo	- 0.4 ± 0.2	- 0.4 ± 0.2
Success[a]	Orgotein	+26.7 ± 6.1	+13.5 ± 9.3
	Placebo	+15.3 ± 10.3	+ 3.1 ± 13.1

[a]Visual analogue scale.

Overall, the results from both the rheumatoid and the osteoarthritis trials
show that orgotein has pronounced anti-inflammatory effects. Once maximal effects
of orgotein therapy have been reached, treatment can be stopped with the beneficial
effects remaining for long periods of time, a behaviour which is very much in contrast to what is seen when non-steroidal anti-inflammatory drugs are used.

Amelioration of Inflammatory Side Effects due to Radiation Therapy. Therapy
of malignancies still strikes a rather fragile balance between efficacy and side
effects, irrespective of whether one is dealing with chemotherapy or radiation
therapy. If one could find drugs of low toxicity which prevent or ameliorate the
side effects, one would be able not only to give the patients a better quality of
life, but also increase the efficacy of therapy by raising the dose with impunity.
A step in this direction has been taken by showing that orgotein is effective and
safe in ameliorating the inflammatory side effects encountered during external irradiation of prostatic and bladder carcinomas.

Our studies were initiated upon the belief that orgotein could reduce inflammatory side effects in patients getting high-dose irradiation for pelvic tumors.
As destruction of malignant cells by high-energy radiation is largely a direct-hit
concentration of SOD in the cytosol, it was felt that interference with the tumorolytic events by extracellular orgotein would not occur. That this is so has been
supported by in vitro as well as by in vivo experiments which document the absence

of an orgotein effect on tumor tissue for a broad spectrum of models. Some representative results are presented in Table 10.

TABLE 10

DOES ORGOTEIN INTERFERE WITH RADIOTHERAPEUTIC EFFECT ON TUMOR GROWTH?

Test System	Species No./Group	Irradiation (250 KV, 15 mA) Rads	Orgotein	Orgotein Effects
KHT Sarcoma[a]	Mice, 8	300 x 10 2 wk	4 & 40 mg/kg, s.c. 5 x/ wk x 4	None
L_1A_2 cell[25] line C₃H mouse lung	Mice, Tissue Culture,[b] 3	200, 400, 600 800, 1000 Resp. to culture	200 mcg (10 ml)/ 300-400 cells, 3 hr before or ½ hr after irradiation	None
Jejunal[25] crypt cells in vivo	Mice, 4	100, 200, 400 600, 800, 1000 Resp. to gut	50 mg/kg, i.p. 2 mg/ml, 2 hr before or ½ hr after irradiation	None
13762 Mammary adenocarcinoma transplants	Rats, 10	None	250 mcg/4 sites intra- tumor, days 13-17, 19-23	None
Walker carcimona	Rats, 6	None	1.6 mg/kg, intratumor x 7 days	None

[a]Hill, R.P., personal communication.
[b]BME (GIBCO C-14) + 10% fetal calf serum.

For the clinical trials we selected two types of pelvic tumors, i.e., urinary bladder carcinoma and prostatic carcinoma, as they provide useful parameters for a protocol evaluating the effect of orgotein and after local irradiation of the tumor (Table 11). Four or 8 mg orgotein or placebo dissolved in about 1 ml saline injection USP were injected deep subcutaneously 15-30 minutes after completion of each daily radiation session. All patients received 6 MV X-rays delivered by linear accelerator, using a three-field technique.[24] The patients received anti-bacterial therapy throughout the trial and were permitted to use one specified anti-diarrheal as needed. No other anti-inflammants in addition to orgotein were permitted.

The effects of the experimental medication were assessed, using parameters such as pain, dysuria, maximum voided volume, interval between voidings during day and night, severity of diarrhea, and amount of medication needed to control diarrhea.

The patients were evaluated at regular intervals after entry into the trial. One visit always coincided with the termination of therapy. Follow-up evaluations were done at about 4 months and 2 years, where possible. Hematology and urinalysis were performed at each visit and clinical chemistry at the beginning and end of the treatment.

TABLE 11

PROTOCOL PARAMETERS OF DOUBLE-BLIND PLACEBO-CONTROLLED STUDY ON THE ORGOTEIN EFFICACY IN AMELIORATING SIDE EFFECTS DUE TO RADIATION THERAPY OF PELVIC TUMORS

Patient Selection	Bladder or prostate tumors T2-T4; malignancy grades 1-3
No. of Patients	38 (bladder); 50 (prostate)
Orgotein or Placebo Dosage	4 mg (bladder) or 8 mg subcutaneously after completion of daily radiation therapy
Radiation Dosage	6400 or 8400 Rads in 8 weeks (bladder) 5400 Rads in 7 weeks (prostate)
Parameters Evaluated	Maximal voiding volume frequency (bowel, bladder) pain, anti-diarrheal medication used

In one double-blind, placebo-controlled trial with 38 patients with urinary bladder carcinoma comparison of orgotein with placebo shows that in 6-7 parameters orgotein was significantly ($P < 0.05$ to < 0.001) more effective in reducing signs and symptoms both in the bladder and the bowel (Table 12).

TABLE 12

BLADDER CARCINOMA EFFICACY OF ORGOTEIN OVER PLACEBO IN AMELIORATING SIDE EFFECTS DUE TO RADIATION THERAPY

Parameters	Level of Statistical Significance (P)
Maximum voided volume >200 ml	< 0.05[a]
Interval between voidings during day	< 0.05[a]
Interval between voidings during night	NS[b]
Severity of signs and symptoms drom bladder	< 0.05[a]
Percent visits with diarrhea	< 0.025[a]
Percent of "diarrhea visits" requiring medication	< 0.001[a]
Dose of anti-diarrheal medication	< 0.0025[c]

[a]Chi-square test.
[b]Not significant.
[c]Student's t-test.

On follow-up at 4 months, the beneficial effects of orgotein over placebo on bladder and bowel, if anything, had become more marked than at termination of thera-py. Thus, 4 of 15 patients in the placebo group still had symptoms of proctitis, while all patients in the orgotein group were free of it. At 2-year follow-up, only 9 of the 38 patients were still alive (5 orgotein, 4 placebo), indicating that orgotein treatment concurrent with radiation therapy did not influence sur-vival time or long-term tumorolytic efficacy.

In another placebo-controlled trial, 8 mg orgotein or placebo per injection was administered to 50 patients receiving radiation therapy (5400 rads during 6 weeks) for poorly differentiated prostate carcinomas (Table 11).

In this study, radiation-induced side effects were seen in considerably fewer patients in the placebo group than expected from prior clinical experience. While orgotein therapy ameliorated these radiation-induced side effects more effectively than placebo (Figure 3), the differences reached statistical significance only occasionally because of the low-overall frequency of serious radiation-induced side effects.

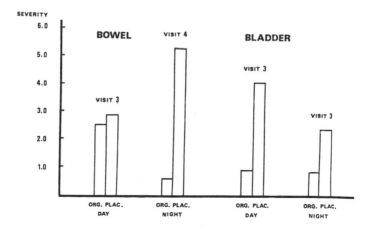

Fig. 3. Prostate carcinoma trial: Maximal deterioration in placebo group compared with change in orgotein group at the same visit.

The results of the above trials and others now underway open the challenging aspect that the use of concomitant orgotein treatment may permit the use of higher radiation doses, and with it the possibility of a better therapeutic result in the treatment of the malignancy. As a first attempt to explore this, a double-blind, placebo-controlled trial in patients suffering from poorly differentiated prostate carcinomas has been started (Table 13). In this trial, a total dose of 6500 rads will be directed at the prostate while a larger area, including the lymph nodes adjacent to the prostate, will be exposed to a total dose of 5000 rads.

TABLE 13

PROTOCOL PARAMETERS OF DOUBLE-BLIND PLACEBO-CONTROLLED STUDY ON THE ORGOTEIN EFFICACY IN AMELIORATING SIDE-EFFECTS DUE TO LARGE FIELD RADIATION THERAPY OF PROSTATE TUMORS AND ADJACENT LYMPHY NODES

Patient Selection	Prostate tumors T2-T4; malignancy grade 1-3
No. of Patients	50
Orgotein or Placebo Dosage	16 mg subcutaneously after completion of daily radiation therapy and during rest period
Radiation	> 6 MV X-rays 5000 Rads in 7 weeks to prostate and lymph nodes; 2 weeks rest, then 1500 Rads in 2 weeks booster to prostate alone
Parameters Evaluated	Maximal voiding volume, frequency (bowel, bladder), pain, anti-diarrheal medication used

Conclusion. The results of the trials described above clearly show the abso-lute need for controlled, blind trials to establish unequivocally (1) the per se usefulness of orgotein in a clinical indication and (2) its degree of efficacy. In our experience, treatments with "open-label" orgotein performed to date by us and an increasing number of other investigators are at best guideposts for subse-quent blind trials using rigid protocols, as in view of the observed placebo effects, the observed "efficacy" too often turns out to be spurious. All the same, there is today no doubt that orgotein, the drug version of bovine Cu-Zn SOD is a proven efficacious drug and not just an active enzyme. The future will have to show how broad its use in the clinic will be. But, as it has two very essential properties for a successful drug – efficacy and low or no toxicity – its future should be promising, indeed.

REFERENCES

1. Menander-Huber, K.B. and Huber, W. (1977) in Superoxide and Superoxide Dis-mutases, Michelson, A.M., McCord, J.M. and Fridovich, I., eds., Academic Press, New York, pp. 537-549.

2.   United States Adopted Names Council (1971) New Names List No. 106. J. Amer. Med. Ass. 218, 1936.

3.   Lund-Olesen, K. and Menander, K.B. (1974) Curr. Ther. Res. 16, 706-717.

4.   Lund-Olesen, K. and Menander-Huber, K.B. 14th Int. Congr. Rheumatol. San Francisco, California, USA, June 36-July 1, 1977, Abstr. 892.

5.   Restifo, R.A., 14th Int. Congr. Rheumatol. San Francisco, California, USA, June 26-July 1, 1977, Abstr. 896.

6.   Commandré, F., 14th Int. Congr. Rheumatol. San Francisco, California, USA, June 26-July 1, 1977, Abstr. 876.

7.   Edsmyr, F., Huber, W. and Menander, K.B. (1976) Curr. Ther. Res. 19, 198-211.

8.   Edsmyr, F., Manander-Huber, K.B. and Huber, W. in Proc. Symp. Active Oxygen and Medicine, Honolulu, Hawaii, January 30-February 2, 1979, Autor, A.P., ed., Raven Press, New York, in press.

9.   Marberger, H., Bartsch., Huber, W., Menander, K.B. and Schulte, T.L. (1975) Curr. Ther. Res. 18, 466-475.

10. Marberger, H., Huber, W., Bartsch, G., Schulte, T.L. and Swoboda, P. (1974) Int. J. Urol. Nephrol. 6, 61-74.

11. Marberger, H., Menander-Huber, K.B., Huber, W. and Bartsch, G. Proc. Symp. Active Oxygen and Medicine, Honolulu, Hawaii, January 30-February 2, 1979, Autor, A.P., ed., Raven Press, New York, in press.

12. Ahlengard, S., Tufvesson, G., Petterson, H. and Andersson, T. (1978) Brit. Equine Vet. J. 10, 122-124.

13. Coffman, J.R., Johnson, J.H., Tritschler, L.G., Garner, H.E. and Scrutchfield, W.L. (1978) J. Amer. Vet. Med. Ass. 174, 261-264.

14. Beecher, H.K. (1955) J. Amer. Med. Ass. 159, 1602-1606.

15. Colton, Th. (1974) Statistics in Medicine, Little, Brown and Co., Boston, pp. 253-270.

16. Huber, W., Menander-Huber, K.B., Saifer, M.G.P. and Dang, P.H-C. (1977) in Perspectives in Inflammation: Future Trends and Developments, Willoughby, D.A., Giroud, J.P. and Velo, G.P., eds., MTP Press, Lancaster, pp. 527-540.

17. O'Brien, Wm.M. 4th Meeting Arthritis Advisory Committee U.S. Food and Drug Administration, Rockville, Maryland, December 11-12, 1975.

18. Rothermich, N.O. (1979) in Clinics in Rheumatoid Arthritis, Vol. 5 No.2, Huskisson, E.C., ed., W.B. Saunders Co. Ltd., London, pp. 631-640.

19. Ehrlich, G.E. (1975) J. Amer. Med. Ass. 232, 157-159.

20. Ali, S.Y. (1977) in Perspectives in Inflammation: Future Trends and Developments, Willoughby, D.A., Giroud, J.P. and Velo, G.P., eds., MTP Press, Lancaster, pp 211-223.

21. Dieppe, P.A. (1977) in Perspectives in Inflammation: Future Trends and Developments, Willoughby, D.A., Giroud, J.P. and Velo, G.P. eds., MTP Press, Lancaster, pp 225-237.

22. Menander-Huber, K.B., Hukisson, E.C. and Huber, W., Int. Congr. Inflammation, Bologna, Italy, October 31-November 3, 1978, Abstr. P2/44.

23. Flohé, L., Biehl, O., Hofer, H., Kadrnka, F., Kölbel, R. and Puhl, W., this volume.

24. Littbrand, B., Edsmyr, F. and Révész, L.W. (1975) Bull. Cancer 62, 241-248.

25. Overgaard, J., Nielsen, O.S., Overgaard, M., Steenholdt, S., Jakobsen, A. and Sell, A., Acta Radiol., in press.

EFFECTIVENESS OF SUPEROXIDE DISMUTASE IN OSTEOARTHRITIS OF THE KNEE JOINT.
RESULTS OF A DOUBLE BLIND MULTICENTER CLINICAL TRIAL.

L. FLOHÉ,[+] O. BIEHL,[++] H. HOFER,[+++] F. KADRNKA,[+] R.KOLBEL[++++] AND W. FUHL[+++++]

[+]Center of Research, Grünenthal GmbH, Aachen, W. Germany; [++]Orthopaedics Department, University of Homburg/Saar, W. Germany; [+++]Orthopaedics Department, State Hospital, Salzburg, Austria; [++++]Orthopaedics Department, University of Berlin, Oskar-Helene-Heim, Berlin, W. Germany; [+++++]Orthopaedics Department, University of Heidelberg, W. Germany

INTRODUCTION

The anti-inflammatory activity of injected superoxide dismutase (SOD) in animal model systems was discovered by Huber et al.[1,2] and could be supported by a variety of clinical observations (for review see Ref. 3). This anti-inflammatory activity became conceivable by the emerging evidence that the substrate of SOD, the O_2^- radical is directly or indirectly detrimental to macromolecules and biomembranes[4-8] and thus may cause or contribute to inflammatory tissue response, if its production is not properly balanced by enzymatic break-down. While intracellular O_2^- formation is adequately counteracted by endogenous SOD, O_2^- released into the extracellular compartment by activated macrophages,[9] neutrophilic[10,11] and eosinophilic[12] leuko-cytes hits a largely unprotected micro-environment. The resulting tissue damage may in turn trigger O_2^- release by the mentioned inflammatory cells and thus the process may become self-sustaining, once a critical level of O_2^- release is reached. Injection of SOD which distributes in the extracellular compartment might be able to interrupt this chain of unfavorable events. It could be anticipated and was established by McCord and Wong[13] that the anti-inflammatory efficacy of injected SOD increases with the time of survival at its presumed site of action.

With these considerations in mind, we designed a randomized placebo-controlled double blind mulicenter trial to establish the safety and efficacy of injected SOD in a clinical condition which meets the theoretical requirements. Osteoarthritis of the knee joint was considered the indication of choice for the following reasons:
(1) Osteoarthritis is considered a degenerative disease with concomitant or inter-mittent inflammation.
(2) A number of subjective and objective parameters indicative of the degree of inflammation can be easily evaluated.
(3) Preliminary investigations had shown a favourable response to SOD (Ref. 14).
(4) The localized character of the disease provides the opportunity to achieve comparatively high and long lasting levels of exogenous SOD, if it is injected

intraarticularly, since SOD is not degraded by synovial fluid in vitro[15] and the articular capsule will limit the leakage of a high-molecular weight compound via diffusion.[16]

MATERIALS AND METHODS

Drugs. Superoxide dismutase was provided in code-labelled single use vials containing 4 mg orgotein (the drug version of copper-sinc SOD prepared from bovine liver; Diagnostic Data, Inc., Mountain View, California plus 8 mg sucrose USP as lyophilized powder. Placebo was provided in identical code-labelled vials containing 12 mg sucrose USP as lyophilized powder. Before injection, the vial contents was reconstituted in 2 ml Sodium Chloride Injection USP, pH 6.5-7.0, without preservative. The solutions were clear and appeared colorless.

Patients. A total of 101 patients with active inflammatory osteoarthritis of the knee joint entered the trial according to the following selection criteria:
(1) Roentgenological evidence of osteoarthritis of the knee;
(2) More than one sign of acute inflammation;
(3) No concomitant medication or disease interfering with the evaluation of therapeutic effects (rheumatoid arthritis, gout, psoriasis, etc.).
Eighty five patients finished the trial and could be evaluated, while 7 had to be disregarded, because the diagnostic criteria were not adequately met, 7 concomitantly took other anti-inflammatory drugs and 2 dropped out after the first or second medication for unknown reasons.

Design. The investigation was performed at four different university hospitals in Germany and Austria according to an identical protocol. The protocol was essentially based on the guide lines for investigations of degenerative joint disease developed by the FDA and the American Rheumatism Association (ARA)

Superoxide dismutase (4 mg) or placebo, respectively, were injected intraarticularly at weekly intervals for eight weeks under strict double blind conditions. Before treatment and at weekly intervals thereafter the following parameters were evaluated:
(1) "Status of the joint" comprising stability of the joint (only used as diagnostic criterion); thickening of the synovia (scale 1-5; weighting factor 1); and hydrops (scale 1-5; weighting factor 1).
(2) Joint dircumference (in cm) was evaluated separately.
(3) Pain was scored (scale 1-4) at night and during day time at rest (weighting factor 2) and upon walking (weighting factor 1).
(4) "Function" composed of "use of walking aids" (range 1-8; weighting factor 1); limp (range 1-5; weighting factor 2); maximum walking distance without stopping (1 = 500 m; 2 = 300-500 m; 3= 100-300 m; 4= 50-100 m; 5 = only indoors; weighting factor 1) and ability to climb stairs (scale 1-6; weighting factor 1).

(5) Global impression (scale 1-5) of the severeness of the disease was rated independently by the investigators and the patients. Adverse effects had to be examined at each visit. Discrepancies between the total number of patients evaluated and the number given in some tables result from the absence of specific symptoms at the beginning of the trial. Informed consent was obtained from all patients entered into the study.

The results were analyzed for statistical significance at the 5% level (predetermined in advance) by two-tailed Student's t-test.

RESULTS AND DISCUSSION

The improvement of the patients' clinical condition after 8 weeks of treatment with orgotein or placebo is presented in Tables 1 to 4. The given parameters and scores are defined under MATERIALS AND METHODS, unless otherwise stated. For obvious reasons the ratings of the clinical symptoms and parameters do not include clinical manifestation which are not liable to be reversed within the period of treatment such as instability of the joint or changes in X-rays. All symptoms related to the inflammatory component of the disease are improved by the end of the treatment and the considerable difference between the treatment groups is evident and significant in each individual parameter despite a fairly high placebo response. The status of the joint is improved by about 50% in the orgotein-treated group (Table 1). Joint circumference is decreased due to orgotein treatment by 1.3 cm, while it remained nearly unchanged by placebo treatment (Table 2). The average pain scores decrease to 35% of the pretreatment period, although orgotein has no intrinsic analgesic properties (Table 3). The functional relevance of these therapeutic effects may be derived from Table 4 summarizing the patients' ability to walk and climb stairs. An average improvement of more than 40% of the initial disability scores is impressive, if one considers the chronic character of osteoarthritis and its tendency to deteriorate. The improvement of the individual parameters is reflected in the global ratings of the clinical improvement by the patients (Table 5) and the physicians (Table 6) which in turn show a high degree of congruence. While 43 of the 45 patients treated with orgotein improved according to both independent ratings, more than 50% of the placebo-treated patients remained unchanged or deteriorated. Again, the superiority of orgotein-treatment over placebo treatment is significant according to both self-rating and physician's rating. Table 7 finally demonstrated the onset of a remarkable improvement according to the physician's opinion. While most of the orgotein-treated patients had responded by the fifth week of treatment and none failed to respond at all, 18 out of 40 patients treated with placebo did not improve, deteriorated or dropped out (seven patients) due to lack of efficacy.

Adverse reactions possibly related to treatment such as irritation of the articular capsule or swelling of the knee joint were observed three times after placebo

TABLE 1

IMPROVEMENT OF CLINICAL STATUS OF THE JOINTS (THICKENING OF SYNOVIA, HYDROPS)
AFTER EIGHT WEEKS OF TREATMENT

Drug	Number of patients	% $(\bar{x} \pm s_{\bar{x}})^a$	Significance
Orgotein	41	46.6 ± 7.4	
			P < 0.05
Placebo	34	21.4 ± 7.3	

[a]% of pretreatment scores

TABLE 2

DECREASE OF JOINT CIRCUMFERENCE AFTER EIGHT WEEKS OF TREATMENT

Drug	Number of patients	% $(\bar{x} \pm s_{\bar{x}})^b$	Significance
Orgotein	41	1.3 ± 0.2	
			P < 0.05
Placebo	40	0.2 ± 0.24	

[a]In some patients either initial or final values were missing.
[b]In cm.

TABLE 3

DECREASE OF PAIN SCORES AFTER EIGHT WEEKS OF TREATMENT

Drug	Number of patients	% $(\bar{x} \pm s_{\bar{x}})^a$	Significance
Orgotein	44	65.5 ± 3.7	
			P < 0.05
Placebo	40	19.8 ± 6.6	

[a]% of pretreatment scores

TABLE 4

IMPROVEMENT OF DISABILITY SCORES AFTER EIGHT WEEKS OF TREATMENT

Drug	Number of patients	% $(\bar{x} \pm s_{\bar{x}})$ [a]	Significance
Orgotein	44	45.3 ± 4.8	
			P < 0.05
Placebo	35	11.1 ± 6.3	

[a]% of pretreatment scores

TABLE 5

PATIENTS' SELF-RATINGS OF OVERALL THERAPEUTIC RESULTS

Rating (Score)	Number of patients	
	Placebo	Orgotein[a]
Much improvement (1)	8	28
(1) to (2)	1	1
Improved (2)	7	14
(2) to (3)	1	–
Unchanged (3)	14	1
Deteriorated (4)	6	1
Much deteriorated (5)	2	–
No comments	1	–
Total	40	45

[a]The superiority of orgotein treatment is significant at the prefixed level (P < 0.05).

and once after orgotein injection. Intestinal pain and an exanthema was seen once each after placebo. None of the adverse reactions required cessation of treatment.

In conclusion, weekly repeated intraarticular injections of SOD (orgotein) are well tolerated and clearly effective in osteoarthritis, as far as the inflammatory symptoms are concerned. Although the present investigation does not yet provide any information on long term therapeutic effects of SOD in this condition, it may be anticipated that SOD treatment does not only suppress the acute symptomatology, but might also delay or prevent irreversible alterations of the joints, since the

the intermittent inflammatory reactions certainly contribute to the final pathological manifestations in osteoarthritis.

TABLE 6

PHYSICIANS' RATINGS OF THE OVERALL THERAPEUTIC RESULTS

Rating	Number of patients	
(Score)	Placebo	Orgotein[a]
Much improved (1)	5	21
(1) to (2)	-	2
Improved (2)	13	20
Unchanged (3)	15	2
Deteriorated (4)	4	-
Much deteriorated (5)	2	-
No comments	1	-
Total	40	45

[a]The superiority of orgotein treatment is significant at the prefixed level
($P < 0.05$).

TABLE 7

ONSET OF THERPEUTIC EFFECT ACCORDING TO PHYSICIANS' RATINGS

Onset of improvement	Number of patients	
after injection indicated	Placebo	Orgotein
1st injection	5	11
2nd "	6	4
3rd "	3	12
4th "	3	11
5th "	3	4
6th "	-	3
7th "	-	-
8th "	-	-
No improvement[a]	18	-
No comments	1	-
Total	40	45

[a]Includes dropouts due to lack of efficacy.

430

SUMMARY

Safety and efficacy of intraarticularly injected SOD (orgotein) in 85 patients with osteoarthritis of the knee joint was investigated in a randomized placebo controlled double blind multicenter trial. Orgotein (4 mg SOD) or placebo was applied at weekly intervals for a total treatment period of 8 weeks. SOD proved to be well tolerated and clearly effective and was significantly ($P < 0.05$) superior to placebo in improving all individual parameters evaluated (e.g. joint circumference, pain and disability scores) and the overall clinical status, as rated by patients and physicians. All 45 patients treated with orgotein improved at the latest after the sixth injection (maximum after the 3rd injection), whereas 18 out of 40 placebo-treated patients failed to show any beneficial effect within 8 weeks or dropped out (7 patients) due to lack of efficacy. It is concluded that intraarticular injection of SOD is a useful treatment of the concomitant or intermittent inflammatory reaction in osteoarthritis and might thereby delay or prevent the manifestation or irreversible changes of the affected joints.

ACKNOWLEDGEMENT

Dr. J. Nijssen, Grünenthal GmbH, Stolberg, W. Germany is gratefully acknowledged for statistical analysis.

REFERENCES

1. Huber, W., Schulte, T.L., Carson, S., Goldhamer, R.E. and Vogin, E.G. (1968) Toxicol. Appl. Pharmacol. 12, 308.
2. Huber, W., Menander-Huber, K.B., Saifer, M.G.P. and Dong, P.H-C.(1977) in Perspectives in Inflammation: Future Trends and Developments, Willoughby, D.A., Giroud, J.P. and Velo, G.P., eds., MTP Press, Lancaster, pp. 527-544.
3. Menander-Huber, K.B. and Huber, W. (1977) in Superoxide and Superoxide Dismutases, Michelson, A.M., McCord, J.M. and Fridovich, I., eds., Academic Press, New York, pp. 537-549.
4. Flohé, L. and Zimmermann, R. (1974) in Glutathione, Flohé, L., Benöhr, H. Ch., Sies. J., Waller, H.D. and Wendel, A., eds., Georg Thieme Publishers, Stuttgart, pp. 245-260.
5. McCord, J.M. (1974) Sciences, 185, 529-531.
6. Michelson, A.M. (1977) in Superoxide and Superoxide Dismutases, Michelson, A.M., McCord, J.M. and Fridovich, I., eds., Academic Press, New York, pp. 245-255.
7. Goldberg, B. and Stern, A. (1976) J. Biol. Chem. 251, 6468-6470.
8. Kellogg, E.W. and Fridovich, I. (1977) J. Biol. Chem. 252, 6721-6728.
9. Lowrie, D.B. and Aber, V.R. (1977) Life Sci. 21, 1575-1584.
10. Babior, B.M., Kipnes, R.S. and Curnutte, J.T. (1973) J. Clin. Invest. 52, 741-744.
11. Johnston, R.B., Keele, B.B., Misra, H.P., Lehmeyer, J.E., Webb, L.S., Boehner, R.L. and Rajagopalan, K.V. (1975) J. Clin. Invest, 55, 1357-1372.
12. Goetzel, E.J., Valone, F.H. Smith, J.A. and Weller, P.F. (1978) Abstr. Int. Congr. Inflammation, Bologna, p. 52.
13. McCord, J.M. and Wong, K. (1979) in Oxygen Free Radicals and Tissue Damage, Ciba Foundation Symposium 65 (New Ser.) Excerpta Medica, Amsterdam, pp. 343-351.
14. Lund-Olesen, K. and Menander, K.B. (1974) Curr. Ther. Res. 16, 706-717.
15. Flohé, L. and Loschen, G. (1979), unpublished results.
16. Zeitlin, I.J., Grennan, D.M., Buchanan, W.W. and Miller, W. (1978) Abstr. Int. Congr. Inflammation, Bologna, P. 170.

Index